Springer Series in

OPTICAL SCIENCES 73

founded by H.K.V. Lotsch

Springer
Berlin
Heidelberg
New York
Barcelona
Hong Kong
London
Milan
Paris
Singapore
Tokyo

Springer Series in
OPTICAL SCIENCES

The Springer Series in Optical Sciences, under the leadership of Editor-in-Chief *William T. Rhodes*, Georgia Institute of Technology, USA, and Georgia Tech Lorraine, France, provides an expanding selection of research monographs in all major areas of optics: lasers and quantum optics, ultrafast phenomena, optical spectroscopy techniques, optoelectronics, information optics, applied laser technology, industrial applications, and other topics of contemporary interest.
With this broad coverage of topics, the series is of use to all research scientists and engineers who need up-to-date reference books.

The editors encourage prospective authors to correspond with them in advance of submitting a manuscript. Submission of manuscripts should be made to the Editor-in-Chief or one of the Editors. See also http://www.springer.de/phys/books/optical_science/os.htm

Frédérique de Fornel

Evanescent Waves

From Newtonian Optics to Atomic Optics

With 277 Figures

Dr. Frédérique de Fornel
Groupe Optique de Champ Proche
Laboratoire de Physique de l'Université de Bourgogne
9, avenue A. Savary BP400
21011 Dijon
France
E-mail: ffornel@u-bourgogne.fr

Library of Congress Cataloging-in-Publication Data

Fornel, Frédérique de, 1953-
Evanescent waves : from Newtonian optics to atomic optics / Frédérique de Fornel.
p. cm. -- (Springer series in optical sciences, ISSN 0342-4111 ; 73)
Includes bibliographical references and index.
ISBN 3540658459 (alk. paper)
1. Optics. 2. Electromagnetic waves. 3. Integrated optics. 4. Atoms. I. Title. II. Springer series in optical sciences ; v. 73.

TA1520 .F67 2000
621.36--dc21

00-022166

ISSN 0342-4111

ISBN 3-540-65845-9 Springer-Verlag Berlin Heidelberg New York

Springer-Verlag Berlin Heidelberg New York
a member of BertelsmannSpringer Science+Business Media GmbH

Printed in Germany

Typesetting by the author using a Springer TEX macro package.
Final typesetting and figure processing: LE-TEX Jelonek, Schmidt & Vöckler GbR, 04229 Leipzig.

Cover concept by eStudio Calamar Steinen using a background picture from The Optics Project. Courtesy of John T. Foley, Professor, Department of Physics and Astronomy, Mississippi State University, USA.
Cover production: *design & production* GmbH, Heidelberg

Printed on acid-free paper SPIN 10652964 56/3141/mf 5 4 3 2 1 0

To Agnès, Gilles, Pierre and Daniel

Preface

That an object of physical investigations can be described as evanescent might at first sight seem paradoxical. This is not caused by the ideas of poetry and mystery which are conveyed by the term evanescent, because examples of optical phenomena to which the same terms are applied can be easily found. But evanescent also suggests the idea of something which disappears and fades away. The strangeness of the phenomena to which the term evanescent has been applied is directly related to this character, in the sense that, although something has been generated, it nevertheless escapes direct measurement. Therefore, the possibility that phenomena described as evanescent could be quantified and thus could become an objet of physical investigations was not at all obvious.

The decisive and in many respects seminal step was made by Newton when he recognized that light can transfer through media where nothing seemed to exist, as, for example, near the surface where total internal reflection occurs. But the existence of the evanescent waves, which ultimately are involved in the phenomena observed by Newton, was proven only at the end of the 19th century, when the first quantitative measurements of these waves were realized, this being true at least for large wavelengths. In spite of this, some of the distinctive properties of evanescent waves were not discovered before the middle of the 20th century.

This being said, we would like to briefly elaborate the reasons behind our interest in these waves, and which have led us to writing this book. Of course, the number and richness of the different fields where these waves are involved would have been a sufficient reason for devoting a book to the description of these waves, to the related experiments and to their analysis on the basis of the laws of electromagnetic fields. However, such a book is also justified because of the large variety of realizations where specific properties of these waves are used. Properties of evanescent waves are used in particular for designing new components and for characterizing them, especially in the optics of nanotechnology. The exploitation of evanescent properties has led to impressive progress in areas as different as atom optics and near-field microscopy, and subsequently to realizations which might have hardly been imagined only a few decades earlier. An example of such a realization is the guiding of atoms using the evanescent field of the modes generated by waveguides with certain characteristics.

In microscopy, the use of distinctive properties of evanescent waves has permitted us to break down the limitation established by Lord Rayleigh on the resolution of microscopes. Again, the enforcement of the Rayleigh limitation on the resolution is an achievement which would have been previously unbelievable. To mention just one further example, evanescent waves are utilized in the realization of optical structures with band gaps, which are therefore similar in this respect to crystals.

While the examples just mentioned all date from the last 10 years, the guiding of light through optical fibers, where the evanescent field plays a crucial part, is an area of research which still attracts interest. Until recently, the phenomena of data transfer, whether at optical or at microwave frequencies, were in general analyzed in terms of propagative waves. The presence and the role of evanescent waves remained implicit in the analyses of these phenomena. At the same time, only a few applications of the properties of these waves to the construction of actual instruments had been implemented, with the exception of some spectroscopy techniques and dark-field microscopy.

Presently, due to the constant necessity to increase the transfer capabilities of systems and to enhance the resolution of the instruments designed for characterizing optical devices, it has become necessary to make full use of the potential of evanescent waves. A deeper understanding of the physics of optical devices, and hence a better utilization of these devices, requires an exhaustive analysis of their near-field.

The near-field of an object extends within an area where the distance to the object remains smaller than the wavelength of the light used for the illumination. The near-field presents both an evanescent part and a propagative part. The structure of the far-field of the object is to a large extent determined by the structure of its near-field, and even weak perturbations arising in the near-field region may have significant effects on the field propagated far from the object. This is one of the reasons for the increasing amount of research investigating the near-field of objects, and especially the evanescent field, which is a part of this field. Further, the development of the techniques of miniaturization of optical devices has led to the possibility of constructing devices where the different elements are located within the near-field regions of each part. This imposes a different analysis of systems consisting of several such components. Indeed, a system of this type cannot be analyzed as a mere juxtaposition of n independent elements, but the ensemble chain needs to be analyzed as a whole.

For these reasons, it has become necessary to understand the physics of the near-field of an object. The intent of this book is to describe the near-field associated with different optical systems. I have chosen here to present the near-field through a description of the role of the evanescent field in different areas of research. Even if the near-field could also have been described without so much emphasis on the role of the evanescent field, it seemed to me that the approach presented here has the advantage of providing a better insight into the role of the evanescent field. Further, it gives prominence to the intimate relation which exists between the propagative and evanescent fields. Indeed, these two fields are not physically separated and independent – only the whole of the propagative and evanescent fields is a physical reality.

The description of the near-field which is presented here will therefore begin with a description of evanescent waves. The theoretical study of these waves, which were first observed during the 17th century, has not really been addressed until recent decades. The very name 'evanescent waves' indicates the strangeness of these waves, which remain confined in the vicinity of the object that has generated them.

An analysis of the optical signal emitted by this object does not provide any direct information on these waves. To obtain such information, the evanescent waves have to be perturbed so that a part of the evanescent field can be transformed into a propagative field. This characteristic of evanescent waves is treated at length within the theoretical section of this book, where some of the optical systems which generate evanescent waves are examined. As the evanescent part of the fields plays a significant part in the guiding of modes, a description of optical waveguides has been included here.

The evanescent field can be perturbed in such a way that a transfer of the energy contained in the initial field arises. Evanescent-field couplers provide examples of systems based on the use of a perturbation of this kind. Fiber-optic couplers, as well as integrated-optical couplers are described in Chaps. 4 and 5, respectively.

Even a slight modification of the object which has generated evanescent waves can significantly modify the fields emitted by this object. Different types of sensors are based on this property of evanescent waves. Moreover, the use of optical fibers permits us to fabricate either localized or delocalized sensors. Another application of great importance concerns the possibility of producing spectroscopic measurements by total internal reflection. The spectrum of elements present in very small quantities can be measured with the techniques developed using these principles. These techniques are assembled under the name 'internal reflection spectroscopy'.

A distinctive feature of the evanescent field is the fact that it presents a high spatial pressure gradient. If certain conditions are satisfied, an atom emerging with a low velocity inside this field will be submitted to a pressure which can be used to modify the path of this atom, to make it rebound, for example. This striking property of the evanescent field is being extensively used in the area of atom optics. In particular, it is the principle behind atom mirrors.

If the perturbation of the evanescent field has a localized character, one obtains local information about the object. This phenomenon is used in particular in near-field microscopy. This property of evanescent waves will be discussed in relation to the description of several local-probe microscopes which have been developed in the last decade.

The main intent of this book is to provide an insight into the role of the information contained within the evanescent fields of different types of objects. As such, the different parts of the book have been conceived in such a way that each of them can be read independently. In the first three chapters the reader will find a theoretical analysis of the structure of the near-field of different optical systems. This description is far from being exhaustive, but might be useful in the analysis of instruments based on the use of the evanescent field.

The next two parts are devoted to the utilization of the evanescent field in different areas. Here the applications based on localized interactions have been separated from those based on delocalized interactions, following a criterion of lateral misalignment. We consider a phenomenon as localized if it arises in an area much smaller than the wavelength of the light used for illumination of the system. The application of this criterion allows us to draw a sharp distinction between the localized measurement of a parameter and the average measurement with respect to the wavelength used.

As a consequence of the very nature of the evanescent field, the analysis of this field is at the boundary between different physical approaches. On the other hand, the properties of the evanescent field are being used in a large variety of areas. Therefore, the choice of the subjects treated in this book necessarily has an arbitrary character, and several other topics could have been included here, such as, for example, evanescent-field holography. Likewise, the case of the evanescent aspect of solid lasers has not been addressed here, since cavity modes are similar to optical guided modes.

Acknowledgements. Some years ago, as I was still graduating in physics, P. Facq explained me that total internal reflection was still actively investigated and in particular that research was being carried out on the measurement of the Goos–Hänchen shift. I did not know at this time that I would describe this research in a book devoted to evanescent waves.

I owe a special debt to all those who helped me in acquiring and preserving a certain curiosity, first to my parents, to my husband and to those with whom I have worked. Without them, I would perhaps never have been led to involve myself in research in the field of evanescent waves. My first thanks are to my children and to my husband, and I dedicate this book to them. Their confidence and support have been essential in writing this book.

I am greatly indebted to P.N. Favennec for his constant support and for all the improvements which he suggested. I also want to acknowledge the referees who read and judged this book, and particularly M. Monerie and J.P. Pochole for their constructive and pertinent suggestions. I would like to express my sincere thanks to all the fellow researchers who have kindly sent me articles dealing with the evanescent field, even if, due to a lack of space and time, I have not been able to include here all the information which I received. I am also grateful to all those who took the trouble to send me their constructive suggestions, and in particular to N. Kallas, H. Lotsch and L. Mathey.

I would like to express my gratitude to my son Pierre de Fornel for his active participation in the translation of this book. I would also like to thank H.J. Kölsch for his valuable advice and J. Lenz for her help during the preparation of this book.

This book, although it is not exactly the result of the 15 years of research which I spent at the CNRS, is closely linked with them, and I am grateful to all of those teachers, researchers, technicians and students who trained me and made me discover different sides of physics. During that time, I had the pleasure to be part of three different laboratories: the IRCOM directed by Y. Garault, the Electronics Laboratory of Southampton University directed by A. Gambling, and the Physics Laboratory of the University of Burgundy directed by H. Berger. I would like in particular to acknowledge J. Arnaud, G. Boutinaud, B. Colombeau, P. Dawson, J.P. Dufour, P. Facq, C. Froehly, A. Hartog, J.D. Love, D. Pagnoux, D.N. Payne, C. Ragdale, A. Rahmani, M. Remoissenet, L. Salomon, P. Teyssier, M. Vampouille and P. Vernier. Of course, this enumeration does not intend to be complete, and I would like to express my gratitude to all those who are not mentioned here.

Dijon, October 2000 *Frédérique de Fornel*

Contents

Symbols and Definitions of Abbreviations Used

AFM	atomic force microscope
BPM	beam propagation method
FTR	frustrated total reflection
LP mode	linearly polarized mode
PSTM	photon scanning tunneling microscope
PTM	photon tunneling microscope
SNOM	scanning near field optical microscope
STM	scanning tunneling microscope
TE mode	transverse electric mode
TIR	total internal reflection
TM mode	transverse magnetic mode

Part I

The Evanescent Field

Introduction to Part I

These first three chapters have essentially a theoretical character and are intended to present the basic theory required for analyzing devices based on the use of properties of the evanescent field. Indeed, before examining the physics of different instruments where the evanescent field of specific elements is involved, it is necessary first to analyze some simple systems which generate evanescent waves. These configurations will be met again in the description of devices like sensors, power splitters or local probe optical microscopes. A few examples of optical systems generating evanescent waves are graphically represented here (Fig. I.1).

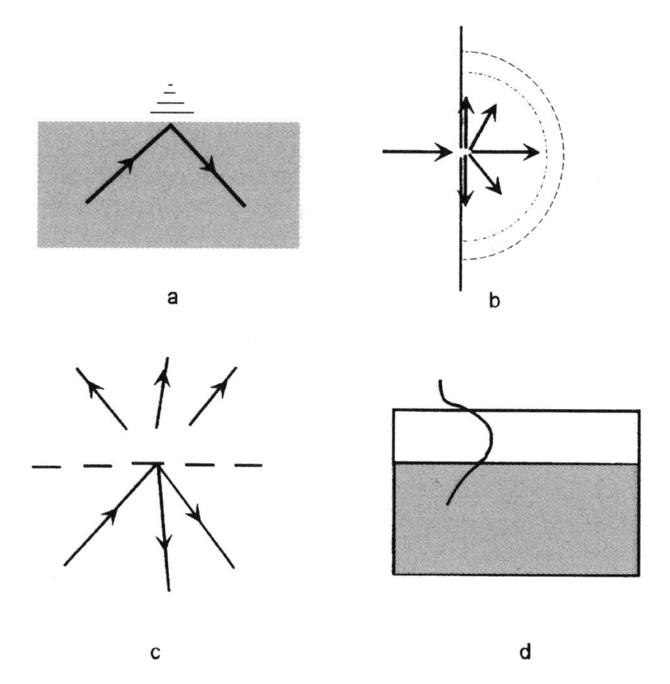

Fig. I.1. Some systems generating an evanescent field (**a**) total internal reflection of a plane wave, (**b**) diffraction of a beam from an aperture, (**c**) diffraction from a grating, (**d**) evanescent field of a guided mode

We address in Chap. 1 the phenomenon of total internal reflection. Even if the discovery of this phenomenon dates back to Newton, it may be useful to recall here the distribution of the field at total internal reflection. The different types of shifts arising at total internal reflection are also described. We then examine the phenomenon of the frustration of a totally reflected wave. Near the end of the first chapter, a brief analysis of the optical tunneling effect is provided, in relation to a description of experimental results on superlattices.

Chapter 2 is concerned with the diffraction of a plane wave from an aperture. In view of the rise of local-probe microscopies in recent years, it seemed to me necessary to review here the analysis of the field in the vicinity of a subwavelength aperture. We first examine the ideal case of an aperture cut in a perfectly thin plane, and then the case of an aperture in a screen with nonzero thickness. The coupling of several apertures is also examined in this chapter.

The physics of a dipole can be regarded as a subject complementary to the diffraction of a plane wave from an aperture. Since a dipole is never isolated in space, we have incorporated in this chapter a description of the effect of the presence of a plane in the vicinity of a dipole. An analysis of the evanescent and propagative fields generated by this dipole is also presented.

The evanescent field has crucial importance in optoelectronics, and so this presentation would not have been complete without a description of the evanescent field of modes of planar waveguides and optical fibers. This is given in Chap. 3. The propagation of light in waveguides is described on the basis of Maxwell's equations and of ray theory. We end this chapter with a description of waveguides presenting distinctive characteristics, like annular waveguides or polarization-preserving bow-tie fibers, where the evanescent field plays an important part. The properties of these waveguides associated with the evanescent field of their modes have led to their utilization in several different applications. Some of these applications are described in later chapters.

These three cases are treated in distinct chapters, each of which can be read separately. The intent of these chapters is to review the basic theory necessary to the analysis of the physics of the optical systems described in the following chapters. Nevertheless, the final two parts of this book, devoted to applications of properties of the evanescent field in different areas, can be read independently.

1. Total Internal Reflection

The intent of this chapter is first to review the conditions of total internal reflection and to provide a description of the associated electromagnetic field. Within this description, the notion of the penetration depth of the evanescent field will be introduced.

When a light beam with a limited extent is totally reflected, a displacement of the beam arises in the plane of the interface where it is reflected. A part of the chapter is therefore devoted to a description and to an analysis of this displacement, referred to under the name of 'Goos–Hänchen shift'.

Near the end of the chapter, we address in detail the phenomenon of the frustration of total internal reflection. The frustration of a totally reflected plane wave from a semiinfinite medium is described and the related experimental facts are displayed. Presently, the structures known as superlattices have a wide range of applications. When they are illuminated under predetermined conditions, these structures present a resonant tunneling effect, which is described at the end of the chapter.

1.1 The Electromagnetic Field at Total Internal Reflection

1.1.1 Snell's Law

Let us consider two media with refractive indices n_1 and n_2 respectively. A plane wave strikes the interface between the two media at an angle of incidence θ_1. If the value of the refractive index of the second medium is higher than that of the refractive index of the first medium, the beam is refracted. In other words, the beam is partially reflected and partially transmitted in the form of plane waves. The directions of propagation of these waves are given by Snell's law

$$n_1 \sin\theta_1 = n_2 \sin\theta_2, \tag{1.1}$$

where θ_2 is the angle of the direction of propagation formed by the refracted beam with the normal to the surface.

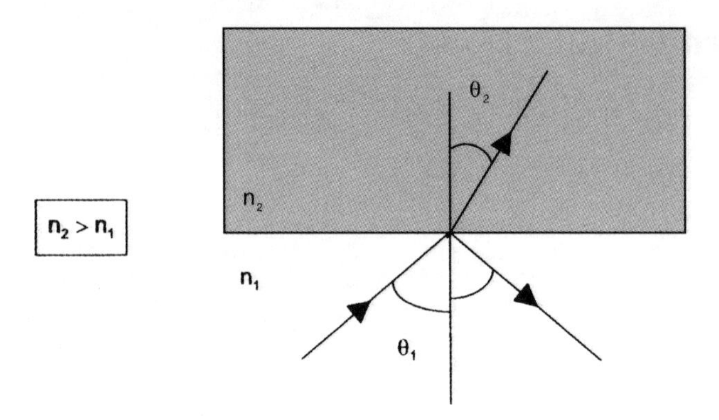

Fig. 1.1. Refraction of a wave from a denser medium towards a rarer medium

When the light beam associated with the plane wave travels from the denser medium into the rarer medium, a reduction of the propagative angle of the beam with respect to the normal to the surface arises (Fig. 1.1).

Let us now examine the inverse case, which is schematized in Fig. 1.2. If the value of the incidence angle is smaller than $\theta_1 = \sin^{-1}(n_2/n_1)$, the angle of the refracted ray can be determined from Snell's law. The value $\theta_1 = \sin^{-1}(n_2/n_1)$ of the incidence angle is referred to either as the 'boundary angle of refraction' or as the 'critical angle', and is usually denoted by θ_c.

If the incidence angle exceeds the value of the critical angle, the light can no longer propagate within the second medium, and is therefore totally reflected. In spite of this, as will be seen later, there are still waves present within the second medium: these waves are referred to as 'evanescent waves'.

The phenomenon of the total internal reflection of light rays was first recognized by Newton. He observed that when two dioptres are moved closer together, and as long as the distance between them remains small, a part

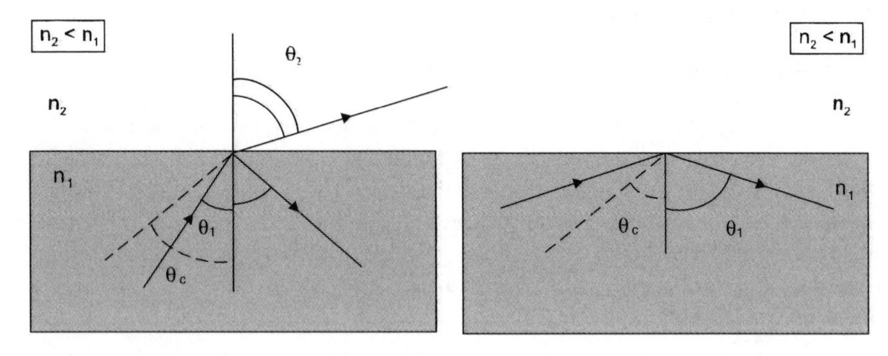

Fig. 1.2. Refraction of a wave from a rarer medium into a denser medium (**a**) $\theta_c > \theta_1$, (**b**) $\theta_c < \theta_1$

of the light can penetrate into the rarer medium before contact is made and travel some distance inside it before re-emerging in the denser medium [Newton 1952].

Quincke in 1966 and Bose in 1897 carried out more precise experiments providing evidence for the dependence of the light transfer into the second medium on the incidence angle as well as on the wavelength of the source [Quincke 1966a, Quincke 1966b, Bose 1897, Hall 1902]. Nevertheless, the full analysis of this phenomenon requires the use of the formalism of electromagnetism. The experiment carried out by Newton will be described in Sect. 1.4. We first examine the phenomenon of total internal reflection as it arises in the simple case of a two-media system.

1.1.2 Analysis of Total Internal Reflection on the Basis of Maxwell's Equations

Let us return to the previous two-media system. In order to analyze this system within the formalism of electromagnetism, it is convenient to assume that an electromagnetic wave reaches the interface between these media.

We hereafter use the following coordinate system for discussing this model: z is to be directed from the more refractive medium towards the less refractive medium, while x and y lie on the interface between the two media (Fig. 1.3). The $y = 0$ plane is the plane of incidence.

We consider the case of total internal reflection, where the incidence angle θ is greater than $\sin^{-1}(n_2/n_1)$. The incident, reflected and transmitted wave vectors are referred to respectively as $\mathbf{K}_\mathrm{i}$, $\mathbf{K}_\mathrm{r}$ and $\mathbf{K}_x$.

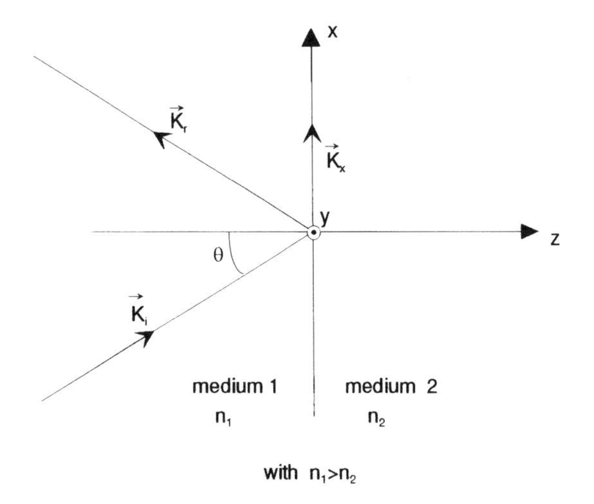

Fig. 1.3. Geometry of the system

Let us consider an incident plane wave whose electric and magnetic fields are referred to as $\mathbf{E}^{\mathrm{i}}$ and $\mathbf{H}^{\mathrm{i}}$ respectively

$$\mathbf{E}^{\mathrm{i}} = \begin{pmatrix} E_x^{\mathrm{i}} \\ E_y^{\mathrm{i}} \\ E_z^{\mathrm{i}} \end{pmatrix} \text{ and } \mathbf{H}^{\mathrm{i}} = \begin{pmatrix} H_x^{\mathrm{i}} \\ H_y^{\mathrm{i}} \\ H_z^{\mathrm{i}} \end{pmatrix}. \tag{1.2}$$

Under these conditions, the field is in general represented in terms of two fields with distinct directions: the p polarized, i.e. parallel to the incidence plane, and s polarized, i.e. normal (in German *senkrecht*) to the incidence plane, electric fields. These fields are expressed as follows

$$\mathbf{E}_{\mathrm{p}} = E_x \mathbf{e}_x + E_z \mathbf{e}_z \text{ and } \mathbf{E}_{\mathrm{s}} = E_y \mathbf{e}_y. \tag{1.3}$$

The p and s polarizations are also referred to as transverse electric (TE) and transverse magnetic (TM) polarizations, respectively.

1.1.3 Components of the Electric Field in the Second Medium in the $z = 0$ Plane

By applying the continuity equations for the fields at the interface to the configuration of two semiinfinite distinct media, we obtain an expression of the field in the case where an incident plane wave is present within the first medium of amplitudes $E_{\mathrm{s}}^{\mathrm{i}}$ and $E_{\mathrm{p}}^{\mathrm{i}}$ contained within the $z = 0$ plane [Axelrod 1992]. In the following equations, we have left out the x dependence and the time dependence, which are of the form $\exp(-\mathrm{j}xn_1(\omega/c)\sin\theta)$ and $\exp(\mathrm{j}\omega t)$, respectively with $\mathrm{j} = (-1)^{1/2}$

$$E_x = \frac{(2\cos\theta)(\sin^2\theta - n^2)^{1/2}}{(n^4\cos^2\theta + \sin^2\theta - n^2)} E_{\mathrm{p}}^{\mathrm{i}} \exp(-\mathrm{j}(\delta_{\mathrm{p}} + \pi/2)), \tag{1.4}$$

$$E_y = \frac{2\cos\theta}{(1 - n^2)^{1/2}} E_{\mathrm{s}}^{\mathrm{i}} \exp(-\mathrm{j}\delta_{\mathrm{s}}), \tag{1.5}$$

$$E_z = \frac{2\cos\theta\sin^2\theta}{(n^4\cos^2\theta + \sin^2\theta - n^2)} E_{\mathrm{p}}^{\mathrm{i}} \exp(-\mathrm{j}\delta_{\mathrm{p}}), \tag{1.6}$$

where $n = n_2/n_1$, while δ_{p} and δ_{s} are solutions of the equations

$$\tan\delta_{\mathrm{p}} = \frac{(\sin^2\theta - n^2)^{1/2}}{n^2\cos\theta} \tag{1.7}$$

and

$$\tan\delta_{\mathrm{s}} = \frac{(\sin^2\theta - n^2)^{1/2}}{\cos\theta}. \tag{1.8}$$

As can be seen in Fig. 1.4, the intensity of the electric field at $z = 0$ depends on the value of the incidence angle. Further, the maximal value for the intensity is reached at a value of the incidence angle equal to the critical angle θ_{c}.

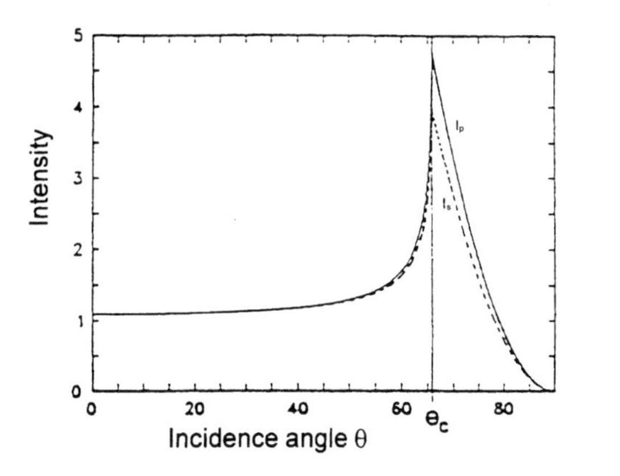

Fig. 1.4. Electric field intensity at the interface of the media as a function of the incidence angle θ, in p and s polarizations respectively. $n_1 = 1.46$ and $n_2 = 1.33$

The z dependence of the evanescent field can be expressed as

$$\mathbf{E}_{\mathrm{p}}(z) = E_{\mathrm{p}}^{\mathrm{i}} \frac{(2\cos\theta)\exp(-z/d_{\mathrm{p}})}{n^2\cos\theta + \mathrm{j}(\sin^2\theta - n^2)^{1/2}} [-\mathrm{j}(\sin^2\theta - n^2)^{1/2}\mathbf{e}_x + \sin\theta\mathbf{e}_z], \quad (1.9)$$

in p polarization, and as

$$\mathbf{E}_{\mathrm{s}}(z) = E_{\mathrm{s}}^{\mathrm{i}} \frac{(2\cos\theta)\exp(-z/d_{\mathrm{p}})}{\cos\theta + \mathrm{j}(\sin^2\theta - n^2)^{1/2}} \mathbf{e}_y = E_{\mathrm{s}}^{t}\mathbf{e}_y, \quad (1.10)$$

in s polarization. In the equation above, E_s^t corresponds to the transmitted field in s polarization.

The amplitude of the electric field decreases exponentially as the distance from the interface increases. The dependence of the amplitude of the field on the distance is represented graphically in Fig. 1.5.

The parameter denoted by d_{p} is referred to as the penetration depth of the evanescent field. The value of d_{p} reflects the decrease of the evanescent field amplitude when the distance to the interface increases. Hence it expresses the confinement of the evanescent field in the vicinity of the interface. The penetration depth d_{p} of the evanescent field is related to the refractive indices of the two media, to the wavelength and to the incidence angle by the equation

$$d_{\mathrm{p}} = \frac{\lambda}{2\pi\sqrt{n_1^2\sin^2\theta - n_2^2}}. \quad (1.11)$$

The penetration depth d_{p} goes from infinity to $\lambda/2\pi\sqrt{n_1^2 - n_2^2}$ as the incidence angle extends from θ_{c} to $\pi/2$. For understanding the confinement of the evanescent field, a few values of the penetration depth for different media illuminated at different angles are reported in Table 1.1.

It is apparent from the values reported here that only a limited part of the second medium actually ‘sees’ the field. We have already seen that the

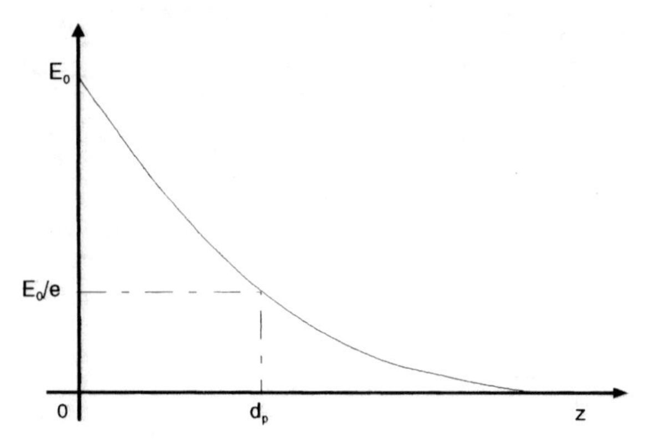

Fig. 1.5. Variation of the amplitude of the evanescent field in the second medium. E_0 is the amplitude value in the $z = 0$ plane and d_p corresponds to the distance where the amplitude is divided by e

amplitude of the field could be very important. Hence, the phenomenon of total internal reflection presents a high sensitivity to the physical state of the interface between the two media.

The fact may be emphasized that when the incident field is s polarized the evanescent field is purely transverse to the direction of propagation. In p polarization, the electric field has two nonzero components with a phase difference equal to $\pi/2$. The extremity of this vector describes an ellipse as time evolves.

The equations for the magnetic components of the evanescent field can be determined in exactly in the same way as for the electric fields. We thus obtain the following equations

$$H_x = \frac{n_1(2\cos\theta)(\sin^2\theta - n^2)^{1/2}}{c\mu_0(1-n^2)^{1/2}} E_\mathrm{s}^\mathrm{i} \exp[-\mathrm{j}(\delta_\mathrm{s} - \pi/2)], \tag{1.12}$$

Table 1.1. Values of the penetration depth of the evanescent field in different configurations.

Wavelength (nm)	Index of the first medium	Index of the second medium	Critical angle (degrees)	Incidence angle θ (degrees)	d_p (nm)
1300	Glass : 1.458	Air : 1	43.3	45	825
1300	Silicium : 3.430	Air : 1	16.9	45	94
633	Glass : 1.458	Air : 1	43.3	45	402
633	Glass : 1.458	Air : 1	43.3	85	96
633	Glass : 1.458	Water : 1.33	65.8	85	173
414	Glass : 1.458	Air : 1	46.3	85	63

$$H_y = \frac{2n_1 n^2 \cos\theta}{c\mu_0(n^4\cos^2\theta + \sin^2\theta - n^2)^{1/2}} E_\mathrm{p}^\mathrm{i} \exp[-\mathrm{j}(\delta_\mathrm{p} - \pi/2)], \tag{1.13}$$

$$H_z = \frac{2n_1 \cos\theta \sin\theta}{c\mu_0(1 - n^2)^{1/2}} E_\mathrm{s}^\mathrm{i} \exp[-\mathrm{j}\delta_\mathrm{s}]. \tag{1.14}$$

In incident p polarization, the magnetic field remains transverse, while the electric component of the field undergoes an elliptic polarization. Conversely, in s polarization, the magnetic field is elliptically polarized.

A rectilinear polarized wave can always be decomposed in terms of the two polarization modes [Imbert 1975, Huard 1997]. The following diagram (Fig. 1.6) summarizes the situation.

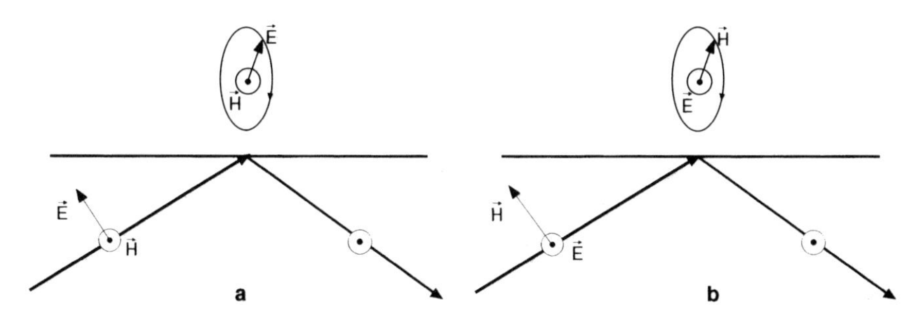

Fig. 1.6. Polarization of the evanescent field at total internal reflection (**a**) in TM polarization, (**b**) in TE polarization

The characterization of an electromagnetic wave requires that the energy it conveys has been determined. In fact, as will be seen immediately after, an evanescent wave does not on average transport any energy into the second medium if the conditions of total internal reflection are fulfilled.

1.2 Flux of the Poynting Vector Associated with the Evanescent Field

The expression of the Poynting vector $\mathbf{P}$, where $\mathbf{E}$ and $\mathbf{H}$ are real values of the complex fields previously described, is

$$\mathbf{P} = \mathbf{E} \wedge \mathbf{H}. \tag{1.15}$$

The expression of the Poynting vector transmitted in the second medium indicates that the y component is always equal to 0. This means that the energy flux along this direction is always zero.

As an example, we shall examine in detail what happens in TE polarization

$$\mathbf{P}_{\mathrm{s}}^{t} = \begin{cases} P_{\mathrm{s}x}^{t} = \dfrac{1}{\mu_0}\dfrac{n_1}{c} E_{\mathrm{s}}^{t\,2} \sin\theta_i \cos^2\left(\omega t - x\dfrac{n_1\omega}{c}\sin\theta + \delta_{\mathrm{s}}\right), \\ P_{\mathrm{s}y}^{t} = 0, \\ P_{\mathrm{s}z}^{t} = \dfrac{1}{\mu_0}\dfrac{n_1}{c}(\sin^2\theta - n^2)^{1/2} E_{\mathrm{s}}^{t\,2} \cos^2\left(\omega t - x\dfrac{n_1\omega}{c}\sin\theta + \delta_{\mathrm{s}}\right). \end{cases} \tag{1.16}$$

Proceeding to the solution of these equations, it turns out that the average flux through the interface over a period is zero. Over a half-period, the average flux is also zero. In contrast, over a quarter-period, the average flux $(P_{\mathrm{s}z}^{t})_{\mathrm{a}}$ has nonzero values

$$\begin{aligned} (P_{\mathrm{s}z}^{t})_{\mathrm{a}} &= \frac{1}{4}\int_0^{T/4} P_{\mathrm{s}z}^{t} \\ &= -E_{\mathrm{s}}^{t\,2}\frac{1}{\mu_0}\frac{n_1}{c}(\sin^2\theta - n^2)^{1/2}\cos^2\left(\omega t - x\frac{n_1\omega}{c}\sin\theta + \delta_{\mathrm{s}}\right). \end{aligned} \tag{1.17}$$

At the next quarter-period, the signs in the equation are reversed. Hence, even if a wave is present in the second medium, the net energy flux through the interface on average is zero. In contrast, there is actually a nonzero net energy flux in the O–x direction.

From these results, it can be seen that the light intensity associated with the waves involved in total internal reflection does not completely characterize evanescent waves. In fact, intrinsic properties of these waves were effectively measured only during this century. In particular, a distinctive characteristic of evanescent waves is the shift of the light beam reflected after total internal reflection.

1.3 Shifts of the Beams at Total Internal Reflection

Let us examine the case of a laterally limited beam totally reflecting at the interface of two distinct media. In this section, we shall first recall that the phenomenon of total internal reflection is accompanied by two distinct shifts, respectively lateral and longitudinal, as illustrated in Fig. 1.7.

The existence of a longitudinal shift had already been predicted by Newton. In 1947, F. Goos and H. Hänchen carried out the first experiments demonstrating the existence of this shift. Since that time, the problem of the measurement of the Goos–Hänchen shift has been extensively studied. The reader interested by this subject will find detailed accounts of these studies in the articles of Lotsch and Imbert [Lotsch 1970a, Lotsch 1970b,

Lotsch 1971a, Lotsch 1971b, Imbert 1975]. Among the theoretical analyses of this phenomenon, the analyses developed by Artmann and Picht, based on a stationary-phase argument, and by Renard, based on an energy-conservation argument, can be mentioned.

The longitudinal shift corresponds to linearly polarized waves, either in p or in s polarization, or in any combination of these two polarizations. As demonstrated by Imbert, when the incident wave is circularly polarized, a lateral shift of the reflected beam arises (Fig. 1.7)[Imbert 1972)].

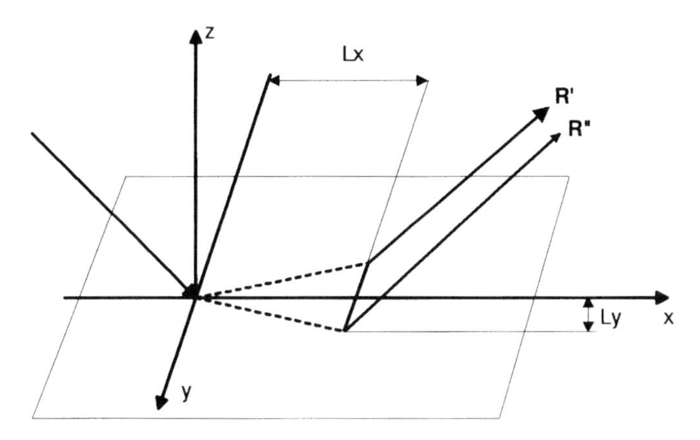

Fig. 1.7. Schematic representation of the two shifts: L_x is the longitudinal shift, L_y the transverse shift

The existence of a lateral shift, and the physics underlying this phenomenon, were foreseen by Fedorov as early as 1955 and experimentally proven by Imbert. The theoretical analysis of the shift can be based on a stationary-phase argument, on a energy-conservation argument or on a description of a beam laterally limited by a superposition of plane waves [Imbert 1975]. Finally, this shift can be viewed as a manifestation of the inertial effect of the proton spin [Costa de Beauregard 1965].

From the principle of energy conservation, L_x and L_y can be written in the following form [Imbert 1975]

$$L_x = \frac{1}{P_z^{\mathrm{r}}} \int_{-\infty}^{0} P_x^{\mathrm{t}} \mathrm{d}z, \tag{1.18}$$

$$L_y = \frac{1}{P_z^{\mathrm{r}}} \int_{-\infty}^{0} P_y^{\mathrm{t}} \mathrm{d}z, \tag{1.19}$$

where P_z^{r} is the z component of the Poynting vector of the totally reflected wave, and P_x^{t} and P_y^{t} are the x and y components of the Poynting vector of the evanescent wave. We thus obtain the following equations

$$L_x^{\mathrm{p}} = K_1^{\mathrm{p}} | E_{\mathrm{p}}^{\mathrm{t}} |^2 \text{ and } L_x^{\mathrm{s}} = K_1^{\mathrm{s}} | E_{\mathrm{s}}^{\mathrm{t}} |^2, \tag{1.20}$$

$$L_y = K_2 \left[E_{\mathrm{p}}^{\mathrm{t}} E_{\mathrm{s}}^{\mathrm{t}*} - E_{\mathrm{s}}^{\mathrm{t}} E_{\mathrm{p}}^{\mathrm{t}*} \right]. \tag{1.21}$$

The experimental arrangement presented in Fig. 1.8 allows a measurement of the longitudinal shift L_x. The shift is measured by comparing the position of the beam reflected by the metallized surface with the position of the totally reflected beam.

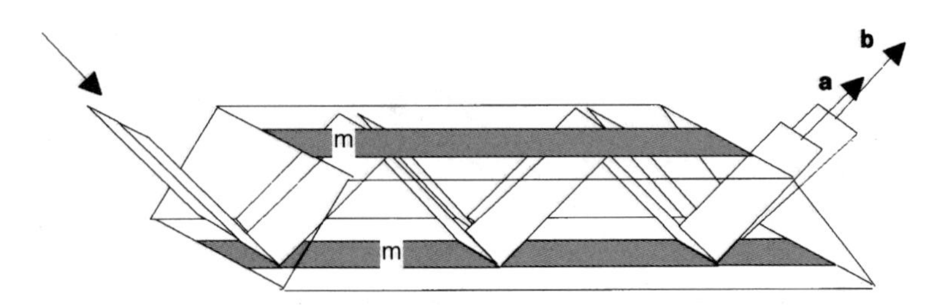

Fig. 1.8. Experimental arrangement used for the measurement of the Goos–Hänchen shift

A prism is partially coated with a metallic film m where the reflection of the beam is, unlike total internal reflection, not accompanied by a shift. After several successive reflections, the shift between parts a and b of the beam becomes detectable.

Equations (1.18) and (1.19) express the dependence of the shift on the polarization. This dependence has been exploited in an experiment reported by Imbert *et al.* for realizing polarization splitters.

The image A′B′ of a metallic wire AB is formed upon a screen E, the source being an unpolarized laser, as represented in Fig. 1.10. Figure 1.11. shows that the image of AB appears in the form of two lines $\mathrm{A}'_{||}\mathrm{B}'_{||}$ and $\mathrm{A}'_{\perp}\mathrm{B}'_{\perp}$, which respectively correspond to the *p*- and *s*-polarized beams. In order to demonstrate this correspondence, an analyzer was placed behind the plate. The rotation of the plate makes each image alternately disappear, these disappearances corresponding to the positions of the analyzer parallel and perpendicular to the incidence plane. Similarly, by polarizing the incident beam, either of the two images can be produced, depending on the direction of the incidence field (parallel or perpendicular to the incidence plane).

Since the L_y shift obtained after a single reflection is very small, it is necessary to amplify it through multiple reflections. The arrangement used for this purpose is represented in Figs. 1.11 and 1.12.

A $\lambda/2$ plate placed across a half of the beam reverses the polarization of the two half-images. By placing a circular light analyzer, consisting of

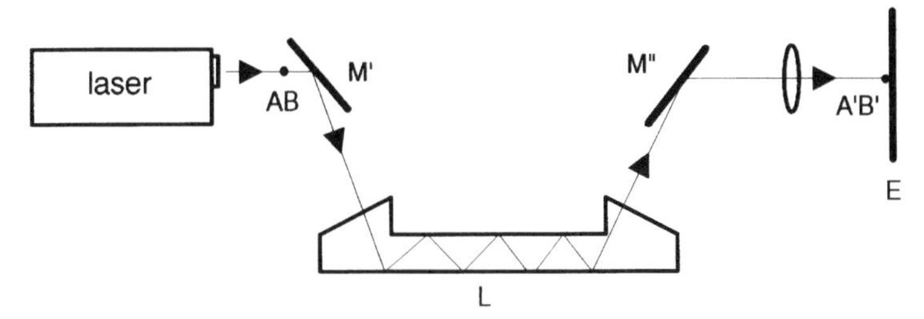

Fig. 1.9. Experimental arrangement, described by Imbert and Lévy, used for carrying out multiple total internal reflections and observing a large L_x shift. A non-polarized laser beam strikes a wire AB perpendicular to the incidence plane defined by the laser beam and the plate L. After 31 multiple total internal reflections, the image of AB is formed upon the screen E [Imbert 1975]

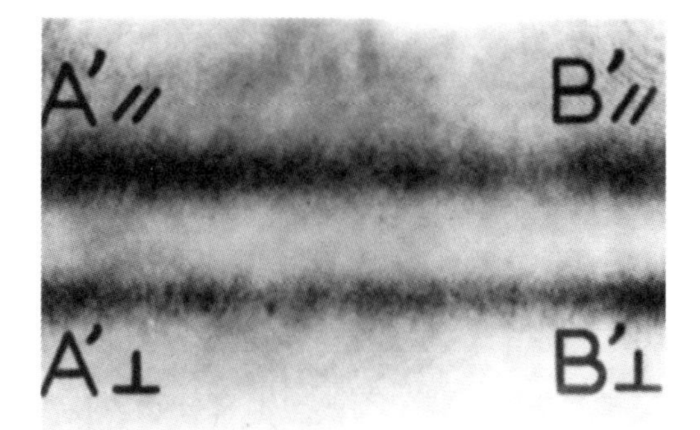

Fig. 1.10. Photograph of the two images displaying the filtering of the rectilinear polarization states parallel and perpendicular to the incidence plane [Imbert 1975]

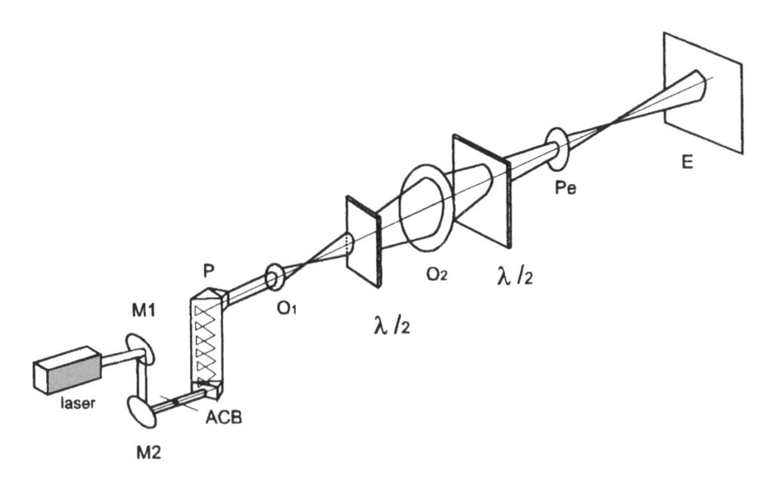

Fig. 1.11. Experimental arrangement used for observing the L_y shift

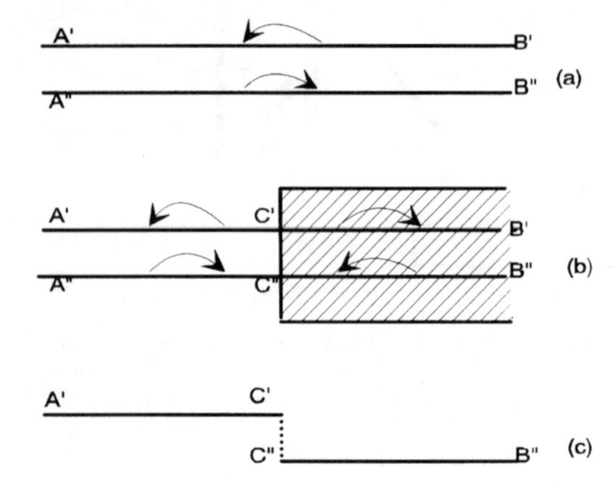

Fig. 1.12. A $\lambda/2$ plate placed across a half of the beam reverses the polarization of the two half-images C′B′ and C″B″. The result is schematized by (**c**) for a position of the Pe polarizer

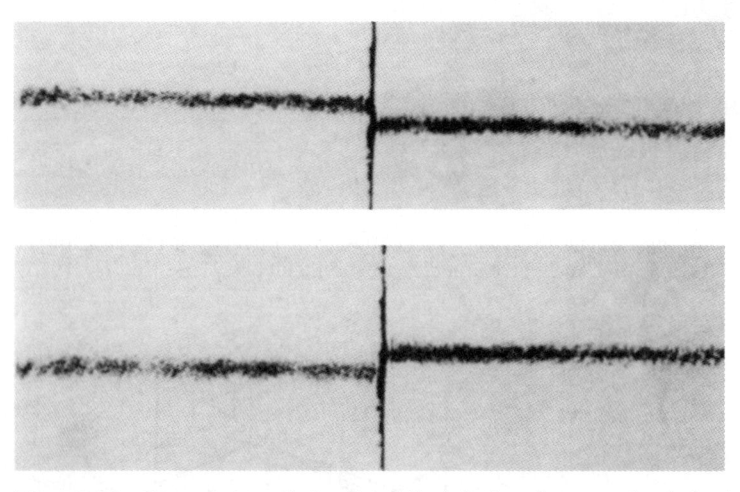

Fig. 1.13. Observation of the L_y shift (**a**) the photograph of the image on the screen E corresponds to the case illustrated in Fig. 1.12, (**b**) this photograph was obtained by turning the Pe polarizer through an angle of 90° [Imbert 1975]

a quarter-wave plate and of a Pe polarizer, the L_y shift can be directly observed. The results obtained with this arrangement are presented in Fig. 1.13.

The longitudinal shift can be amplified by making the incident beam reflect on a prism coated with several thin layers. The arrangement used to this end is illustrated in Fig. 1.14.

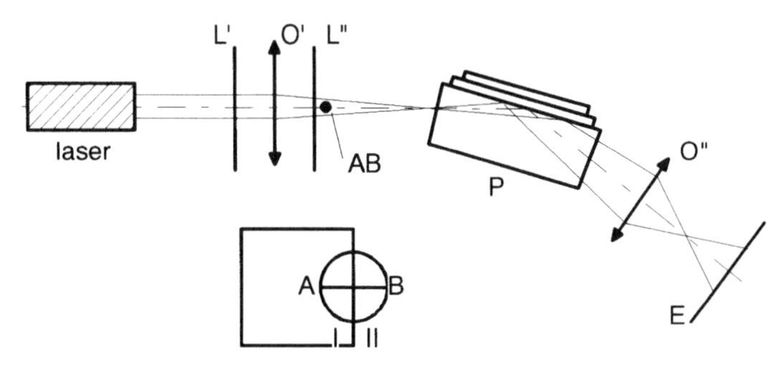

Fig. 1.14. Experimental arrangement used for amplifying the L_y shift [Imbert 1975]

We shall not examine in detail here all the measurements reported by Imbert [1975]. The remarkable images of the shifts obtained show that with convenient arrangements it is possible to observe shifts of the order of about ten microns. In his article, Imbert stresses the fact that the measurement of one of the shifts prevents the measurement of the other.

Other measurements of the Goos–Hänchen shift have been achieved in the following years. As an example, we may mention here results presented by Bretenaker. A prism is inserted within a laser cavity (Fig. 1.15) [Bretenaker 1992]. For measuring the longitudinal Goos–Hänchen shift, the optical signal is modulated from the displacements of a knife which acts as an obstacle, these displacements being controlled from a piezo-electric device. This induces a modulated signal, which directly depends on the polarization of the beam. Figure 1.16 shows that these results are consistent with the values derived from Artmann's formulas [Huard 1997].

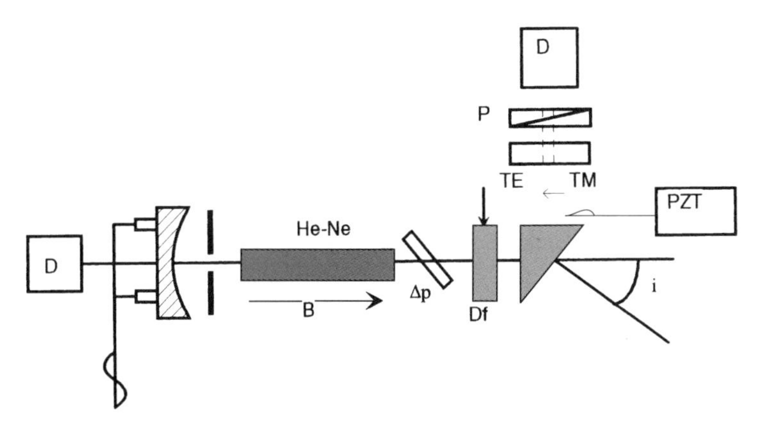

Fig. 1.15. Experimental arrangement used for measuring the difference between the Goos–Hänchen shifts in TE and TM modes inside a laser cavity [Bretenaker 1992]

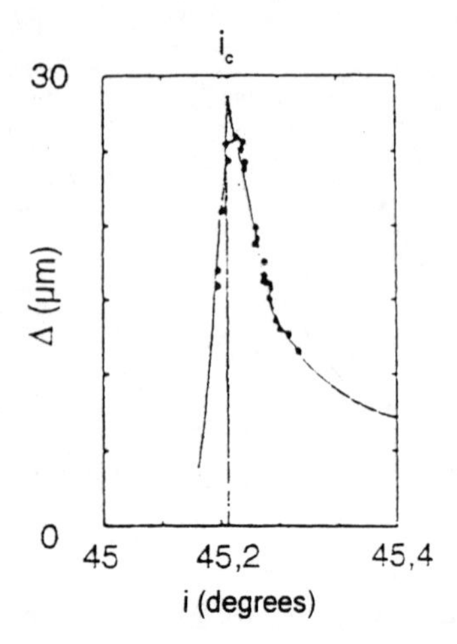

Fig. 1.16. Difference between the shifts in TE and TM polarizations as a function of the incidence angle of the Gaussian beam [Bretenaker 1992]

The Goos–Hänchen shift is still an object of researches, directed either at a theoretical understanding of the physics of this phenomenon or at the collection of experimental results [Kallas 1997]. The very nature of evanescent waves renders the determination of their properties difficult. Whereas the Goos–Hänchen shift can be measured without having to perturb total internal reflection, a more complete understanding of evanescent waves is likely to require measurements that might modify the system where these waves have originated. In order to prevent a modification of the nature of the evanescent waves, the perturbation induced must be as slight as possible. The case of the interaction of an electron or an atom placed in the vicinity of the interface, described by Vigoureux, will not be examined here, and we shall content us with describing the interaction of the evanescent wave with a semiinfinite medium [Vigoureux 1974, Vigoureux 1975a].

1.4 Frustrated Total Internal Reflection

The average energy transported by the evanescent field is zero. Therefore, for determining the value of the field at a given time, it is necessary to perturb the system in such a way that a part of the evanescent wave will be transformed into a propagative wave, which, unlike evanescent waves, can be detected. This basically was what Newton achieved in the well-known experiment schematized in Fig. 1.17. This experiment consisted in placing a prism against a lens with a very large radius of curvature. Newton observed that the light intensity transferred onto the lens was located on an area of the lens larger than the point of contact.

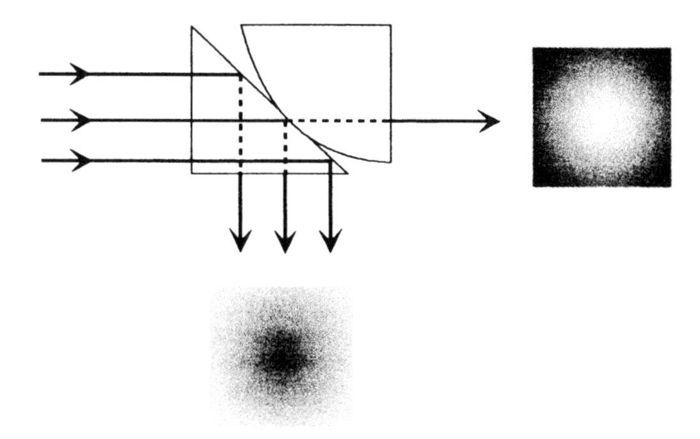

Fig. 1.17. Scheme of Newton's experiment. A prism is illuminated in total internal reflection and a lens with a large radius of curvature is brought into contact with the prism. Along the axis of the incident beam, a luminous area larger than the point of contact is observed

This experiment suggests that a light transfer can arise from the first medium into the second medium, even if the surfaces of the lens and of the prism are not in contact, but provided that the distance between them is very small, namely far below the value of the half-wavelength of the source.

In order to examine the transfer of light onto the lens, we consider a simple model, consisting of two semiinfinite media separated by an air gap with variable thickness. This three-media model is represented in Fig. 1.18.

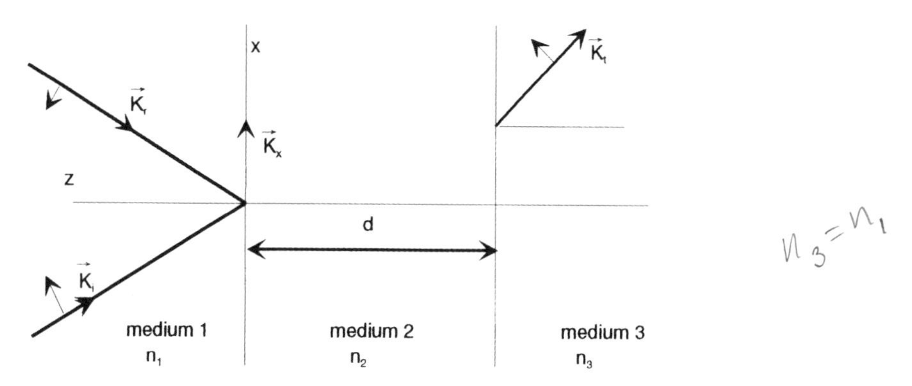

Fig. 1.18. Model with three media of refractive indices n_1, n_2 and n_3 respectively

In this model, the third medium has the same refractive index as the first medium. It will be assumed here that the wave re-emerges in the third medium with the same direction of propagation as the incident wave. The presence of this third medium requires us to express the field within the intermediate region in terms of two exponential functions, respectively increasing and decreasing [Salomon 1991b].

The equation for the amplitude of the field at the interface of the second and third media is therefore

$$E_3 = \frac{\exp\left(\mathrm{j}n_1(\omega/c)d\cos i\right)}{\cosh Kd + \mathrm{j}\sinh Kd\cot 2\varphi}E_i. \tag{1.22}$$

The field intensity at the interface between air and the third medium depends on the wavelength of the incident wave, on the incidence angle i and on the thickness of the intermediate medium. E_i here denotes the amplitude of the incident electric field. Unlike the dependence of the evanescent field on the distance between the two outer media, the dependence of the intensity of the field on this distance is not exponential (Fig. 1.19). If this variation is plotted as a function of the distance between the two media, one obtains a curve which decreases exponentially only for large distances between the two media, namely, when the distance between them is greater than the wavelength. Further, the slope at the origin is zero.

The case where the two media are sufficiently far from each other, and where the amplitude of the field decreases exponentially, is referred to as the low-coupling regime. The similarity of the field variation in this case with the variation of the evanescent field in the absence of a third medium, reflects the fact that the third medium does not, or at least not much, perturb the total internal reflection.

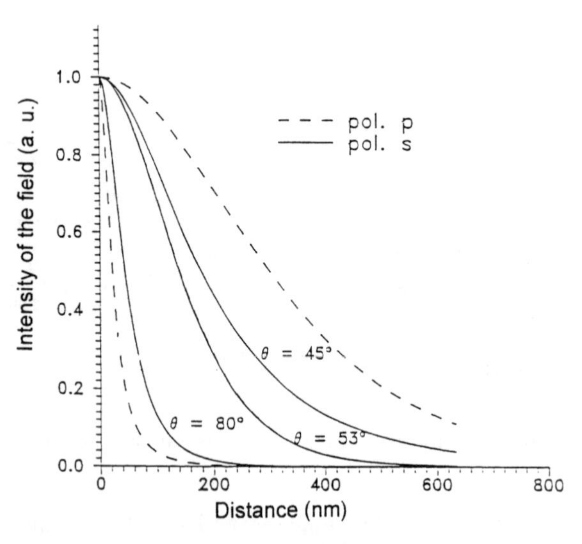

Fig. 1.19. Intensity of the field in the third medium as a function of the distance between the outer media [Salomon 1991]

The observation of the phenomenon of the frustration of evanescent waves in the field of centimetric waves does not encounter any serious difficulty, because the necessary displacements of the elements of the experimental arrangement can be mechanically controlled [Albiol 1993]. The first experimental evidence of the frustration of the evanescent wave by a third medium was achieved by Bose at the end of the nineteenth century. These experiments were carried out with one of the first available sources of centimetric waves [Bose 1897].

In optics, the displacements of the elements in the experimental arrangement are to be carried out at a micrometric scale. Therefore, experiments on the frustration of the evanescent wave from a third medium are much more recent in this field than for centimetric frequencies [Zhu 1986].

Bose demonstrated that the transmitted energy decreases very rapidly when two prisms are very far from each other. He also recognized the fact that, from a certain distance, the presence of the second prism does not perturb the total internal reflection. Further, this distance increases with the wavelength. These discerning remarks can be easily verified with the instrumentation presently available.

As an example, we present one of the first curves of the intensity frustrated by the second prism, in the case of illumination by a visible source with wavelength $\lambda = 546.1$ nm [Zhu 1986]. Zhu *et al.* report that, with the

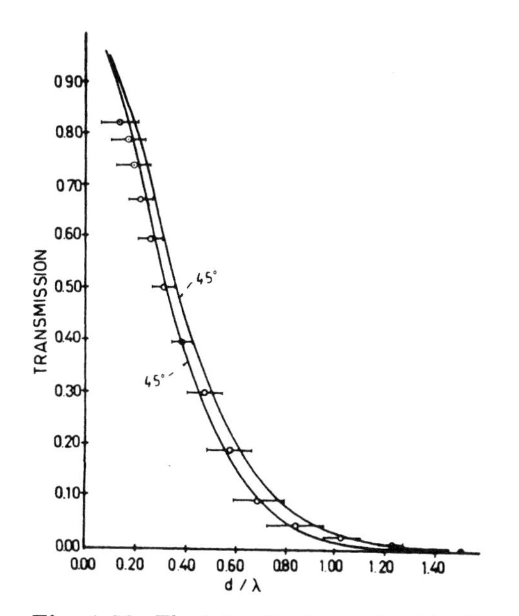

Fig. 1.20. The intensity transmitted in the second medium as a function of the distance between the two prisms (*dotted lines*): comparison with a theoretical model calculated for two incidence angles (*represented by the two solid curves*) [Zhu 1986]

exception of the measurements carried out by Coon in 1966 in the case of distances between 3.5λ and 8λ for the mercury line at 546.1 nm, they were the first to perform a measurement of the variation of the frustrated intensity in the case of distances between the two prisms comprised between 0.3λ and λ [Zhu 1986].

These experimental curves agree quite well with the theoretical curves (Fig. 1.20). The slight difference between the curves may be due to contamination on the surfaces of the prisms. For this experiment, the authors have used Pellin–Broca prisms, as represented in the scheme of the experimental arrangement (Fig. 1.21).

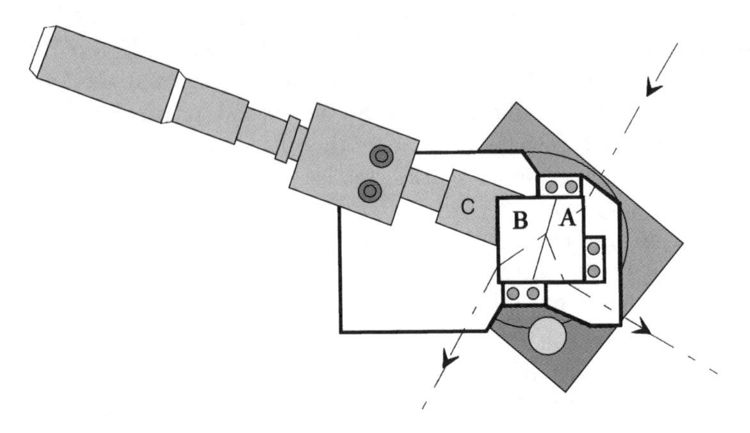

Fig. 1.21. Experimental arrangement used for the measurement of the coupling ratio between the two prisms

The distance between the two prisms was measured by using an interferometer for calibrating the vernier of the displacements of the prisms. In order to determine exactly the distance separating the two prisms, it was essential to determine in advance the position corresponding to the contact ($d = 0$). Depending on the distance between the two prisms, any ratio for the transferred light can be obtained. Such a system can be used as a beam splitter, as can be seen from Figs. 1.22 and 1.23 [Hecht 1974].

Despite this, the high wavelength sensitivity of these components restricts the field of their applications.

1.5 Resonant Tunneling Effect

The analysis of frustrated total internal reflection was developed in the foregoing with a model comprising only three media. In the case of more complex systems, three-media models may be insufficient and it might be necessary to introduce models with additional media.

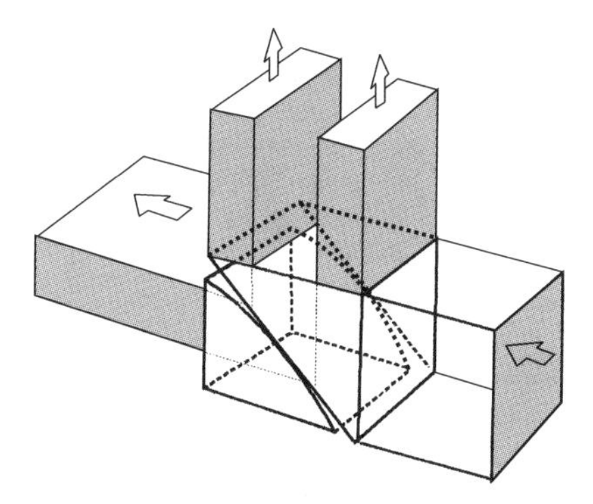

Fig. 1.22. Beam splitter [Hecht 1974]

Fig. 1.23. Beam splitter [Melles Griot]

In the description of the phenomenon of frustrated total internal reflection, we have seen that the quantity of light that is frustrated from the third medium decreases in proportion to the distance between this medium and the interface where total internal reflection arises. The introduction of a fourth medium might significantly modify this behavior. As shown by different authors, the frustrated intensity is a periodic function of the thickness of the second medium [Salomon 1992]. We discuss in detail the case of a four-media system (Fig. 1.24).

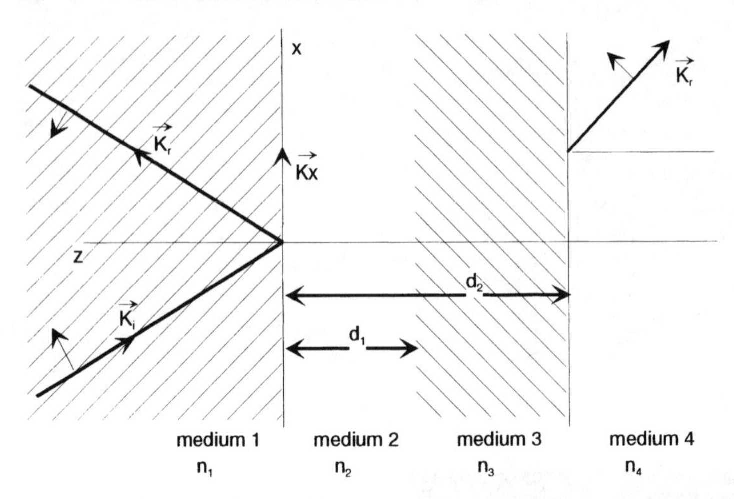

Fig. 1.24. Four-media model

The transmissivity in the fourth medium satisfies the following relation

$$T_i = \frac{U_i}{(P_i \sinh z \sin z' + Q_i \cosh z \cos z')^2 + (R_i \sinh z \cos z' + S_i \cosh z \sin z')^2}, \tag{1.23}$$

where i corresponds either to s or p polarization, while $z = B_s(d_1 - d_2)$, and $z' = A_s d_1$, with d_1 and d_2 as defined on the previous schematic. The variables in the expression of T_i are given by the equations

$$P_i = \frac{a_i A_i C_i - B_i^2}{2A_i B_i}, \quad Q_i = \frac{A_i + a_i C_i}{2A_i}, \quad R_i = \frac{a_i B_i^2 - A_i C_i}{2A_i B_i},$$

$$S_i = \frac{a_i A_i + C_i}{2A_i}, \quad a_s = \frac{n_2 \cos\theta_2}{n_1 \cos\theta_i}, \quad a_p = a_s \frac{n_1^2}{n_2^2},$$

$$A_s = n_2 \frac{\omega}{c} \cos\theta_2, \quad A_p = \frac{A_s}{n_2^2}, \quad C_s = n_4 \frac{\omega}{c} \cos\theta_4,$$

$$C_p = \frac{C_s}{n_4^2} \quad B_s = \frac{\omega}{c}(n_2^2 \sin\theta_2 - n_3^2)^{1/2}, \; B_p = \frac{B_s}{n_3^2},$$

$$U_s = \frac{n_4 \cos\theta_4}{n_1 \cos\theta_i}, \quad U_p = U_s \frac{n_1^2}{n_4^2}.$$

From these relations, it may be noted that, if the second and fourth media have the same refractive index, the maximal value of the transmissivity in the latter medium equals 1. If the intensity frustrated by the fourth medium is represented as a function of the thickness of air, the form of this function depends on the thickness of medium 2. A maximal value of transmissivity does not necessarily correspond to an air gap with a zero thickness.

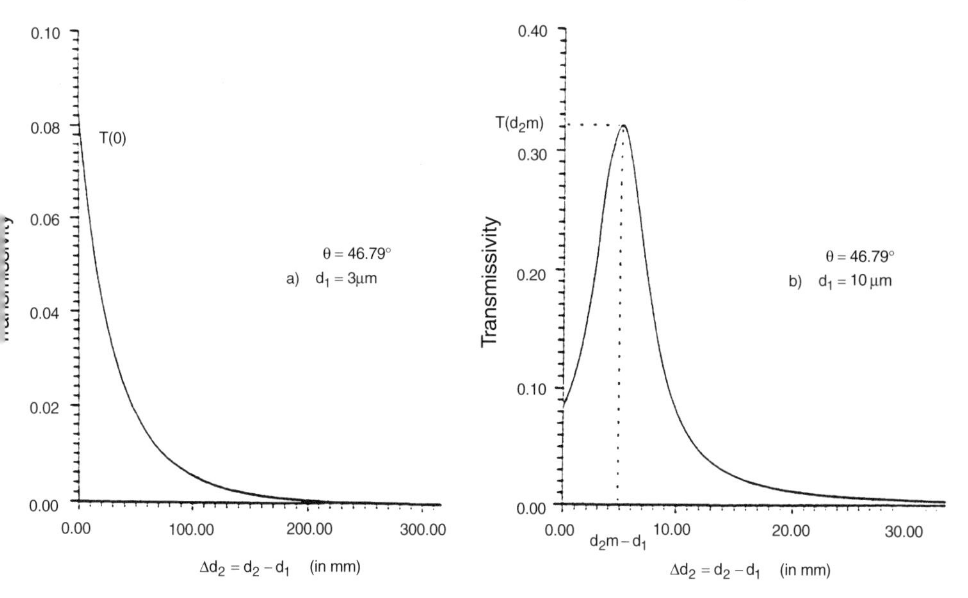

Fig. 1.25. The transmissivity as a function of the thickness of the air gap, for two values of the thickness d_1 [Salomon 1991a]

Furthermore, the maximal value $T_{\max}$ of the transmissivity is a periodic function of the thickness of medium 2. The curve represented in Fig. 1.26 exhibits such a variation, which corresponds to a resonance of the system.

At resonance, the transmissivity reaches 1, and the decrease of the intensity goes on for more than a hundred nanometers, against ten nanometers only outside the resonance. This very rapid decrease was measured by means of a scanning tunneling optical microscope.

These results, which suggest the phenomenon of resonance, can be compared with results formerly presented by Yeh on superlattices [Yeh 1985]. A superlattice is a structure obtained by depositing very thin layers, usually monocrystalline, of materials with alternately high and low indices. Figure 1.27 provides a schematic representation of a superlattice.

GaAs-GaAlAs superlattices are the most extensively studied and have numerous applications in optics as well as in optoelectronics. The index profile of a superlattice can be expressed as a function of the thickness, as follows

$$\begin{cases} n(x) = n_2 & \text{if } x < -a,\ 0 < x < b \text{ or } N\Lambda < x, \\ n(x) = n_1 & \text{if } -a < x < 0, \\ \text{where } n(x+\Lambda) = n(x) & \text{if } 0 < x < N\Lambda. \end{cases} \tag{1.24}$$

The different coefficients in the equation are as defined in Fig. 1.27. The modes propagating in such a structure are of the form

$$\mathbf{E} = \mathbf{E}(x)\exp[\mathrm{j}(\omega t - \beta z)], \tag{1.25}$$

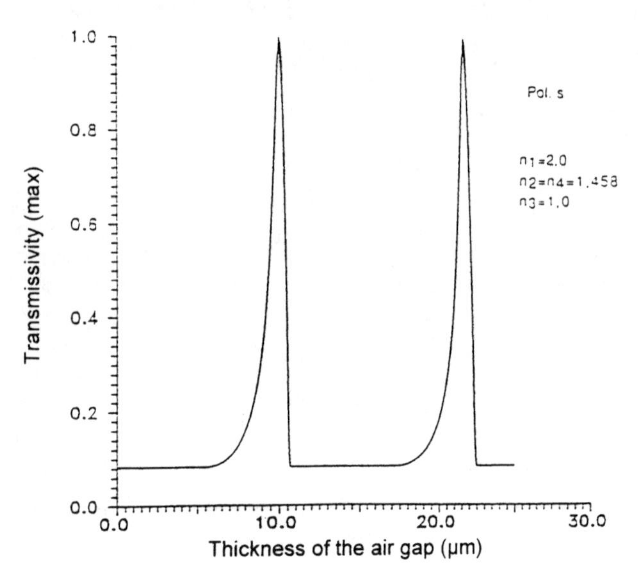

Fig. 1.26. Maximal value of the transmissivity within the second medium, as a function of the thickness of medium 2, with $n_1 = 2$, $n_2 = n_4 = 1.458$ and $n_3 = 1$ in s polarization [Salomon 1991a]

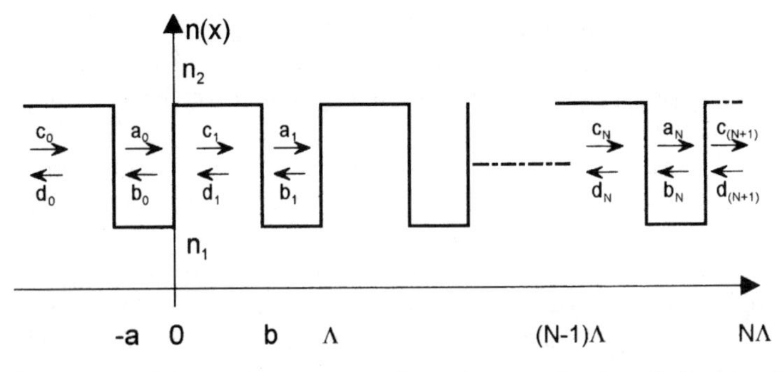

Fig. 1.27. Schematic representation of a superlattice. Refractive indices and reflection coefficients in the structure

where β is the propagation constant associated with the fields propagating within the structure. The amplitude of the field is given by the equation

$$E(x) = \begin{cases} c_0 \exp[-\mathrm{j}p(x+a)] + d_0 \exp[\mathrm{j}p(x+a)] & \text{if } x < -a, \\ a_0 \exp(-qx) + b_0(qx) & \text{if } -a < x < 0, \\ c_1 \exp[-\mathrm{j}p(x-b)] + d_1 \exp[\mathrm{j}p(x-b)] & \text{if } 0 < x < b, \\ a_1 \exp[-q(x-\Lambda)] + b \exp[q(x-\Lambda)] & \text{if } b < x < \Lambda, \\ \cdot & \\ \cdot & \\ a_N \exp[-q(x-N\Lambda)] + b_N \exp[q(x-N\Lambda)] & \text{if } N\Lambda - a < x < N\Lambda, \\ c_{N+1} \exp[-\mathrm{j}p(x-N\Lambda-b)] & \text{if } N\Lambda < x. \end{cases} \quad (1.26)$$

We consider only fields which present an evanescent behavior in media with low refractive index and therefore satisfy the equations

$$p = [(n_2\omega/c)^2]^{1/2},$$
$$q = [\beta^2 - (n_1\omega/c)^2]^{1/2}. \tag{1.27}$$

The transmissivity in such a system can easily be calculated by means of a matrix method. In s polarization, the value of the transmissivity is given by the equation

$$T_{N+1} = \left|\frac{c_{N+1}}{c_0}\right|^2 = \frac{\left[\dfrac{\sin K\Lambda}{\sin[(N+1)K\Lambda]}\right]^2}{[\exp(jpq)\frac{j}{2}(\frac{p}{q}+\frac{q}{p})\sinh qa]^2 + \left[\dfrac{\sin K\Lambda}{\sin[(N+1)K\Lambda]}\right]^2}, \tag{1.28}$$

where $K\Lambda$ is given by the relation

$$\cos K\Lambda = \cosh qa \cos pb + 1/2(p/q - q/p)\sinh qa \sin pb. \tag{1.29}$$

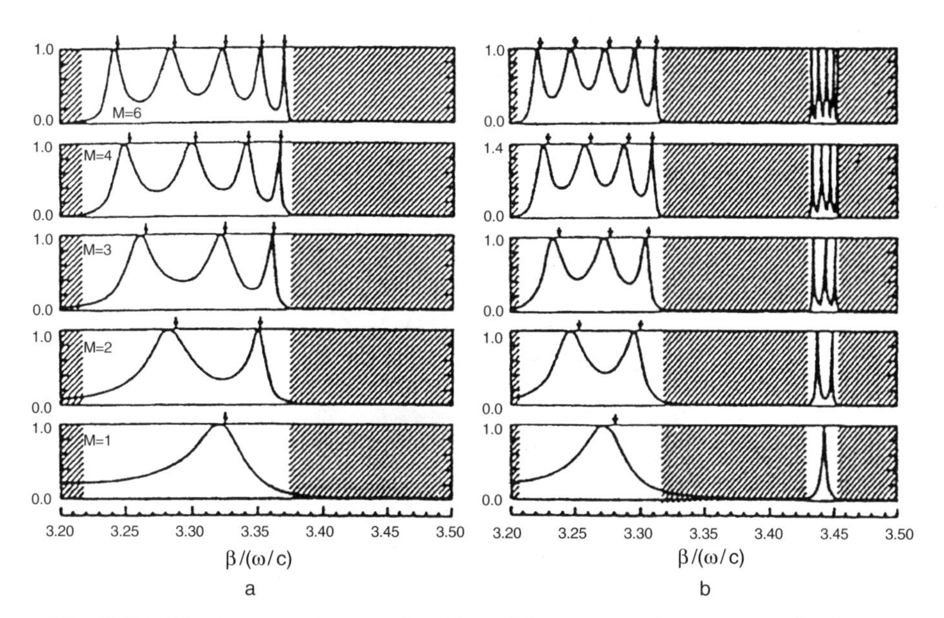

Fig. 1.28. The transmission as a function of the propagation constant β. The arrows indicate the propagation constant of the guided modes. The structures consist of layers of GaAs and AlGaAs, with $n_1 = 3.2$, $a = 0.20$, $n_2 = 3.5$. (**a**) $b = 0.20$ and (**b**) $b = 0.55$

K is known as the 'Bloch wave number'. The phenomenon of resonant optical tunneling arises when

$$\frac{\sin[(N+1)K\Lambda]}{\sin K\Lambda} = 0. \tag{1.30}$$

If these conditions are fulfilled, the Bloch wave propagates throughout the whole structure.

Calculating the eigenvalues associated with the modes guided in a structure bounded by two low-index media, thereby ensuring the guiding, we obtain values close to the values found in the case of the superlattice, as shown in Fig. 1.28. In the second case, it may be noted that a second band appears [Yeh 1985].

Superlattices, which are also known under the names of 'photonic crystals', by analogy with solid crystals, and 'band gap photonics', have potential applications in the field of light sources [Berger 1996]. Indeed, the production of superlattices should permit the fabrication of sources emitting with very high directivity, as well as the classical realization of filters. Different groups are presently working on the production of superlattices from semiconductors.

We present here a type of superlattice different from those mentioned above. This superlattice is a fiber obtained by the collective fusion of a bundle of n capillaries. The illumination of this fiber under special angles reveals the existence of privileged directions of illumination [Knight 1996]. This fiber is represented in Fig. 1.29.

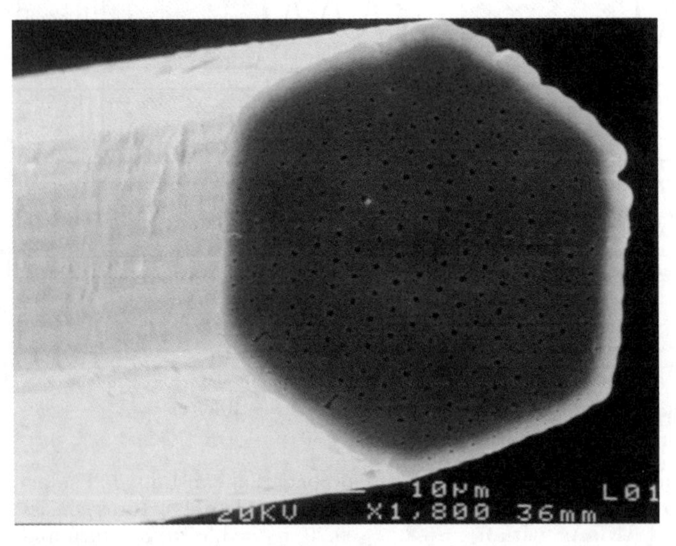

Fig. 1.29. Image obtained by scanning electronic microscopy of the photonic superlattice fiber [Knight 1996]

Researches on photonic crystals are still going on, for these devices have potential applications in a number of areas [Joan 1995, Tayeb 1997, Painter 1999, Broeng 1999, Centeno 1999]. The production of these superlattices usually involves high technologies. However, the theoretical models developed thus far still need to be experimentally confirmed.

1.6 Conclusion

The phenomenon of the total internal reflection of a plane wave at the interface of two media with different refractive indices has been described in this chapter. Even if this model is unreal, it presents the advantage of being rigorously treatable on the basis of Maxwell's equations. In particular, this model permits an exhaustive analysis of the field in the vicinity of the interface where total internal reflection arises. This analysis shows that if on average there is no energy transfer into the second medium, the field near the interface might have a very high amplitude. The phenomenon of total internal reflection can therefore be used for achieving an amplification of the electromagnetic field amplitude, in particular at the interface where it arises.

The amplification of the electromagnetic field depends on different parameters, such as the wavelength of the source, the indices of the different media, or the angle of incidence of the incident beam. We shall later see that a judicious choice of the values of these parameters theoretically enables an amplification of the field which might reach an order of 10^4.

When a laterally limited beam is totally reflected, a shift of this beam occurs. It seemed to us interesting to devote a part of this chapter to this phenomenon, which was experimentally proven at the beginning of the twentieth century. The fact that this shift can be analyzed on the basis of classical optics indicates that the evanescent properties of the electromagnetic field are at the boundary of different physical approaches.

The measurement of the totally reflected beam does not provide any direct information on the evanescent field itself. Therefore, it is necessary to induce a perturbation on the system. The simplest way for perturbing the evanescent field is by introducing a third medium with such a refractive index that the total internal reflection is partially or totally frustrated. The frustration of total internal reflection reflects a coupling between the different media. The coupling depends on the value of the evanescent field between the first two media. In the case of a weak perturbation, the value of the evanescent field can be determined directly. Since the coupling involves several different parameters, like the wavelength, the angle of incidence or the refractive index, a large range of components are based on it.

2. Diffraction from an Aperture and Dipolar Radiation

Among the different systems that generate evanescent waves, the illumination of an aperture by a plane wave is of particular importance for the recently developed near-field microscopies. In the present chapter, we shall review procedures developed by different authors for analyzing the field generated by the diffraction of a plane wave from an aperture with extent smaller than the wavelength.

The existence of evanescent waves in the vicinity of apertures was first revealed by researches carried by Toraldo di Francia [Toraldo di Francia 1942a,b]. Experimental results for microwave frequencies were presented a few years later [Schaffner 1949]. Similar results were obtained for antennas by Woodward in 1948. The latter subject can be seen as equivalent to the diffraction of light from an aperture with finite extent and will not be detailed here [Schaffner 1949, Toraldo di Francia 1952].

We end this chapter by recalling the description of the field emitted by a dipole in free space and in the vicinity of an interface. The simplest element to be described here is the dipole itself. The field generated by the dipole depends on several parameters. In particular, the medium in which the dipole lies plays an important part. Until recently, fluorescent labels were essentially used by biologists for accurately investigating molecular interactions. In order to be able to realize components from materials doped with fluorescent molecules, for example with erbium, it is essential to have a thorough understanding of the interaction of the fluorescent element with its environment, and this can be achieved through the description of the field emitted by a dipole.

2.1 Analysis of the Propagation of Light Through an Aperture

Let us consider a plane wave illuminating an aperture. The simplest model, presented in Fig. 2.1, is the case of a single aperture in an opaque screen.

The limitations of ray optics, where light is described as propagating along straight lines, first became apparent in the seventeenth century, when it was recognized by Grimaldi that the light which has propagated through

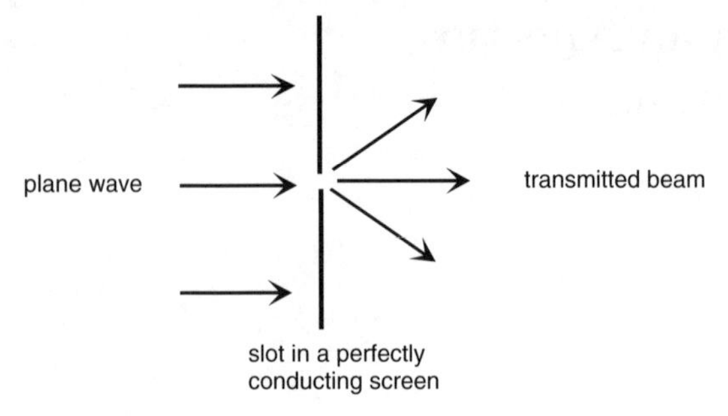

Fig. 2.1. Slot with width larger than the wavelength illuminated by a plane wave

a slot is not entirely confined (Fig. 2.2) [Grimaldi 1665]. From this discovery of diffraction phenomena therefore arose the necessity of developing a new approach to optics.

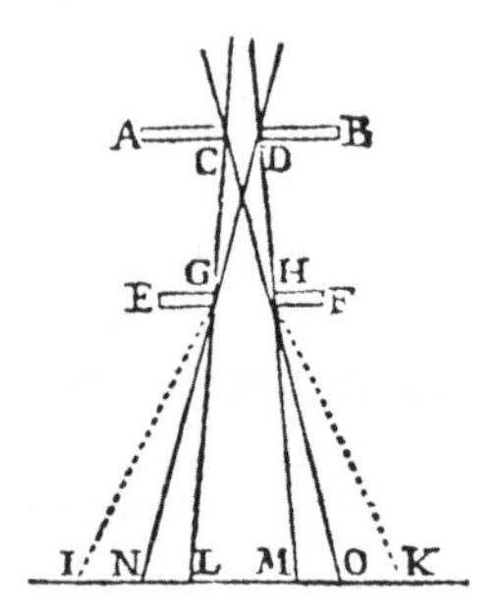

Fig. 2.2. Diffraction of light from a circular aperture [Grimaldi 1665]

The exhaustive analysis of the diffraction of light from an aperture really began with Fresnel [Maitte 1981].

Let us consider an electromagnetic wave propagating normally to the z plane where the aperture is located. At a distance greater than the half-wavelength, the field can be expressed in terms of a superposition of elementary waves

$$\begin{aligned} E(x,y,z) = \frac{E_0}{(2\pi)^2} \iint f^*(k_x,k_y) \\ \times \exp[-\mathrm{j}(\omega t - k_x x - k_y y - k_z(k_x,k_y)z]\ \mathrm{d}k_x \mathrm{d}k_y, \end{aligned} \tag{2.1}$$

where $f^*(k_x,k_y)$ is the Fourier transform of $f(k_x,k_y)$. The integration domain of this function is interrelated with the dimensions of the aperture by the equations $\Delta x \Delta k_x \simeq 2\pi$ and $\Delta y \Delta k_y \simeq 2\pi$. The expression of the field involves its decomposition into elementary waves. Each wave vector of these waves satisfies the relation

$$\frac{\omega^2}{c^2} = k_x^2 + k_y^2 + k_z^2. \tag{2.2}$$

If the width of the slot is much smaller than the wavelength of the source, Δk_x remains small in comparison with $k_z = 2\pi$. Light propagates without undergoing any significant deviation, and remains confined within a cone whose numerical aperture is as large as the size of the aperture is small. The waves which propagate behind the aperture are propagative waves.

The smaller the extent of the aperture, the larger the aperture of the emerging beam (Fig. 2.3). If the width of the slot is equal to $\lambda/2$, the emerging beam fills the entire half-space. Elementary waves like $k_z = 0$ and ${k_x}^2 + {k_y}^2 = \omega^2/c^2$ are in this case generated within the half-space.

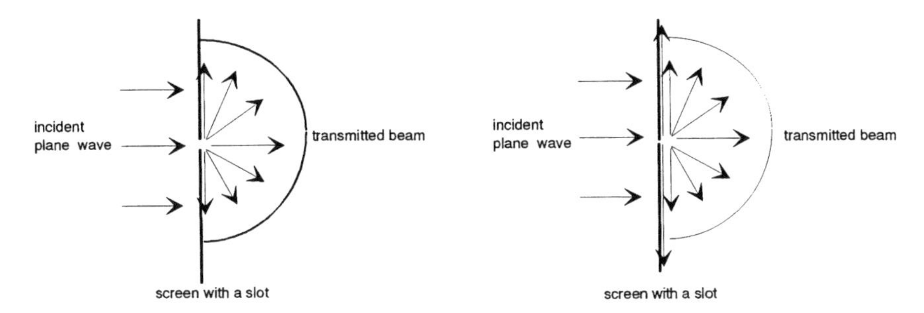

Fig. 2.3. Schematic representation of the diffraction of light from an aperture with width a

Let us now assume that the width of the aperture is smaller than $\lambda/2$. In this case, elementary waves of wave vectors, like ${k_x}^2 + {k_y}^2 > \omega^2/c^2$, are diffracted from the slot. The dispersion relation has to be satisfied and restricts k_z to be purely imaginary. The field of these waves is thus expressed as

$$E = E_0 \exp(-z/d_{\mathrm{p}}) \exp(\mathrm{j}\omega t - k_x x - k_y y), \tag{2.3}$$

where d_{p} is the penetration depth associated with the evanescent waves

$$d_{\mathrm{p}} = \left(k_x^2 + k_y^2 - \frac{\omega^2}{c^2}\right)^{-1/2} . \tag{2.4}$$

These waves do not propagate along the z axis, but remain confined within the $z = 0$ plane. They are related to high spatial frequencies of the slot [Wolf 1985].

These direct calculations indicate the existence of evanescent waves in the vicinity of apertures with dimensions smaller than half the wavelength. Little experimental evidence for the presence of evanescent waves in the vicinity of subwavelength apertures is presently available. Only in the field of microwave frequencies have some measurements demonstrated the existence of evanescent waves in the vicinity of such apertures.

In the following section, we examine in more detail the case of the diffraction of light from a circular aperture.

2.2 Diffraction of Light from a Circular Aperture

The theoretical analysis of the diffraction of light from an aperture with an extent smaller than the wavelength was developed only in the 1940s, with the researches carried out by Bethe [1944]. In the next decade, the analysis developed by Bethe was resumed by Bouwkamp and was developed into a more complete description of the problem [Bouwkamp 1950b].

Evidently, the field can be determined from numerical simulations. We base our presentation on works carried out by Leviatan at Cornell University (New York) [Leviatan 1986].

2.2.1 Diffraction from an Aperture in an Infinitely Thin Plane

Figure 2.4 illustrates the parameters of the aperture and the illumination conditions. An aperture with a limited extent is cut in an infinitely thin perfectly conducting plane. The system is analyzed in different stages. The situation in the initial problem is first divided into two equivalent systems: in each half-space, the presence of the aperture is assumed to be equivalent to a system consisting of a screen not containing any aperture. Further, it is assumed that sources of magnetic current $-M$ and M are present at the location of the aperture, as illustrated in Fig. 2.5.

In Fig 2.5, **n** stands for the unit vector normal to the aperture. From the equation of the field emitted by the sources and from the continuity conditions of the fields at the aperture, the form of the field can be determined in relation to the distance from the aperture, for different values of the radius of the aperture (Figs. 2.6 and 2.7).

In the case of the system of a single aperture, three regions can be distinguished, as is illustrated in Figs. 2.6 and 2.7. In the vicinity of the aperture, the field remains constant (proximity regime); it then very rapidly decreases

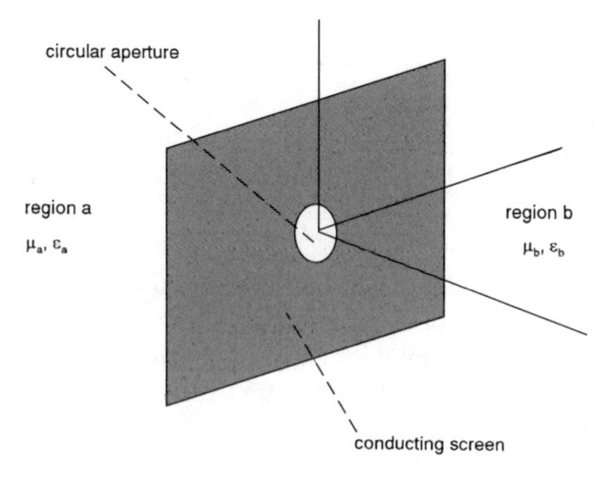

Fig. 2.4. The two regions coupled by the aperture

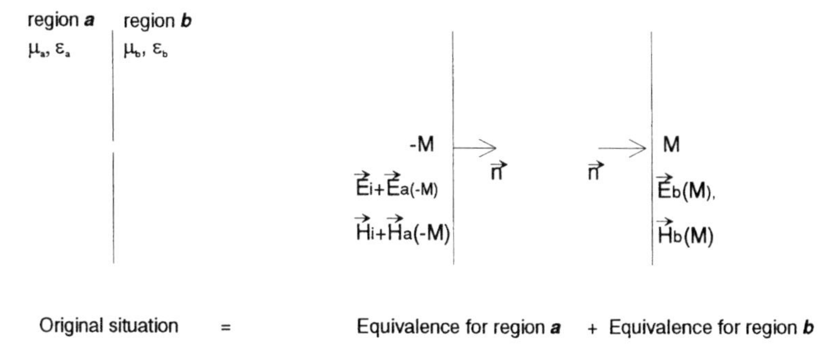

Fig. 2.5. Decomposition, as described by Leviatan, of the system of an apertured screen into two distinct systems

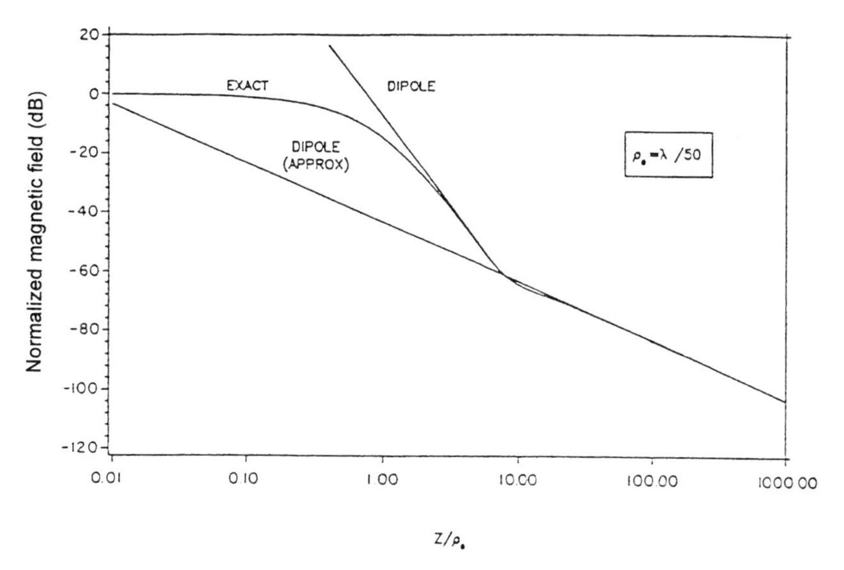

Fig. 2.6. The magnetic component of the field as a function of the distance from the circular aperture of radius $\rho_0 = \lambda/50$. Three regions characterizing diffraction can be observed: proximity regime, near-field regime and far-field regime [Leviatan 1986]

(near-field regime), and finally becomes similar to the field of a dipole at a great distance (far-field regime).

The dependence of the field on the distance from the dipole is graphically represented in the curves presented in Figs. 2.6 and 2.7. The field generated by a dipole will be described near the end of the chapter. As soon as the far-field is reached, a very good agreement between the two curves can be observed. At great distance, the aperture behaves like a dipole.

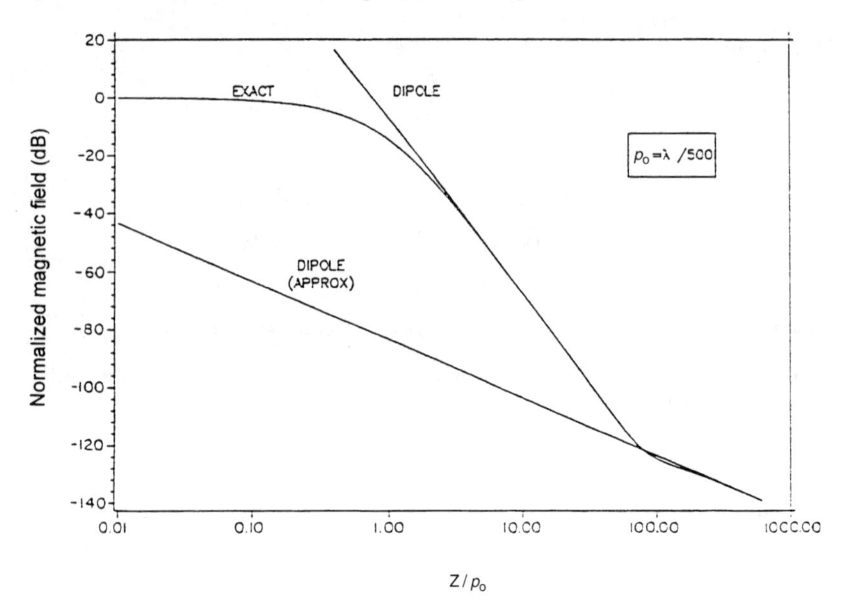

Fig. 2.7. The magnetic component of the field as a function of the distance from the circular aperture with radius $\rho_0 = \lambda/500$ [Leviatan 1986]

The average intensity of the energy flux of the field in the vicinity of the aperture has also been determined by Leviatan. He thus obtained the equation

$$S_z|_{r=0} = \mathrm{Re}\{[E^b(M_1)xH^{b*}]u_z\,|_r\} = 0. \tag{2.5}$$

From Fig. 2.8 it can be noted that the average flux of the Poynting vector decreases exponentially as the distance from the aperture increases.

Evidently, the model of an infinitely thin plane remains ideal. The problem of the diffraction from an aperture in a screen with a nonzero thickness has thus been studied by A. Roberts. These results will be reviewed in the next subsection.

2.2.2 Diffraction from a Circular Aperture in a Thick Screen

The model we now describe applies to the diffraction of a plane wave from an aperture in a screen with non-negligible thickness [Roberts 1987, Roberts 1989]. The field is decomposed outside the aperture on the basis of plane waves, and inside it on the basis of the modes guided by the aperture. The geometry of the system is illustrated in Fig. 2.9.

The field outside the aperture can be written in the form

$$R(\alpha, \gamma, x, y, z) = \frac{k_0}{2\pi}[\mathrm{j}k_0(\alpha x + \beta y + \gamma z)], \tag{2.6}$$

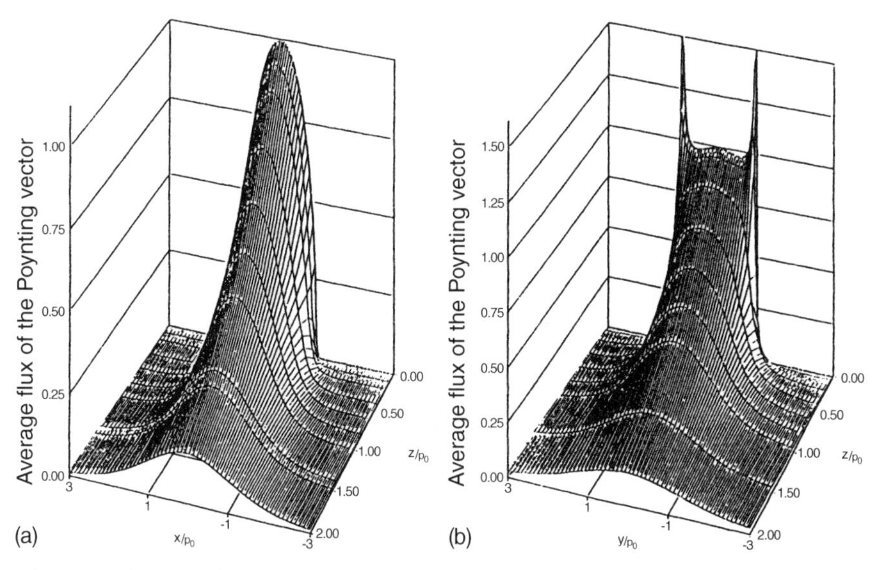

Fig. 2.8. Average flux of the Poynting vector near the aperture, as a function of the normalized distance z/ρ_0 from the aperture, with respect to the x axis (**a**) and y axis (**b**) respectively [Leviatan 1986]

where α and γ are real terms and are bound to β by the following equations

$$\beta(\alpha,\gamma) = \begin{cases} \sqrt{1-\alpha^2-\gamma^2}, & \text{if } \alpha x + \beta y + \gamma z \leq 1, \\ \mathrm{j}\sqrt{\alpha^2+\gamma-1}, & \text{if } \alpha x + \beta y + \gamma z > 1. \end{cases} \tag{2.7}$$

The equation for the transmitted field is

$$\hat{R}(\alpha,\gamma,x,y,z) = \frac{k_0}{2\pi}\exp[\mathrm{j}k_0(\alpha x - \beta y + \gamma z)]. \tag{2.8}$$

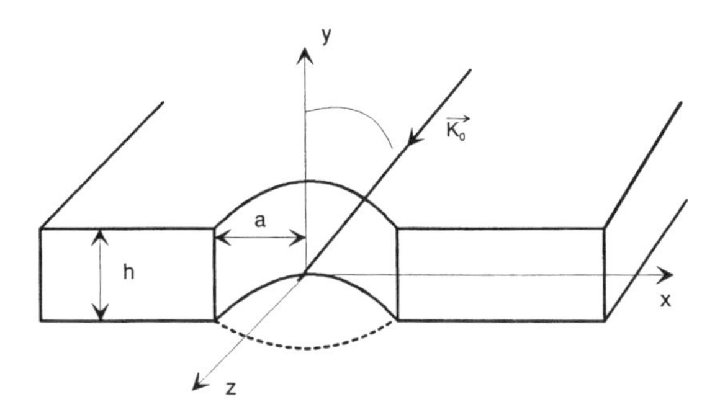

Fig. 2.9. Parameters used in the description of the aperture, of the screen and of the incident field

The transverse component of the electric field within the aperture can be determined from the expression of the modes of the aperture

$$E_t = \sum_{s=1}^{2} \sum_{n=0}^{\infty} \sum_{m=1}^{\infty} \sum_{l=1}^{2} [a_{snml} \sin(k_0 \nu_{snml} y) + b_{snml} \cos(k_0 \nu) \Psi_{snml}(\rho, \theta)]. \tag{2.9}$$

Since the aperture presents an azimuthal symmetry, the eigenmodes of the aperture can be expressed in terms of Bessel functions. The equations for the modes $\Psi_{snml}(\rho, \theta)$ and for the related variables a_{snml}, b_{snml} and v_{snml} are quite heavy. The interested reader is referred to the paper by A. Roberts, where these calculations are presented in full [Roberts 1987]. The solution for the problem of the diffraction of a plane wave by an aperture is deduced from the continuity equations of the field at the different interfaces. The solution of these equations leads to long developments, which can be found in the article mentioned above.

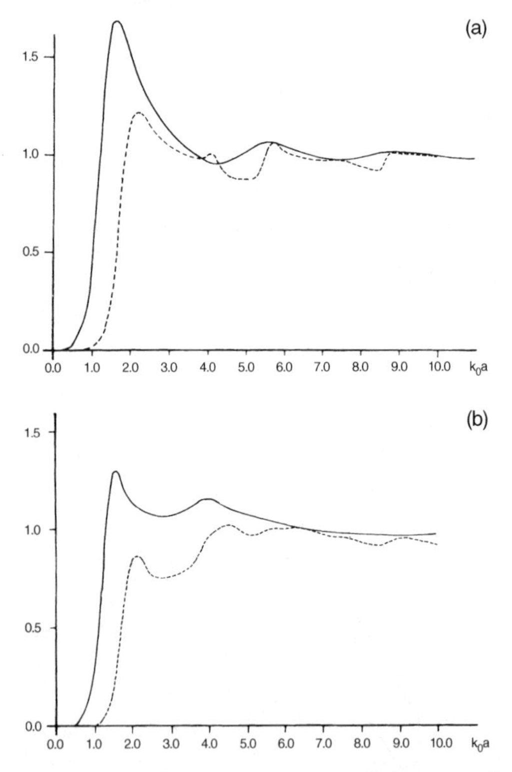

Fig. 2.10. The transmission coefficient as a function of the $k_0 a$ parameter for two values of the screen thickness ($h = 0$ for the *solid curve* and $h = a$ for the *curve in dashed lines*) (**a**) normal incidence, (**b**) the incident beam is 30° from the normal to the aperture in TE polarization [Roberts 1987, with permission of the Optical Society of America]

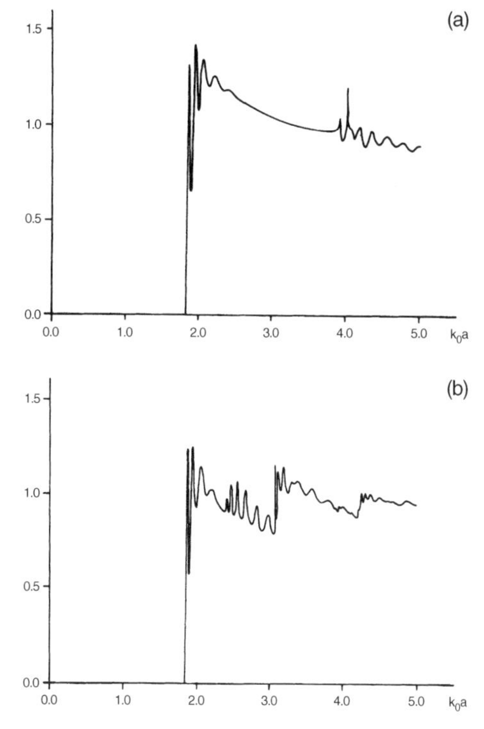

Fig. 2.11. The transmission coefficient as a function of the k_0a parameter with $h = 10a$ (**a**) normal incidence, (**b**) the incident beam is 30° from the normal to the aperture in TM polarization [Roberts 1987, with permission of the Optical Society of America]

The dependence of the transmission coefficient T on the radius of the aperture, as well as the effect of the incidence angle of the incident wave on the value of the transmission coefficient, were determined in this way. These dependences are graphically represented in Figs. 2.10 and 2.11 respectively.

An abrupt fall of the transmitted signal arises when the value of the k_0a parameter becomes smaller than 1.8. Below this value, the modes propagated by the aperture are purely evanescent. Since the screen has a nonzero thickness, these modes are totally attenuated as they propagate through the screen. A value of T higher than 1 reflects the presence of the evanescent field. If the width of the aperture rises, the coefficient of the evanescent field falls and T tends towards a value of the order of 1.

The intent of the calculations performed by A. Roberts was primarily to provide a better evaluation of the resolution power of microscopes where an aperture with very small extent is used as a source. Incidentally, the resolving power of these microscopes does not seem to be enhanced by the use of a screen with nonzero thickness.

2.3 Coupling Between Several Apertures

In practice, actual optical systems, whether they be designed for data transfer or for microscopy, generally include several coupled apertures. In order to characterize these systems, it is therefore necessary to analyze the phenomenon of coupling between different apertures.

If the diffraction of a plane wave by an aperture generates evanescent waves, the diffraction of the evanescent waves correspondingly creates propagative waves. Leviatan has investigated the problem of the coupling between two slots perforated in two separated screens. The parameters of this coupling are illustrated in Fig. 2.12 [Leviatan 1988].

The space is first divided into different parts. Proceeding in the same way as in the case of a single aperture, the system is separated into different equivalent systems.

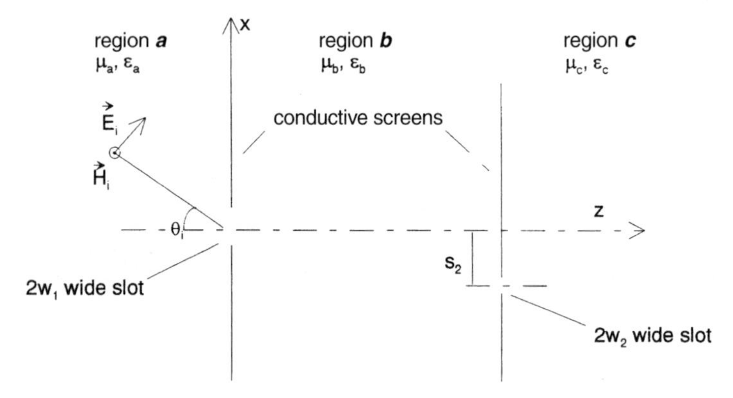

Fig. 2.12. Illustration of the coupling between two apertures in two conductive screens

The calculations of the field will not be presented in full here. Instead, we shall discuss some of the results that were obtained with this method.

Leviatan was able to determine the dependence of the transmission coefficient on the distance separating the two screens. Figure 2.13 displays the variation of the transmission coefficient for different values of the distance. The curves presented in this figure exhibit resonance peaks corresponding to values of the distance that are multiples of the half-wavelength. Further, the rapid decrease which can be observed between 0 and $\lambda/2$ reflects the coupling with the evanescent waves generated by the first aperture.

As illustrated in Fig. 2.14, the transverse shift obviously modifies the value of the transmission coefficient. For very small distances between the screens, the value of the transmitted intensity falls very abruptly when the slots are shifted. In contrast, if the screens are far from each other, the transmitted energy does not tend to a zero value, but remains constant. This phenomenon

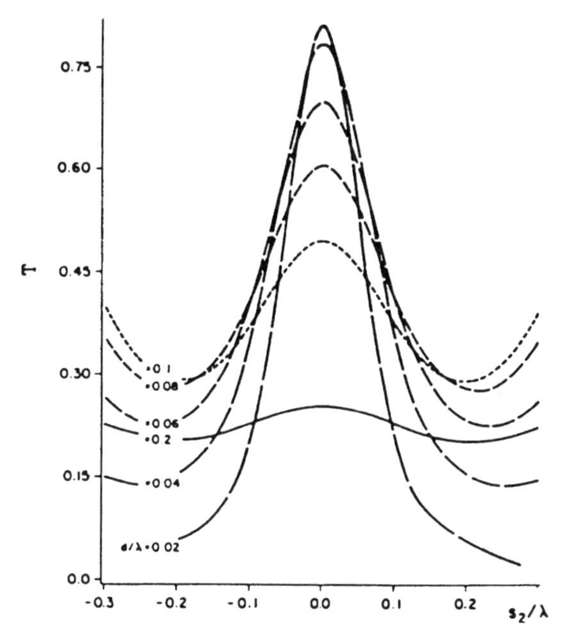

Fig. 2.13. Effect of the distance between the slots on the value of the transmission coefficient [Leviatan 1988, with permission of IEEE]

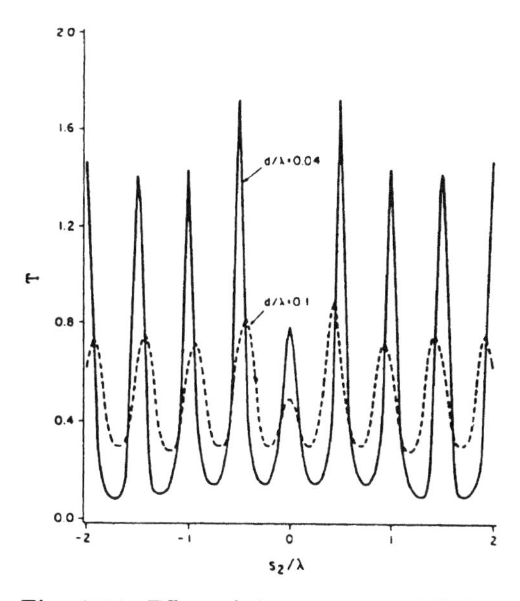

Fig. 2.14. Effect of the transverse shift between the slots on the transmission coefficient [Leviatan 1988, with permission of IEEE]

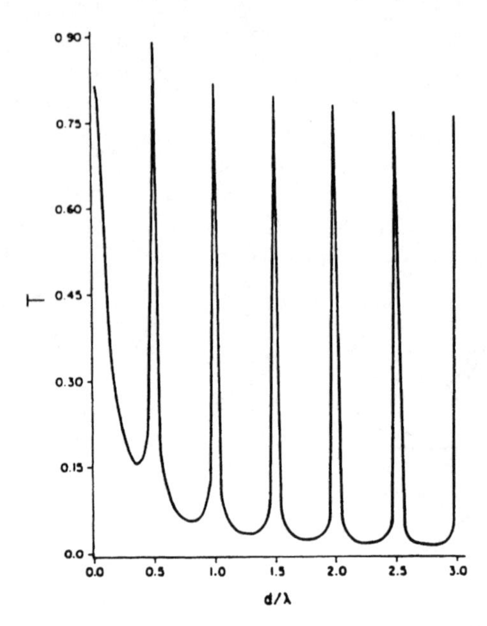

Fig. 2.15. The transmission coefficient as a function of the transverse shift between the slots and of the distance between the slots [Leviatan 1988, with permission of IEEE]

is related to the fact that the light emitted from a small aperture remains confined only in the vicinity of the near-field of the aperture but spreads out beyond. In this case, irrespective of the value of the distance separating the slots, a part of the light can be collected by a second slot, as can be seen in Fig. 2.15.

2.4 Dipolar Emission

The calculations of the field present in the vicinity of an aperture with a subwavelength extent suggest that at great distances the variation of the field is similar to the emission of an oscillating dipole [Bouwkamp 1950b]. We shall hereafter return in more detail to the case of the emission of a dipole. The importance of understanding the physics of the field emitted by a dipole is related with the fact that the development of near-field optics have led to the proposal of several models for describing the scattering of an electromagnetic wave by different structures. An example of an elementary object often encountered is that of a subwavelength-sized spherical particle. A simple method for calculating the electromagnetic response of such a sphere is by considering a dipole with polarisability as given by the Clausius-Mossotti relation. More detail on this subject can be found in recent articles [Xiao 1996, Novotny 1997, Chaumet 1998].

2.4.1 Expression of the Dipolar Field

The description presented here will be based on the analysis of the field present in the vicinity of an aperture which is developed in the book by Born and Wolf [Born 1959]. Figure 2.16 defines the parameters used in these calculations. It is assumed that the dipole is located at the origin of the coordinate system.

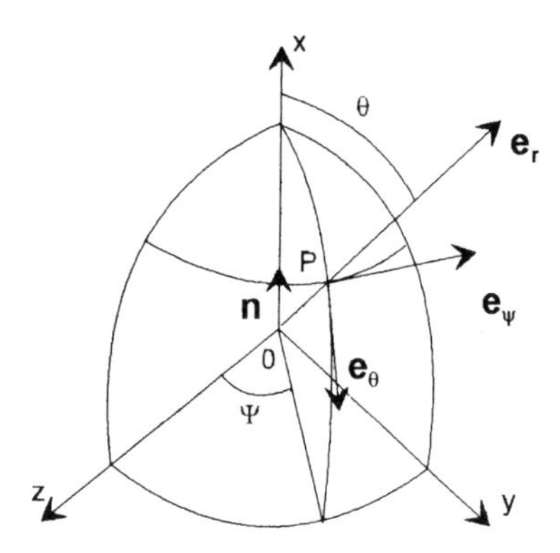

Fig. 2.16. Schematic of the dipole with momentum parallel to the z axis

We shall resume in the following the calculations described by Born and Wolf [Born 1959].

Let us consider a linear dipole vibrating along the unit vector $\mathbf{n}$. The electric polarization satisfies the equation

$$\mathbf{P}_e(\mathbf{r},t) = p(t)\delta(\mathbf{r} - \mathbf{r}_0)\mathbf{n}, \tag{2.10}$$

where $\delta()$ is the Dirac distribution and $p(t)$ the polarizability, the latter being time-dependent.

We skip a few steps in these calculations, which are presented in full in Born [1959]. The equations for the components of the electromagnetic field generated by the dipole are

$$\begin{cases} E_r = 2\left(\dfrac{p}{r^3} + \dfrac{\dot{p}}{cr^2}\right)\cos\theta, \\[2ex] E_\theta = \left(\dfrac{p}{r^3} + \dfrac{\dot{p}}{cr^2} + \dfrac{\ddot{p}}{c^2 r}\right)\sin\theta, \\[2ex] E_\psi = \left(\dfrac{p}{r^3} + \dfrac{\dot{p}}{cr^2}\right)\cos\theta, \end{cases} \tag{2.11}$$

where $\dot{p}$ and $\ddot{p}$ denote the first and second derivatives of $p(t)$ versus t.

In this form, the expression of the components does not allow us to separate the terms corresponding to propagative waves and the terms corresponding to evanescent waves. In order to differentiate the evanescent part of the field, it is necessary to assume that the dipole is surrounded by a closed surface, and then to determine the flux of the Poynting vector through this surface.

2.4.2 Energy Emitted by a Dipole

The Poynting vector $\mathbf{P}$ can be expressed as follows:

$$\mathbf{P} = \mathbf{E} \wedge \mathbf{H}, \tag{2.12}$$

$$\mathbf{P} = E_\theta H_\psi \mathbf{e}_r - E_r H_\psi \mathbf{e}_\theta . \tag{2.13}$$

Proceeding to the integration of $\mathbf{e}_r \cdot \mathbf{P}$ upon a sphere of radius r and retaining the real terms, we determine the average value of the energy passing through the surface of the sphere. The power emitted by the dipole in a medium with dielectric constant ϵ and magnetic constant μ is equal to

$$W = \frac{\omega^4}{12\pi} \mu \sqrt{\mu\epsilon} \mid p \mid^2, \tag{2.14}$$

where W is expressed in CGS units.

The smaller the wavelength, the higher the emitted power. It is apparent that only the $1/r$ components present a nonzero contribution, which corresponds to the $1/r^2$ variation of the energy of a spherical wave. The remaining $1/r^2$ and $1/r^3$ components represent the evanescent waves. For very small distances from the dipole, the amplitude of the evanescent field is very large and extends largely over the amplitude of the propagative field.

Everything happens as if the energy associated with the evanescent waves was leaving the source and periodically returning inside it, without being ever lost by the system. This character of evanescent waves has already been noted in the preceding chapter.

Evanescent waves can be detected only if they are perturbed and therefore are partially transformed into radiative waves. A possible approach for achieving this perturbation is by frustrating the evanescent waves from a semiinfinite medium. The first demonstration was achieved near the beginning of the twentieth century by Sélényi [Sélényi 1913].

The experiment carried out by Sélényi basically consisted in illuminating a fluorescent material placed at the surface of a semicylindrical prism, as illustrated in the schematic of Fig. 2.17.

The fact that the presence of light can be observed when θ has a value higher than the critical angle proves the existence of evanescent waves associated with the fluorescence of the material deposited at the surface of the prism. Indeed, only if evanescent waves are present within the fluorescent medium can waves with directions of propagation comprised between θ_c and $\pi/2$, and between $-\theta_c$ and $-\pi/2$ be generated inside the prism.

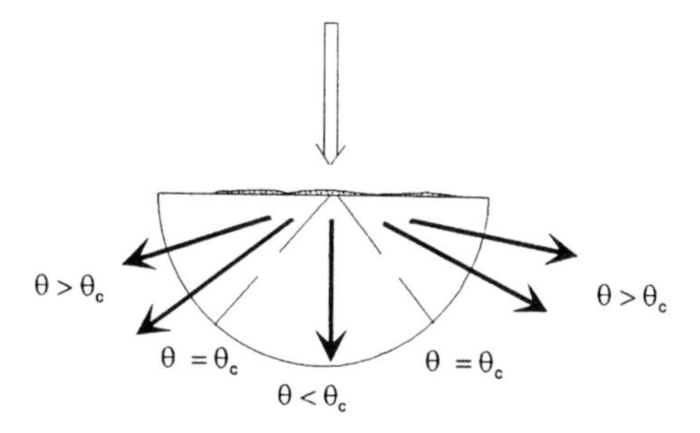

Fig. 2.17. Principle of the Sélényi experiment [Sélényi 1913]

More accurate experiments based on the same principle were carried out by Carniglia *et al.* The arrangement used for these experiments is represented in Fig. 2.18. The experimental results and the theoretical curves are presented in Fig. 2.19. The calculations on which the latter are based are detailed in the article by Carniglia [1972].

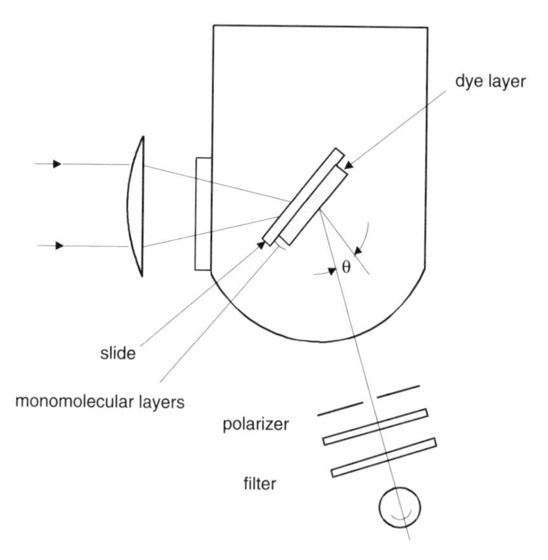

Fig. 2.18. Arrangement used for observing the light emission with respect to the angle of measurement. The plate is kept fixed while the detector rotates about the axis of the cylindrical window [Carniglia 1972]

It may be noted that the experimental measurements are quite consistent with the numerical simulation. This agreement quantitatively demonstrates the existence of evanescent waves emitted by the dipole.

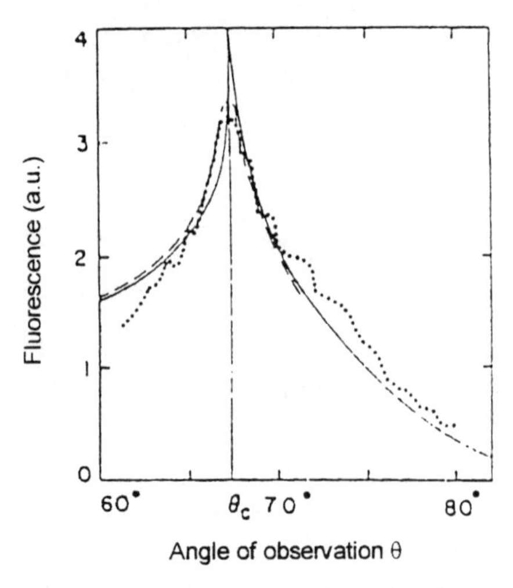

Fig. 2.19. Experimental results for an incident plane wave. The *solid curve* is based on theoretical calculations. The curve in *dashed lines* takes into account the effect of the finite size of the aperture [Carniglia 1972]

The authors have demonstrated in the same way, and with a similar arrangement, that if a dipole can generate evanescent waves, it can correspondingly absorb evanescent photons [Carniglia 1972].

We now return in detail to the effect of the presence of a second medium on the dipolar emission.

2.5 Dipolar Emission in the Vicinity of a Surface

The presence of a surface in the vicinity of a dipole modifies the nature of its emission. In 1909, Sommerfeld recognized that the Earth acts on the evanescent part of the waves emitted by an antenna, and that the energy emitted by the antenna is absorbed by the Earth. In the presence of a medium near the dipole, the dipolar emission has the same form.

In Fig. 2.20 are illustrated the orientations that the dipole may assume with respect to the interface. It will be assumed hereafter that the dipole is located within the medium 1 of refractive index n_1, and that it lies at a distance d from the second medium of refractive index n_2. The dipole is denoted by $\|$ or $\perp$ depending if its momentum is parallel or perpendicular to the surface of the second medium.

We have seen earlier that in the case where the dipole lies within a homogeneous isotropic infinite medium, the total power emitted could be expressed by (2.14). The emitted power corresponds to the propagative waves emitted by the dipole. If the field generated by the dipole is perturbed by the presence of the second medium, a part of the evanescent waves can be

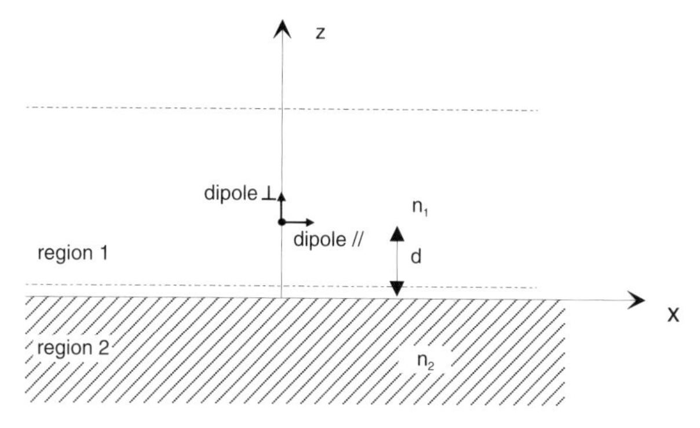

Fig. 2.20. Schematic of a dipole in the vicinity of an interface

transformed into propagative waves. This can be demonstrated with a variety of formalisms. Among the descriptions of this interaction, those developed by Chance and Lukosz can be mentioned [Kuhn 1970, Chance 1975a, Chance 1975b, Lukosz 1977, Nha 1996, Lamouche 1999]. In order to show the role of evanescent waves in the interaction of the dipole with the second medium, we here examine results presented by Lukosz [1977].

Lukosz calculated the energy emitted by the dipole in the presence of the second medium. The following expression yields the energy emitted by the dipole with an arbitrary orientation. The emitted energy is normalized with respect to the energy emitted by the dipole in the presence of a single medium

$$W(d)/W = \cos^2\theta[W(d)/W]_\perp + \sin^2\theta[W(d)/W], \tag{2.15}$$

where

$$W(d)/W = 1 + \frac{p \cdot \mathrm{Im}[E_r(x_0)]}{p \cdot \mathrm{Im}[E(x_0)]}. \tag{2.16}$$

Figure 2.21 describes the variation of the emitted power when the second medium is of higher index than the medium in which the dipole lies. Figure 2.22 describes the inverse case.

Different conclusions can be drawn from the observation of these curves. Let us first consider the case where the dipole is far from the surface ($d > \lambda/2$). We can observe here the presence of oscillations with a period of $\lambda/2$. These oscillations are present in particular when the dipole is parallel to the interface. This phenomenon is due to the dipolar emission which arises normally to the direction of the dipole [Stratton 1941]. In this case, the light emitted by the dipole in the form of propagative waves is reflected at the interface and interferes with the emitted light. Depending on the distance between the dipole and the surface, these interferences are either constructive or destructive.

We now turn our attention to the curves obtained in the case where the surface lies within the near-field of the dipole. If the second medium is

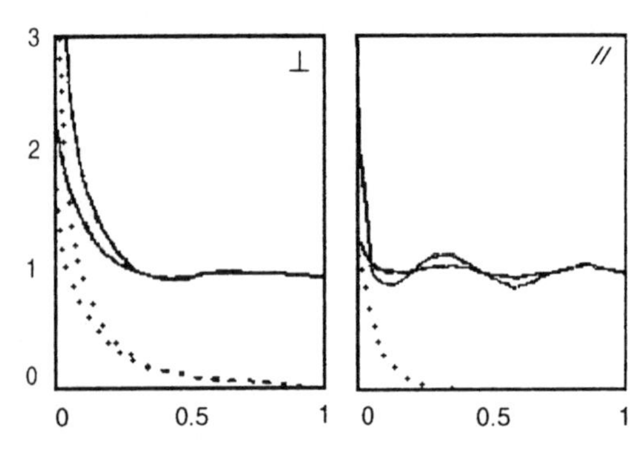

Fig. 2.21. Normalized power emitted by a parallel or perpendicular dipole as a function of d/λ, in the case where $n > 1$. d is the distance between the dipole and the interface, λ is the wavelength and $n = n_1/n_2$ [Lukosz 1977]

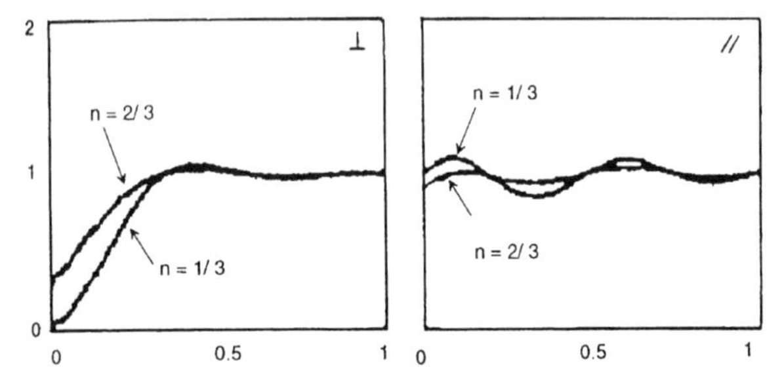

Fig. 2.22. Normalized power emitted by a parallel or perpendicular dipole as a function of d/λ, in the case where $n < 1$. d is the distance between the dipole and the interface, λ is the wavelength and $n = n_1/n_2$ [Lukosz 1977]

of higher refractive index than the medium in which the dipole is located, this medium frustrates the evanescent waves, thus transforming them into propagative waves. This induces a rise of the radiation of the dipole.

If the index of the second medium is smaller than the index of the medium containing the dipole, the waves emitted by the dipole can be totally reflected by the second medium. Evanescent waves are then generated inside the second medium, while the radiation is reduced.

Another approach is based on the consideration of the lifetime τ of the dipolar emission [Drexhage 1970, Drexhage 1974, Lukosz 1977]. The lifetime is related to the energy emitted in the entire space by the following equation [Lukosz 1977]

$$\tau(d)/\tau = [W(d)/W]^{-1}, \tag{2.17}$$

where τ is the lifetime associated with the dipole in free space. Hence, the measurement of the lifetime is equivalent to a measurement of the normalized power emitted in the entire space of the dipole. We end this section with an example which has applications in the realization of optical amplifiers.

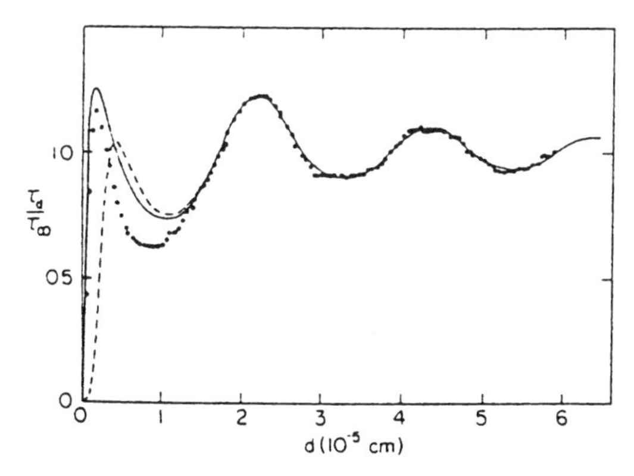

Fig. 2.23. Normalized lifetime of Eu^{3+} ions as a function of the distance to a silver film. The *solid curve* corresponds to the measurements. The dipole is assumed to be parallel to the film with an efficiency of 0.85. $n_1 = 1.5$, $n_2 = 0.06 \pm 0.02$ and $k_2 = 4.11 \pm 0.007$ [Chance 1974]

Figure 2.23 represents the lifetime of erbium molecules as a function of the distance separating the Eu^{3+} ions and a silver film [Chance 1974].

The measurements presented in the Fig. 2.23 were carried out by Drexhage [1968]. As in the case of a dipole in front of a dielectric, the curve exhibits oscillations which indicate the existence of interference phenomena. Further, an important fall of the lifetime can be observed when the metallic film lies within the near-field of the dipole. This fall expresses a nonradiative transfer from the Eu^{3+} ions to the metal.

2.6 Conclusion

For a long time, the diffraction of light from an aperture was analyzed without taking into account the presence of evanescent waves. However, when the aperture is of subwavelength extent, the phenomenon of diffraction requires more complete studies for its analysis. In fact, the rigorous analysis of the field in the vicinity of subwavelength apertures really started being developed with the researches carried out by Bethe [1944] and Bouwkamp [1950a, 1950b].

As described by Leviatan, when an aperture with a subwavelength extent is illuminated by a plane wave, the space beyond the aperture can be divided into different regions with distinct propagation characteristics [Bouwkamp 1950a, Bouwkamp 1950b]. Within the first region, referred to as the proximity region, the field remains constant and confined. The dimension of this region is very small: a fraction of the wavelength. The second region, or region of the near-field of the aperture, extends over a half of the wavelength. In this region, the field decreases very rapidly. Lastly, within the third region, or far-field region, the field decreases inversely with respect to the distance from the aperture. This region extends indefinitely.

More realistic calculations, taking into account the thickness of the screen in which the aperture is located, have vindicated the previous results. Further, as actual systems usually include several successive slots, it was of interest to discuss here some results on the coupling between slots. Here again has appeared the effect of the near-field on the coupling between two slots.

The results of these calculations have several direct applications, related in particular to 'near-field microscopy', both in the case of highly localized sources and for the analysis of the field diffracted by subwavelength objects. These applications will be described in the last part of this book.

Evanescent waves can be generated from the total internal reflection of plane waves or from the light emission of a dipole. This field remains confined near the dipole and does not transport any energy at great distances from the dipole. In order to obtain information about the near-field, it is therefore imperative to have it perturbed, so as to transform a part of the evanescent field into a propagative field. This can be achieved by using a semiinfinite, either dielectric or metallic medium, or else a second dipole.

The light emission of the dipole can be characterized by its probability of emitting a photon. In other words, the emitted power can be related to the notion of lifetime. This possibility is being widely used in spectroscopy. It seemed to us that this needed to be mentioned, even if there are few applications related to dipolar emission described in this book.

3. The Evanescent Field in Guided Optics

The development of lasers during the 1960s has led to a renewal of interest in optics. This trend was amplified in the 1970s, when it was recognized that silica fibers could guide light with losses of the order of dB/km. This was realized first with the sources available at the time, for example with lasers, and later with laser diodes, which present the advantage over lasers of being extremely compact. Another advantage of optical fibers is that the raw material used in their fabrication is silica, which is found in great quantities on Earth. It was soon theoretically demonstrated that perceptible enhancements of the transmission properties of optical fibers could be obtained by modifying their characteristics. Experimental results rapidly confirmed these theoretical predictions.

The evanescent field itself is not actively involved in the guiding of the light. Rather, the evanescent field can be regarded as reflecting the fact that the light is being guided. This chapter begins by describing the basic theory of guided modes, which will serve in the analysis of the characteristics of the evanescent field associated with these modes. We then describe planar and confined waveguides respectively, and finally treat the case of optical fibers.

3.1 The Evanescent Field in Planar Optics

3.1.1 Analysis of Planar Waveguides

Let us consider two semiinfinite media with refractive indices n_2 and n_3 respectively. It will be assumed that these two media are separated by a medium of index n_1. If the conditions of total internal reflection are satisfied at the 1-2 and 1-3 interfaces, a plane wave will be guided between the second and third media (Fig. 3.1).

A wave guided between these media satisfies the following conditions

$$\begin{aligned} &\theta_\mathrm{i} > \sin^{-1}(n_2/n_1), \\ &\theta_\mathrm{i} > \sin^{-1}(n_3/n_1). \end{aligned} \tag{3.1}$$

The simplest description of the propagation of light in a waveguide is in terms of ray-optical arguments. This however requires that the very notion of

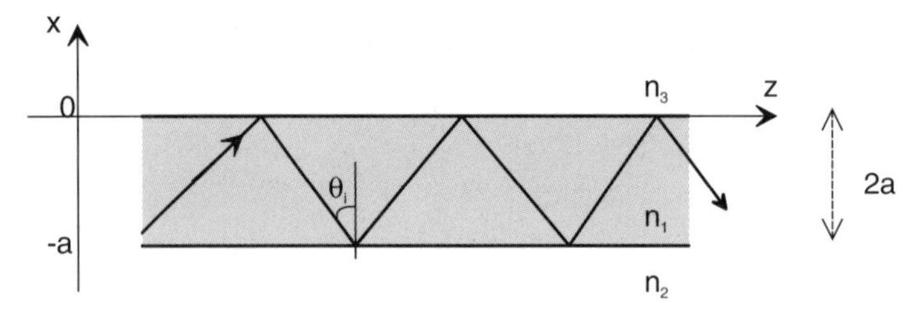

Fig. 3.1. Guiding of light between the two media

'ray' has a physical meaning for such a system, namely that the dimensions of the guiding area, in particular its thickness, are larger than the wavelength. In the following, it will be assumed that these conditions are satisfied.

Let us consider now a more complicated model: instead of a single layer, as before, it will be assumed that several layers are inserted between the second and third media. If the indices of the internal layers are higher than the indices of the two semiinfinite media, and provided that the previous relations hold at the two interfaces, the ray is guided by this stratified system (Fig. 3.2). In this case, the path of the ray is formed by segments of straight lines. If the number of layers is very high, the path of the rays is a curve which directly depends on the index profile of the waveguide.

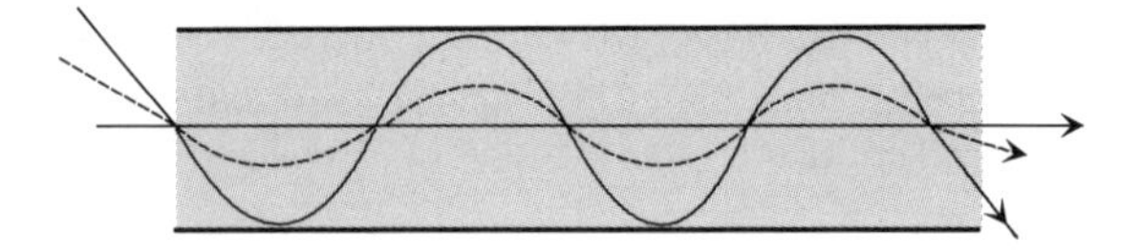

Fig. 3.2. Propagation of rays in a stratified system with parabolic index profile

It may be noted that the guiding, i.e. the fact that the ray returns back into the center of the core, does not necessarily take place at the last interface. If the incidence angle at an internal interface is high enough, the guiding might arise at this interface. In this case, the evanescent field associated with the different plane waves is present not only within the last medium, but also within the internal layers.

The representation of the path followed by the rays in Fig. 3.1 is not entirely exact, for it does not take into account the Goos–Hänchen shift which arises at each total internal reflection (Fig. 3.3). As observed by Adams, the effective propagation of the rays in the waveguide is equivalent to the propagation inside a waveguide with width w.

The propagation in these waveguides can be analyzed on the basis of Maxwell's equations and of the boundary conditions [Adams 1981].

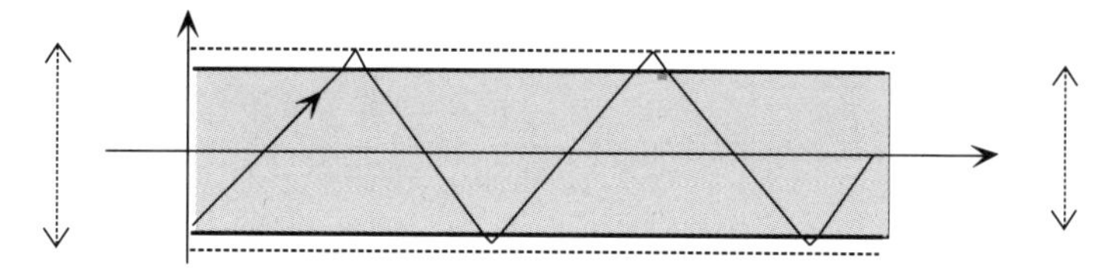

Fig. 3.3. Effective width of the waveguide resulting from the Goos–Hänchen shift

For TE modes, the solution of Maxwell's equations yields

$$E_y = \begin{cases} A\exp(-rx) & \text{if } x \geq 0, \\ A\cos qx + B\sin qx & \text{if } 0 \geq x \geq -2a, \\ (A\cos 2aq - B\sin 2aq)\exp[p(x+2a)] & \text{if } -2a \geq x, \end{cases} \tag{3.2}$$

and

$$H_z = \frac{-\mathrm{j}}{\omega\mu_0} \begin{cases} -A\exp(-rx) & \text{if } x \geq 0, \\ -q(-A\sin qx + B\cos qx) & \text{if } 0 \geq x \geq -2a, \\ p(A\cos 2aq - B\sin 2aq)\exp[p(x+2a)] & \text{if } -2a \geq x, \end{cases} \tag{3.3}$$

where p, q and r are bound to the propagation constant by the equations

$$\begin{aligned} q^2 &= n_1^2k^2 - \beta^2, \\ p^2 &= \beta^2 - n_2^2k^2, \\ r^2 &= \beta^2 - n_3^2k^2, \end{aligned} \tag{3.4}$$

where $k = \omega(\mu_0\epsilon_0)^{1/2}$. The eigenvalue equation derived from the continuity conditions is

$$\tan(2aq) = \frac{q(p+r)}{q^2 - pr}. \tag{3.5}$$

For TM modes, the components of the fields can correspondingly be written

$$H_y = \begin{cases} C\exp(-rx) & \text{if } x \geq 0, \\ C\cos qx + D\sin qx & \text{if } 0 \geq x \geq -2a, \\ (C\cos 2aq - D\sin 2aq)\exp[p(x+2a)] & \text{if } -2a \geq x, \end{cases} \tag{3.6}$$

and

$$E_z = \frac{-\mathrm{j}}{\omega\epsilon_0}\begin{cases} \dfrac{rC}{n_0^2}\exp(-rx) & \text{if } x \geq 0, \\[2ex] \dfrac{q}{n_0^2}(-C\sin qx + D\cos qx) & \text{if } 0 \geq x \geq -2a, \\[2ex] \dfrac{p}{n_0^2}(C\cos 2aq - D\sin 2aq)\exp[p(x+2a)] & \text{if } -2a \geq x. \end{cases} \tag{3.7}$$

The eigenvalue equation is thus

$$\tan(2aq) = \frac{(n_0^2 p + n_2^2 r)n_1^2 q}{n_2^2 n_3^2 q^2 - n_1^4 pr}. \tag{3.8}$$

The field inside the waveguide is a sine or cosine function. In contrast, within the cladding, i.e. within the medium of index n_2 or n_3, the variation of the field is exponential [Ostrowsky 1979, Adams 1981, Hunsperger 1985]. As an infinite field does not have any physical meaning, we restrict ourselves here to the solution of the case of an exponentially decreasing field. Here again, the form of the evanescent field turns out to be similar to the form of the evanescent field generated by the total internal reflection of a plane wave.

3.1.2 Production of Step-Index Planar Waveguides

Step-index planar waveguides are produced by depositing a layer upon a substrate. Different methods are available for the production of the guiding layer. One of these methods is the technique of epitaxial growth. This technique imposes several conditions on the preparation of the sample. The structure has to be effectively guiding, and the materials used must be compatible, in order to adhere to one another and for ensuring the stability of the resulting structure.

A more recent method uses the 'sol–gel' technique: a material in the form of a gel is deposited upon a substrate and submitted to an overheating intended to solidify it. Lastly, the guiding layer can be formed by the deposit of a polymer material.

Each material has a specific transmission band and other distinctive properties. Table 3.1 reproduces the characteristics of step-index planar waveguides fabricated with different techniques.

The value of the optical losses in these waveguides is intimately related with the state of the surface. By adding a supplementary layer at the surface of the waveguide, the losses can be reduced, so that the amplitude of the evanescent field becomes zero at the interface between the waveguide and air.

Ion-implanted and diffused waveguides have an index profile which directly depends on the technique used for their fabrication [Hunsperger 1985,

Table 3.1. Comparison of different fabrication techniques of planar waveguides.

Material	Substrate	Transmission band (μm)	Loss (dB/cm)	Distinctive characteristics
Semiconductor Si, SiGe	Silicon	1–10	0.1	Si technique, microelectronic integration, heavy technology
Semiconductor III-V (binary, ternary, quaternary, superlattice,...)	GaAs, InP	0.5–10	0.1	Heavy and expensive technique adjustable spectral band
Sol–gel Silicon (either doped or pure), $BaTiO_3$, $LiNbO_3$...	Any	0.2–10	Some dB/cm (immature technique)	Simple and inexpensive technique, easily achieved doping
PZG	Glass, CaF_2...	0.4–2	1	Recently developed and relatively simple technique
$LiNbO_3$	$LiNbO_3$		0.1	Ion implantation
Polymer Pure or doped organic material	Any	0.2–10	2	Relatively simple technique, thermal and time stability?

Favennec 1993, Fogret 1994]. As an example of a waveguide with specific characteristics, we may mention waveguides fabricated by the diffusion of silver in glass [Zolatov 1978]. The characteristics of these waveguides are summarized in Table 3.2.

Diffusion techniques are not really adapted to the production of buried waveguides, except in some rather special cases: waveguides produced by the diffusion of silver in certain glasses under an electric field are but an example. For the fabrication of buried waveguides, high-energy ion-diffusion techniques may be used. If the implantation is accompanied with a decrease of the refractive index, a confinement of the field limited to the non-implanted

Table 3.2. Comparison between two fabrication techniques of doped waveguides.

	Technique	
	Ion implantation	Ion diffusion
Implanted elements	Any element in ionized form	Elements compatible with the host material
Host material	Any material	Silicium, silica, ZBLAN, glasses...
Transmission band (μm)	0.2–10 depending on the bombarded material	0.2–10 depending on the bombarded material
Loss (dB/cm)	1	0.1
Distinctive characteristics	This method requires the use of a convenient implanter. In the case where the implanted area is guided, the thickness of the waveguide is limited	Simple and rather inexpensive technique, which can be completed with a diffusion under an electric field

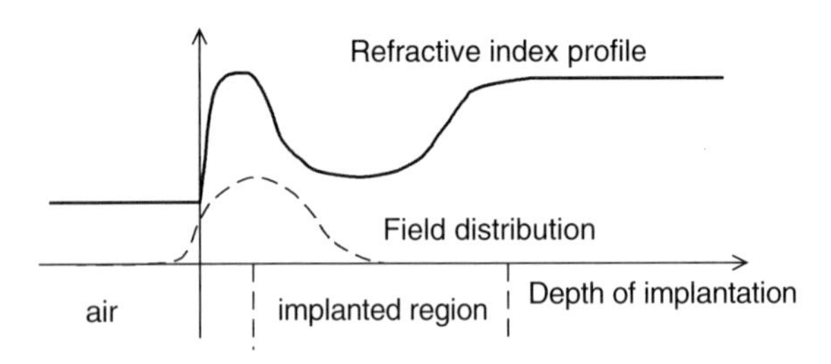

Fig. 3.4. Example of the profile of a lithium niobate plate implanted with a proton beam

zone can be obtained. In this case, the implanted zone acts as an optical cladding (Fig. 3.4).

3.2 Confined Waveguides

The field present inside the waveguide actually needs to be laterally confined. Several different techniques can be used for the production of confined waveguides. Waveguides fabricated with different methods are represented in Fig. 3.5.

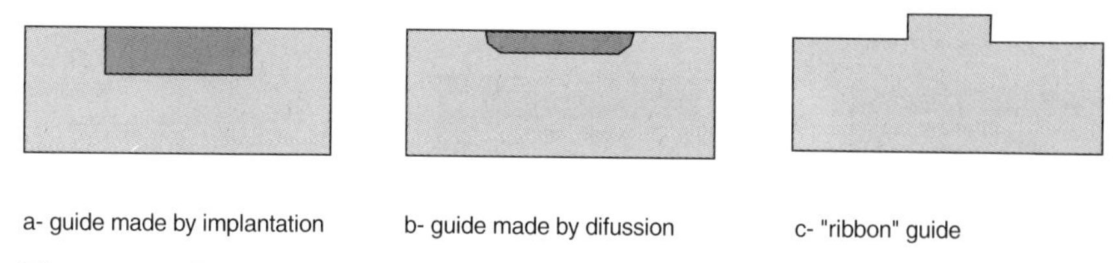

Fig. 3.5. Different types of confined waveguides (**a**) ion-implanted waveguide, (**b**) ion-diffused waveguide, (**c**) strip-loaded waveguide produced by lithography and ion etching

The production of the first type of waveguides represented in Fig. 3.5 proceeds as illustrated in Fig. 3.6. The fabrication technique for strip-loaded waveguides is shown in Fig. 3.7.

The determination of the propagation of the modes in confined waveguides is more complex than in the case described in the preceding paragraph. Further, there is not any general analytical solution for this problem. The analysis of these waveguides can be achieved by using either the finite elements method [Yeh 1975] or the beam propagation method (BPM), developed by Feit and Fleck [Feit 1978, Feit 1980]. However, to detail here at length all the calculations necessary for the analysis of these waveguides would extend largely beyond the scope of this book.

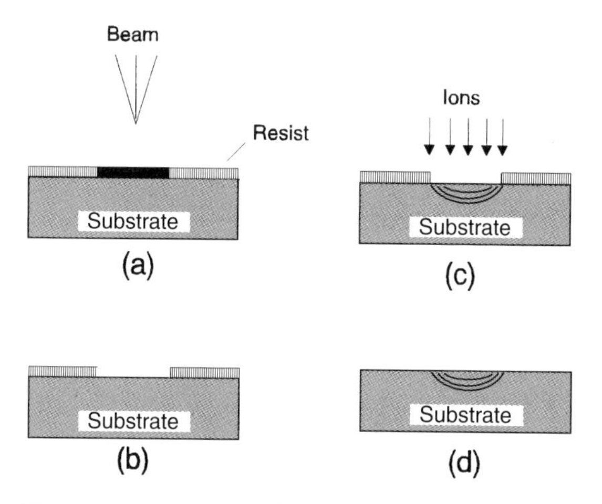

Fig. 3.6. Description of the fabrication process of an ion-implanted waveguide (**a**) deposit, irradiation and developing of an electroresist or a photoresist, (**b**) etching of the unexposed parts for the production of the mask, (**c**) diffusion or implantation through the mask, (**d**) withdrawal of the mask

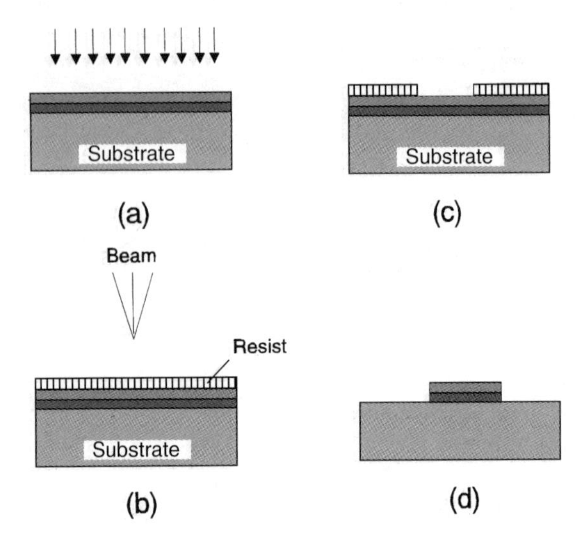

Fig. 3.7. Fabrication process of a strip-loaded waveguide (**a**) deposit by epitaxial growth of the different layers, (**b**) deposit of the sensitive resist and local sensitization, (**c**) developing of the resist, (**d**) selective etching

3.3 Optical Fibers

3.3.1 Ray-Optical Analysis of the Propagation in Optical Fibers

The propagation of light inside an optical fiber can be analyzed in terms of ray-optical arguments [Arnaud 1976]. The relative variations of indices inside the fibers are generally smaller than one, and therefore the theory of paraxial rays can be applied here.

The path followed by the rays is deduced from the following two equations [Arnaud 1977]

$$\begin{aligned} \mathrm{d}^2x/\mathrm{d}z^2 &= -\partial U(x,y)/\partial x, \\ \mathrm{d}^2y/\mathrm{d}z^2 &= -\partial U(x,y)/\partial y, \end{aligned} \tag{3.9}$$

where x and y are z dependent. $U(x,y)$ is related to the index profile by the equation

$$U(x,y) = 1 - n(x,y)/n_0, \tag{3.10}$$

where n_0 is the refractive index along the axis of the fiber.

Representing in such a way the propagation of light in fibers, the presence of fluctuations of the diameter of the fiber as well as microbending can be readily incorporated into the model [Grignand 1980, de Fornel 1983]. Further, this method permits us to analyze the propagation of the rays inside a fiber whatever the index profile of the fiber is.

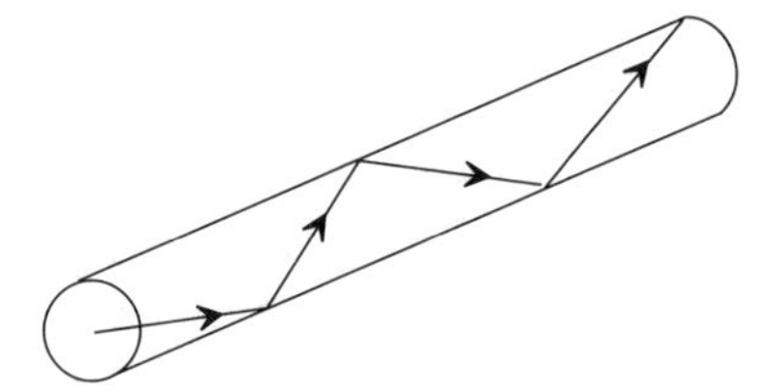

Fig. 3.8. Propagation of a light ray inside an optical fiber

The equations describing the propagation of rays can be easily solved using Euler's method. The light rays which correspond to the source are injected in the $z = 0$ plane. The path followed by the rays is then calculated segment by segment (Fig. 3.8). If a perturbation arises, the only condition to be satisfied is at the interface between the core and the cladding. If the angle of the ray and the interface is less than the critical angle, the ray is not guided any more and the calculation of the propagation of the ray can be ended.

The amount of the losses resulting from perturbations can therefore be determined from the number of rays that reach the location z, by comparing it with the number of rays that were injected (Fig. 3.9). The time of propagation of the rays can be determined in the same way. If all rays were injected at a time t_0 in the form of an impulse, the value of the scattering of this impulse at the end of a given length of fiber can thus be determined. The numerically determined value of the steady-state length of different fibers

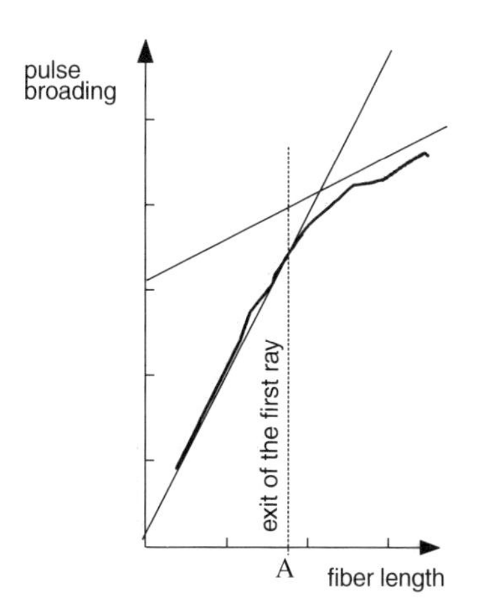

Fig. 3.9. Variation of the spreading of a light impulse injected in a fiber subjected to microbending. The transition to the equilibrium regime appears distinctly on the curve

subjected to random microbending was found in agreement with experimental measurements.

The notion of 'ray' does not correspond exactly to an eigenmode of the fiber and it must therefore be cautiously used. In fact, only a congruence of rays can be said to be equivalent to an eigenmode of the fiber.

Different authors, see for example Marcuse, Adams, Snyder or Joindot have given comprehensive descriptions of the propagation of light in waveguides [Marcuse 1972, Adams 1981, Snyder 1983, Joindot 1996]. In the following sections, we review the theory of optical waveguides and provide an account of the basic principles which are necessary for understanding the physics of devices containing optical fibers. We begin with the simplest case: step-index fibers.

3.3.2 Modes of Step-Index Fibers

A step-index fiber consists simply of a silica cladding and of a doped core with refractive index higher than that of the cladding. This geometry and the associated optical characteristics permit an analytic treatment of these fibers.

We assume here that modes of propagation constant β and azimuthal order ν can propagate inside a structure of this type. β is the constant of propagation of the mode in the direction along which the waveguide is uniform, while ν expresses the azimuthal symmetry of the system. The expression of the field is thus

$$\begin{aligned}\mathbf{E}(r,\theta,z,t) &= [E_r(r)\mathbf{e}_r + E_\theta(r)\mathbf{e}_\theta + E_z(r)\mathbf{e}_z]\exp\mathrm{j}(\omega t - \nu\theta - \beta z),\\ \mathbf{H}(r,\theta,z,t) &= [H_r(r)\mathbf{e}_r + H_\theta(r)\mathbf{e}_\theta + H_z(r)\mathbf{e}_z]\exp\mathrm{j}(\omega t - \nu\theta - \beta z).\end{aligned} \tag{3.11}$$

From Maxwell's equations we derive the equations of propagation in the two media, i.e. in the core and in the cladding respectively

$$\begin{aligned}&\frac{\mathrm{d}^2E_z(r)}{\mathrm{d}z^2} + \frac{1}{r}\frac{\mathrm{d}E_z(r)}{r\mathrm{d}z} + \left(\omega^2\epsilon\mu_0 - \beta^2 - \frac{\nu^2}{r^2}\right)E_z(r) = 0,\\ &\frac{\mathrm{d}^2H_z(r)}{\mathrm{d}z^2} + \frac{1}{r}\frac{\mathrm{d}H_z(r)}{\mathrm{d}z} + \left(\omega^2\epsilon\mu_0 - \beta^2 - \frac{\nu^2}{r^2}\right)H_z(r) = 0.\end{aligned} \tag{3.12}$$

The expression of the z component of the electromagnetic field is given by the following relations

$$\begin{aligned}E_z(r) &= E_0 J_\nu(ur/a) && \text{if } 0 < r < a,\\ E_z(r) &= E_0\frac{J_\nu(u)}{K_\nu(wr/a)} && \text{if } r > a,\end{aligned} \tag{3.13}$$

with u and w defined by the equations

$$u = a\sqrt{k_0^2 n_1^2 - \beta^2},$$
$$w = a\sqrt{\beta^2 - k_0^2 n_2^2}, \tag{3.14}$$

where n_1 and n_2 are the refractive indices of the core and of the cladding respectively. Here J_ν is the Bessel function pf the first kind of order ν and K_ν is the modified Bessel function of the first kind [Abramowitz 1964].For $H_z(r)$, we obtain similar expressions. The other components of the field can be directly derived from these relations on the basis of Maxwell's equations.

The part of the field which extends outside the core corresponds to the evanescent part of the mode. Due to the simple geometry of step-index fibers, the localization of the evanescent part of the field can be easily achieved.

Figures 3.10, 3.11 and 3.12 indicate respectively the variations of the E_z and H_z longitudinal components of the TM_{01}, TE_{02} and EH_{21} modes. V denotes the normalized frequency. In the case of a step-index fiber, the equation for the normalized frequency is

$$V = (u^2 + w^2)^{1/2} = \left[\frac{2\pi a}{\lambda}\sqrt{n_1^2 - n_2^2}\right]. \tag{3.15}$$

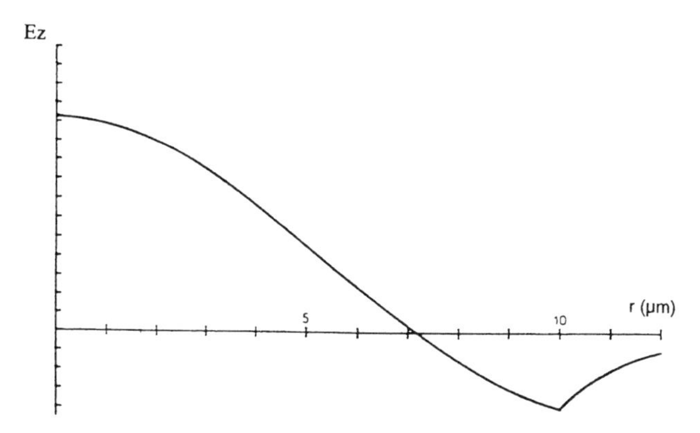

Fig. 3.10. Radial dependence of $E_z(r)$ for the TM_{01} mode of a fiber with normalized frequency $V = 7.2$

The modes of an optical fiber are sensitive to all the perturbations that the evanescent part of the field undergoes. It might be instructive to express the relative ratio of the power propagated in the core η as a function of the normalized frequency of the fiber (Fig. 3.13)

$$\eta = \frac{P_{\text{core}}}{P_{\text{core}} + P_{\text{cladding}}} = 1 - \frac{u^2}{\nu^2}\left[(w^2 + l^2)^{1/2} + \frac{u^2}{u^2 - l^2}\right]^{-1}. \tag{3.16}$$

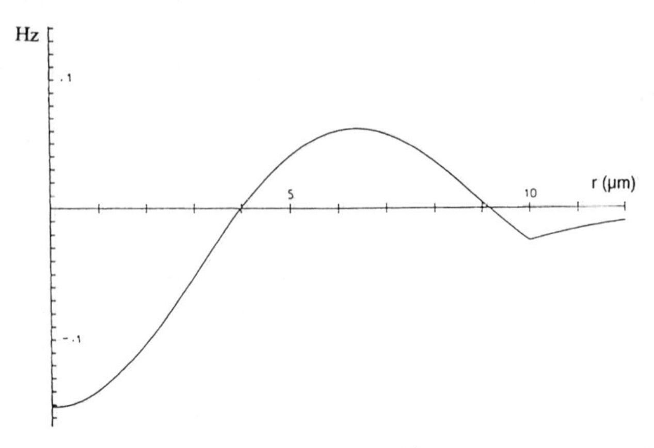

Fig. 3.11. Radial dependence of $H_z(r)$ for the TE_{02} mode of a fiber with normalized frequency $V = 7.2$

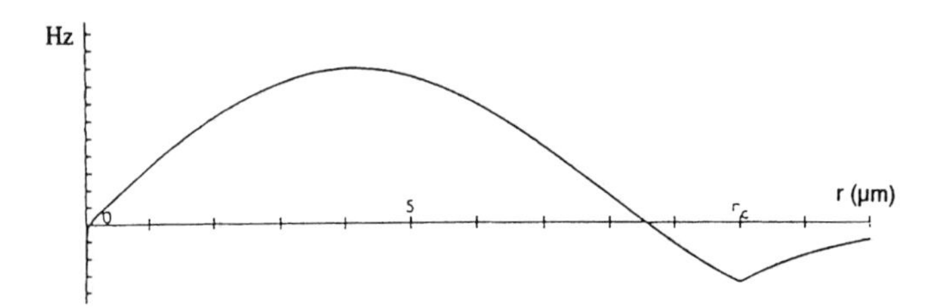

Fig. 3.12. Radial dependence of H_z(r) for the EH_{21} mode of a fiber with normalized frequency $V = 7.2$

For the fundamental mode ($l = 0$), we obtain the curve represented in Fig. 3.13. This curve indicates that the field extends very rapidly inside the cladding when the normalized frequency decreases and becomes less than a predetermined value. Below this value, the relative importance of the evanescent part of the field makes the field very sensitive to any perturbation, for example to bending or to variations of the refractive index.

It is apparent that when the mode draws nearer to its cutoff frequency, the energy of the mode is contained within the evanescent part of the field. In this case, the mode is less guided. The mode is highly sensitive to defects in the fiber as well as to bending. For producing fibers where the light will be effectively guided, it is preferable to proceed under conditions limiting as much as possible the evanescent part of the field, but preventing the distribution inside the core from being too confined. This will permit the avoidance of problems which might arise when components are being connected.

Perturbations induced on the evanescent part of the mode might cause a loss of a part of the energy of the mode. Likewise, the modes can also be excited by coupling the light to the evanescent part of the mode. The first

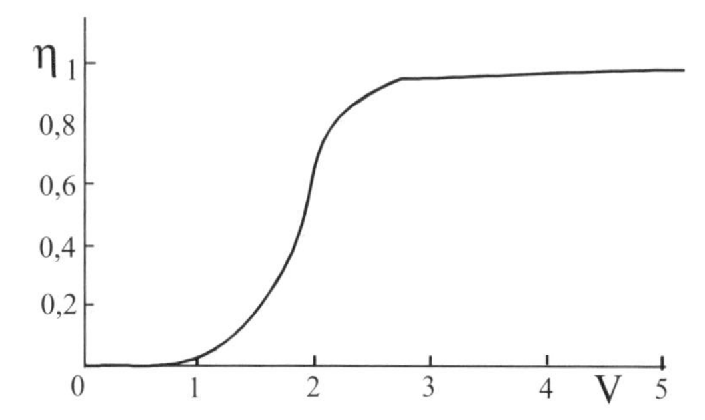

Fig. 3.13. The dependence of the normalized power guided inside the core on the total power as a function of the normalized frequency

experiments in this direction were carried out by Stewart and, a few years later, by Szczepanek [Szczepanek 1978, Facq 1980].

The experimental arrangement represented in Fig. 3.14 was used for exciting and examining the modes. Figure 3.15 represents one of the modes excited under these conditions.

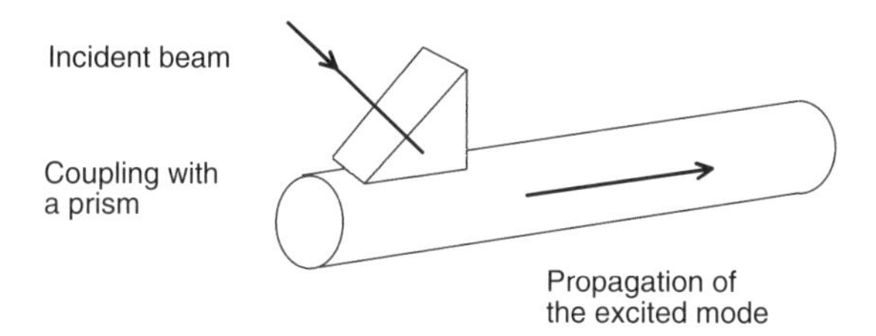

Fig. 3.14. Arrangement used for the excitation of the modes

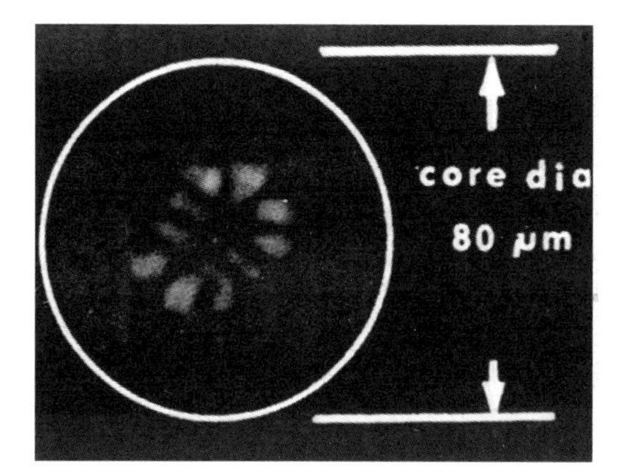

Fig. 3.15. Observation of the near-field of a mode excited inside a parabolic graded-index fiber. The diameter of the core is equal to 80 μm. The orders of the mode are $\mu = 0$ and $\nu = 5$ [Szczepanek 1978]

As shown by Stewart, modes of a high order can be excited using this technique [Facq 1980]. Figure 3.15 corresponds to the excitation of the mode of orders $\mu = 0$ and $\nu = 5$, obtained by Szczepanek in the case of a fiber with core of diameter 80 μm.

An inherent disadvantage of this technique is that it can be used only for the excitation of modes of high order (Fig. 3.15). This technique has therefore been replaced by methods based on the direct excitation of the modes [Facq 1980, de Fornel 1983, Facq 1984].

3.3.3 Modes of Inner-Cladding Fibers

Enhancement of the guiding properties of step-index fibers can be obtained by creating an index depression between the core and the cladding. This leads to the production of fibers with three index levels. The index profiles of two different fibers of this type are schematized in Fig. 3.16. We shall discuss hereafter in general terms this class of fibers, referred to as 'inner-cladding fibers'.

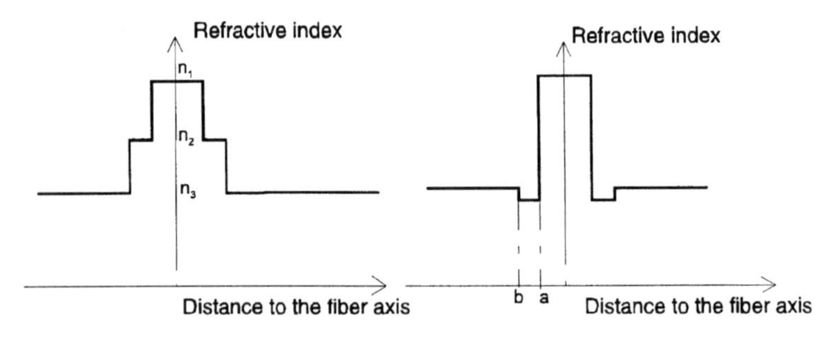

Fig. 3.16. Index profile of raised and depressed inner-cladding fibers

The thorough analysis of structures of this type is quite cumbersome. As the relative variations of the refractive index between the various parts are very small ($|1-n_i^2/n_j^2| \ll 1$), Monerie was able to determine the expressions of the field of the modes of these fibers [Monerie 1982]. The role of each interface obviously depends on the characteristics of the guiding (wavelength, modal order, refractive indices, dimensions, etc.). Depending on the values of these parameters, the evanescent field will be generated either from the first or from the second interface.

Under these conditions, the radial dependence of the transversal components of guided modes can be expressed in the following simple form

$$\psi(r) = \begin{cases} A_0 J_N(ur/a) & \text{if } r < a, \\ A_1 J_N(u'r/b) + A_2 Y_N(u'r/b) & \text{if } a < r < b, \\ A_2 K_N(wr/b) & \text{if } b < r, \end{cases} \tag{3.17}$$

if $k_0 n_3 < \beta < k_0 n_2$, and

$$\psi(r) = \begin{cases} A'_0 J_N(ur/a) & \text{if } r < a, \\ A'_1 I_N(w'r/b) + A'_2 K_N(w'r/b) & \text{if } a < r < b, \\ A'_2 K_N(wr/b) & \text{if } b < r, \end{cases} \tag{3.18}$$

if $k_0 n_2 < \beta < k_0 n_1$. Here J_N and Y_N are Bessel functions of the first and second kinds and of order N, and K_N and I_N are modified Bessel functions of the first and second kinds.

Recall that N corresponds to the azimuthal order of the mode $LP_{N,L}$, where L is the Lth root of the dispersion equation. u, u', w and w' are bound to the propagation constant by the following relations

$$\begin{aligned} u &= a\sqrt{k_0^2 n_1^2 - \beta^2}, \\ u' &= b\sqrt{k_0^2 n_2^2 - \beta^2}, \\ w &= b\sqrt{\beta^2 - k_0^2 n_3^2}, \\ w' &= a\sqrt{\beta^2 - k_0^2 n_2^2}. \end{aligned} \tag{3.19}$$

The coefficients A_i and A'_i deduced from the equations expressing the continuity of $\psi(r)$ and its derivative at each interface are given by Monerie [1982].

The eigenvalue equation of the modes can then be deduced from the continuity equations of the field and its derivative

$$\frac{[\hat{J}_N(u) - \hat{Y}_N(u/S)][\hat{K}_N(w) - \hat{J}_N(u')]}{[\hat{J}_N(u) - \hat{J}_N(u/S)][\hat{K}_N(w) - \hat{Y}_N(u')]} = \frac{J_{N+1}(u'/S)Y_{N+1}(u')}{J_{N+1}(u')Y_{N+1}(u'/S)}, \tag{3.20}$$

if $k_0 n_3 < \beta < k_0 n_2$, and

$$\frac{[\hat{J}_N(u) - \hat{K}_N(w'/S)][\hat{K}_N(w) - \hat{I}_N(w)]}{[\hat{J}_N(u) - \hat{I}_N(w'/S)][\hat{K}_N(w) - \hat{K}_N(w')]} = \frac{I_{N+1}(w'/S)K_{N+1}(w')}{I_{N+1}(w')K_{N+1}(w'/S)}, \tag{3.21}$$

if $k_0 n_2 < \beta < k_0 n_1$. The different terms in these equations are given by the relation

$$\hat{Z}_N(x) = \frac{Z_N(x)}{x Z_{N+1}(x)}.$$

The parameters of the mode being determined, it might be instructive to represent graphically the dependence of the distribution of the field in the core and the cladding on the different characteristics of the waveguide. For this purpose, we plotted the ratio η of the flux of the Poynting vector inside the core to the flux of the Poynting vector in a plane normal to the fiber axis.

The z component of the Poynting vector is

$$P_z(r) = [E_r(r)H_\theta(r) - E_\theta(r)H_r(r)]/2. \tag{3.22}$$

Hence the equation for η is

$$\eta = \frac{\displaystyle\int_0^a P_z(r)\mathrm{d}r}{\displaystyle\int_0^\infty P_z(r)\mathrm{d}r}. \tag{3.23}$$

The dependence of η on the normalized frequency of these waveguides is represented in Figs. 3.17 and 3.18, for two sets of values of the parameters S and R. The normalized frequency is defined by the equation

$$V = (2\pi a/\lambda_0)\sqrt{n_1^2 - n_3^2}. \tag{3.24}$$

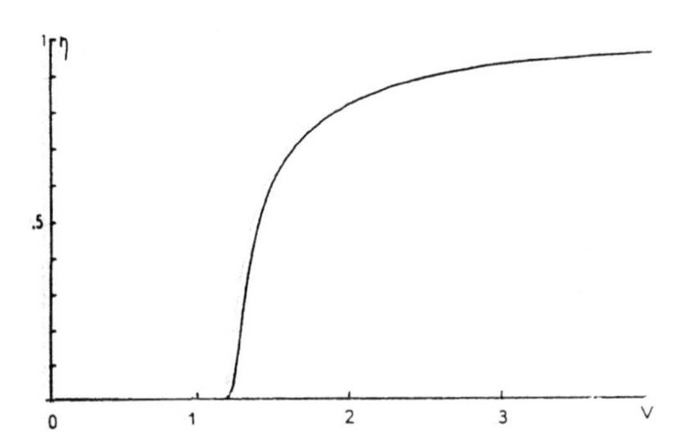

Fig. 3.17. Energy inside the core as a function of the normalized frequency V, where $S = b/a = 2$ and $R = (n_2 - n_3)/(n_1 - n_3) = -0.3$

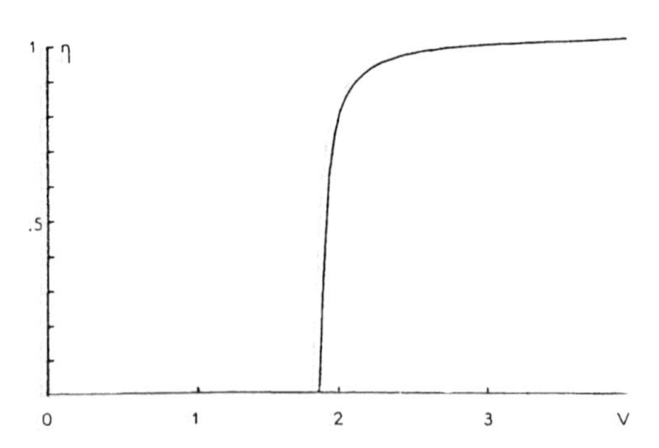

Fig. 3.18. Energy inside the core as a function of the normalized frequency V, where $S = b/a = 1.084$ and $R = (n_2 - n_3)/(n_1 - n_3) = -0.9$

The influence of the size of the depression on the guiding of the mode can also be graphically represented. The curve drawn in Fig. 3.19 corresponds to a value of the normalized frequency equal to 2.

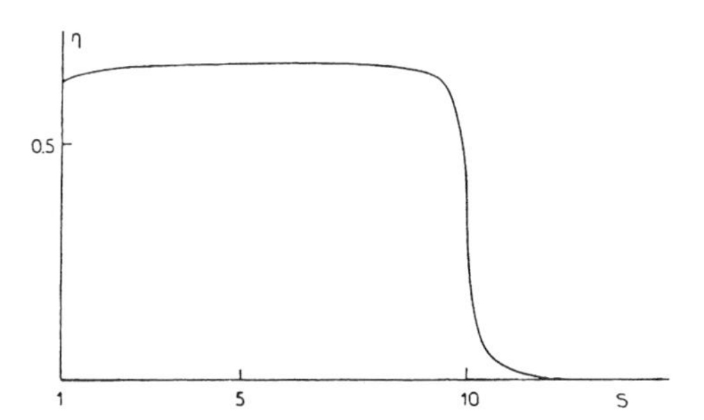

Fig. 3.19. Power inside the core as a function of the size of the depression, where $V = 2$ and $R = -0.1$

3.3.4 Modes of Annular-Core Fibers

In the waveguides previously described, the evanescent field extended outside the waveguide and therefore had a delocalized character. By creating a central index depression in the core of the waveguide, as schematized in Fig. 3.20, an evanescent field localized inside the fiber can be generated. Therefore, one obtains a ring-shaped guiding area, which fulfills the same function as the core in a step-index fiber. This type of guiding is known under the name of annular waveguiding.

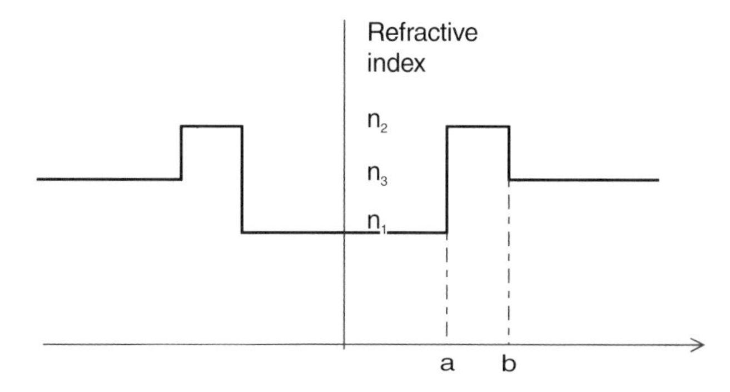

Fig. 3.20. Index profile of an annular-core fiber

The transverse dependence of the field of the guided modes of this fiber on the radial distance is given by the equations

$$\psi(r) = \begin{cases} A_0 I_N(w_1 r/a) & \text{if } r < a, \\ A_1 J_N(u_2 r/b) + A_2 Y_N(u_2 r/b) & \text{if } a < r < b, \\ A_2 K_N(w_3 r/b) & \text{if } b < r, \end{cases} \tag{3.25}$$

where

$$\begin{aligned} w_1 &= b\sqrt{\beta^2 - k_0^2 n_1^2}, \\ u_2 &= b\sqrt{k_0^2 n_2^2 - \beta^2}, \\ w_3 &= a\sqrt{\beta^2 - k_0^2 n_3^2}. \end{aligned} \tag{3.26}$$

Resuming the procedure used by Monerie, the eigenvalue equation is determined in the following form

$$\frac{[\hat{J}_N(u_2c) - \hat{I}_N(w_1)][\hat{Y}_N(u_2) - \hat{K}_N(w_3)]}{[\hat{Y}_N(u_2c) - \hat{I}_N(w_1)][\hat{J}_N(u_2) - \hat{K}_N(w_3)]} = \frac{Y_{N+1}(u_2c)J_{N+1}(u_2)}{Y_{N+1}(u_2)J_{N+1}(u_2c)}, \tag{3.27}$$

where $\hat{J} = J_n'/J_n$ and so on.

Ito *et al.* have excited the fundamental mode of such a structure with an efficiency of 10% [Ito 1995]. These experiments were primarily carried out with the construction of atom waveguides in view. The project of using the radiation pressure exerted by evanescent fields for constructing atom waveguides will be treated more extensively in Chap. 8. Figure 3.21 represents an example of a mode guided inside this structure.

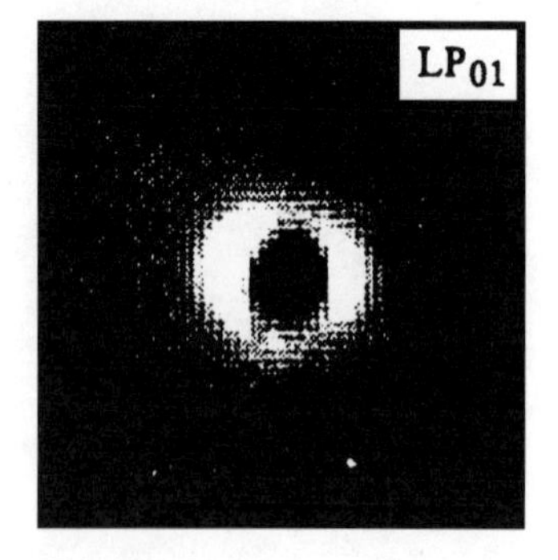

Fig. 3.21. LP_{01} mode excited inside an annular-core fiber [Ito 1995]

3.3.5 Modes of Graded-Index Fibers

In the case of step-index fibers, the region where the evanescent field is present can be readily determined, whatever the number of index depressions in the fiber is. The question which arises then is that of the possibility of localizing the evanescent field inside a graded-index fiber.

Figure 3.22 represents the paths followed by different rays inside a graded-index waveguide with parabolic index profile. Consider the optical path of the first ray: from A_0 to A, the ray is just refracted, and then is totally reflected at point A. A diferent ray will be totally reflected at another value of the distance to the axis. Therefore the field is not evanescent within a unique region of the waveguide. Indeed, the field will depend on the entire set of rays, each of these rays being characterized by a specific direction and optical path.

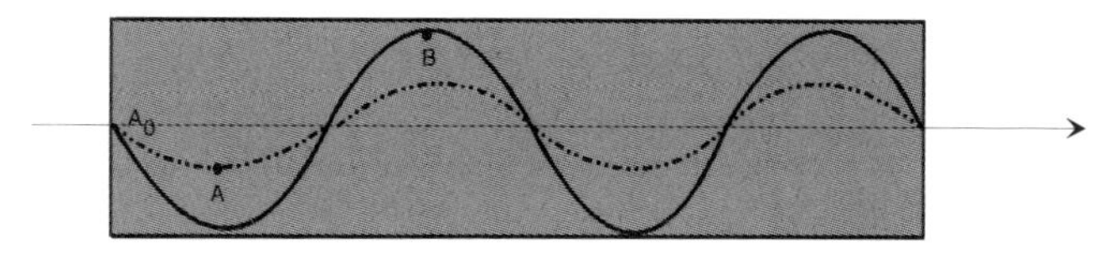

Fig. 3.22. Paths followed by rays inside a waveguide with parabolic profile

3.3.6 Modes of Polarization-Preserving Fibers

The perturbation of the evanescent part of a mode can result in a modification of the properties of this mode. As an example, the polarization generally induces a degeneracy of the modes in fibers presenting a symmetry of revolution. This degeneracy can be removed by perturbing the evanescent part of the mode. This can be achieved either by inducing birefringence in the core or in the cladding, or by creating variations of the indices, in order to break the symmetry of revolution of the waveguide.

There are several different types of polarization-preserving fibers: elliptic-core fibers, panda fibers or bow-tie fibers are a few examples of fibers of this type [Snyder 1983]. We shall detail hereafter the case of fibers where the perturbation arises within the evanescent part of the field. Figure 3.23 represents the index profile of a bow-tie fiber. An effect of the presence of an index depression inside the cladding is that the two normally polarized modes do not have the same propagation constant.

From the curves presented in Fig. 3.24 it can be seen that the two polarized modes of a bow-tie fiber both have a nonzero cutoff frequency. Further, these polarized modes do not have the same cutoff frequency and therefore the extinction ratio of the fiber can be high (Fig. 3.25).

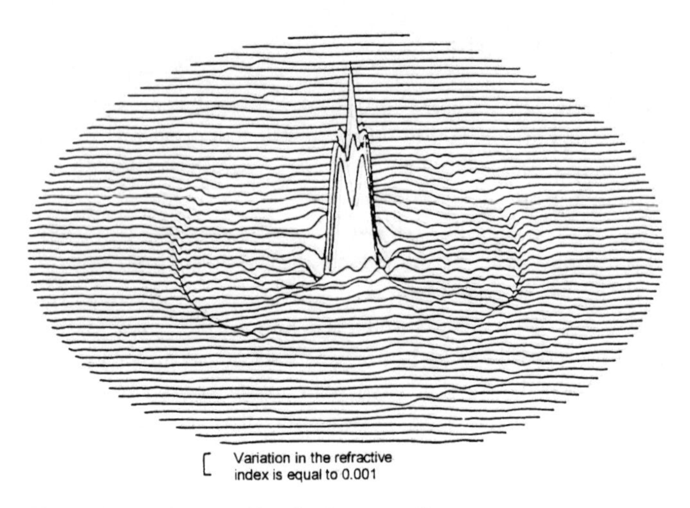

Fig. 3.23. Index profile of a bow-tie fiber

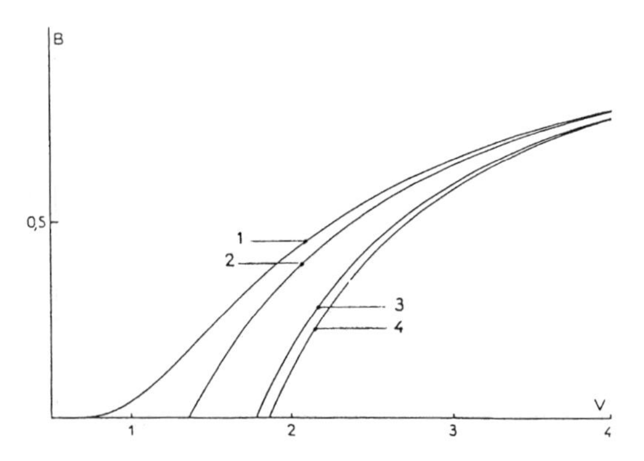

Fig. 3.24. Dispersion curves (**a**) mode of a step-index fiber with no local depression inside the cladding, (**b**) and (**c**) polarized modes of a bow-tie fiber, (**d**) mode of a depressed inner-cladding fiber with a ring-shaped depression

3.4 Whispering-Gallery Modes

The modes that we examined in the preceding subsections were propagating within the core of the waveguide. Therefore, these modes remained relatively well confined. This type of propagation is adapted to applications related to information transfer.

The modes that will be described in this section are referred to under the name of whispering-gallery modes. The concept of whispering-gallery mode dates back to the nineteenth century, and was introduced first by Lord Rayleigh in the study of the propagation of acoustic waves. It is well known that whispers in the gallery of the cathedral of St. Paul's in London can be

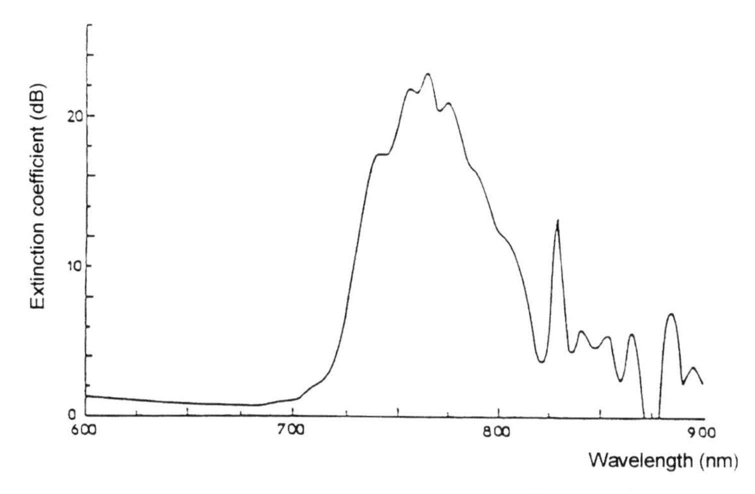

Fig. 3.25. Value of the extinction coefficient of a bow-tie fiber as a function of the wavelength

heard from other places in the gallery without any loss of information. The term 'whispering-gallery mode' refers to this phenomenon, for which Rayleigh developed a solution in terms of Bessel functions.

From the 1960s, the use of whispering-gallery modes has extended from acoustics to the study of electromagnetic waves. In 1961, Arnaud suggested an application of whispering-gallery modes to microwaves [Vedrenne 1982, Cros 1990b]. As described by Garrett in 1961, whispering-gallery modes can be applied to laser sources [Garret 1961, Sandoghdar 1998]. As an example, a CaF_2:Sm_2^+ sphere immersed in liquid hydrogen and optically pumped exhibited a tangential stimulated emission inducing the oscillation of a whispering-gallery mode.

We shall not engage here in an exhaustive description of the whispering-gallery modes. Figure 3.26 represents the geometrical construction associated with the whispering-gallery mode, in the case of a circular propagation of the rays [Lomer 1992].

If the incidence plane is not normal to the axis, the rays propagate helicoidally inside the waveguide. The photograph reproduced in Fig. 3.27 is a lateral view of a silica cylinder where a whispering-gallery mode has been excited. The helicoidal propagation which is associated with whispering-gallery modes can be observed in this figure.

The near-field of the extremity of a fiber where a whispering-gallery mode has propagated is shown in the photograph presented in Fig. 3.28. The ray associated with the whispering-gallery mode can be distinctly recognized here. The excitation of the whispering-gallery mode was achieved by using a coupling prism [Etourneau 1996].

The structure of the whispering-gallery modes was used for determining the value of the losses caused by defects present at the surface of a plastic fiber

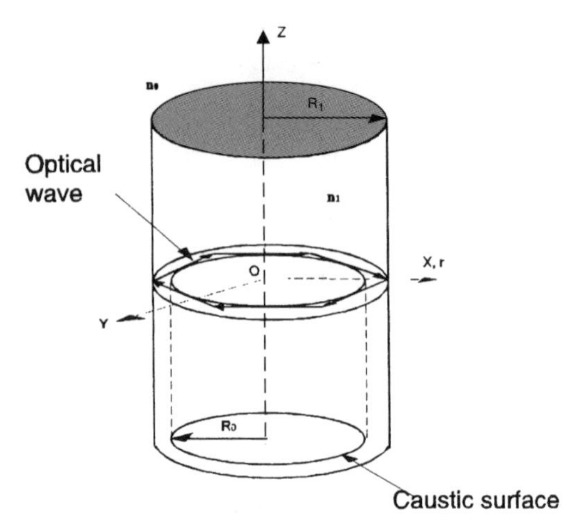

Fig. 3.26. Geometrical construction associated with the whispering-gallery mode for a circular propagation of rays

Fig. 3.27. Lateral view of a cylinder where a whispering-gallery mode is excited, displaying the helicoidal propagation of the rays [Lomer 1992]

[Etourneau 1996]. Researches on whispering-galery modes of microspheres are still progressing [Mukaiyama 1999].

3.5 Band-Gap Photonics Waveguides

Before ending this chapter, it might be of interest to mention the existence of a markedly different type of waveguides [Knight 1996, Broeng 1999, Centeno 1999]. These waveguides are produced from band-gap photonics structures. The particular example represented in Fig. 3.29 is fabricated from

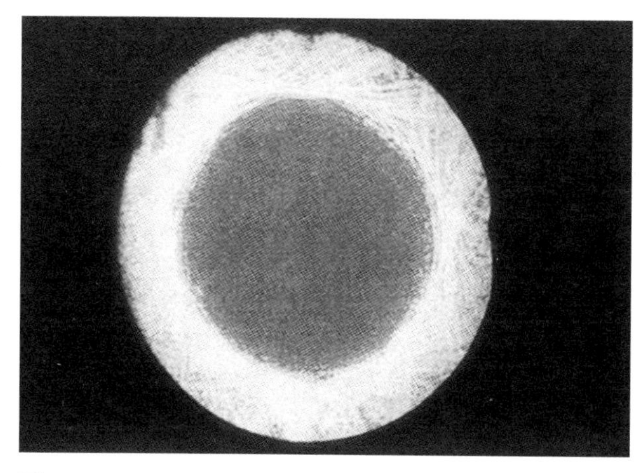

Fig. 3.28. Near-field of a plastic fiber where helicoidal rays are propagating [Etourneau 1996]

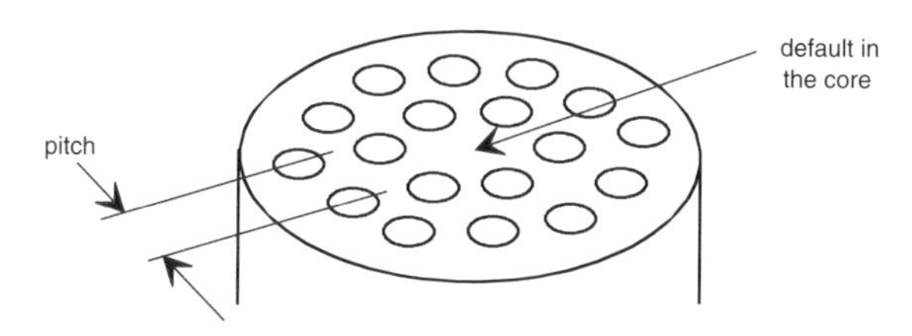

Fig. 3.29. Schematic representation of a band-gap photonics waveguide

a bundle of fused-tapered capillaries. The structure thus produced consists of a set of hexagonally distributed holes, except at the center of the waveguide.

The analysis of these waveguides has demonstrated that they remain single-mode at all wavelengths. The fabrication of these waveguides is quite difficult, and the field of applications of these fibers has not been entirely determined.

3.6 Conclusion

In this chapter we have recalled that light propagates inside a confined medium only if totally reflected along the entire waveguide. The evanescent field is a proper part of the mode, and the properties of the mode are intimately related to the evanescent field, and depend on all the physical parameters of the waveguide, namely the geometry of the waveguide and the index profiles of the cladding and of the core. An example is the possibility of removing the degeneracy of the mode that results from the polarization, simply by breaking the symmetry of the refractive index, thereby inducing a perturbation of the evanescent part of the mode. This phenomenon is related to the fact that modes with distinct polarizations do not have the same propagation constant.

By abrading the cladding of the waveguide, the evanescent part of a mode can be easily reached. This leads to the possibility of producing highly sensitive sensors. The sensors developed on these principles are described in Chap. 6. The applications of hollow-core fibers belong to the field of atom optics and will be described in the chapter devoted to the utilization of the evanescent field in this area.

Conclusion of Part I

These first three chapters are essentially devoted to the description from a theoretical standpoint of the physics of the evanescent field in different optical systems. Beginning with the configuration described by Newton, we first presented the form of the evanescent field generated by the total internal reflection of a plane wave. Even if this case could at first sight seem simpler than some of the other systems which generate evanescent waves, researches on the evanescent field at total internal reflection have continued until recent years. An example is the Goos–Hänchen shift, which was not completely analyzed until the 1960s.

In the analysis of the field present in the vicinity of a simple dipole, the separation between the propagative and the evanescent parts of this field has turned out to be not a trivial matter. Here again, although the study of this phenomenon dates back quite a long time, researches on fluorescence have recently led to a renewal of interest in dipolar resonances.

In the last part of this book, some of the recently developed local-probe microscopies will be described. For understanding the principles of these microscopies, it is necessary to recall first the analysis of the evanescent field generated in the vicinity of subwavelength apertures. For this reason, a chapter has been devoted here to the analysis of this configuration.

Lastly, the description of the evanescent field presented in these three chapters would have been incomplete if the role of the evanescent field in the area of integrated optics had not been examined. Therefore we were led to include here a presentation both of the characteristics of guided modes and of the importance of the evanescent field associated with these modes, in optics as well as in the field of microwaves.

The relation between the evanescent field present in some configurations and the physical parameters of the media involved in these configurations has been presented in the first three chapters. In the rest of this book, different types of devices will be described, the developement of which was a consequence of researches on the evanescent field.

The localized or delocalized character of the interaction with the evanescent field will be used as a criterion for separating these applications into two classes. Part II deals with devices where the interaction with the evanescent field has a laterally delocalized character, while the devices where this interaction is localized are examined in Part III. For some devices, the application of this criterion might not be self-evident. This will be indicated hereinafter every time the case arises.

Part II

Delocalized Interaction with the Evanescent Field

Introduction to Part II

As described in the first three chapters, an evanescent field can be generated when a plane wave either reaches an interface where the conditions of total internal reflection are satisfied or illuminates a structure with subwavelength extent. The detection and analysis of evanescent waves requires the complete or partial transformation of the evanescent field into a propagative field. To this end, it is necessary to induce a perturbation on the evanescent field.

The evanescent field generated by the total internal reflection of a plane wave and frustrated from a plane presents the same character of lateral delocalization as the propagating mode does. In contrast, the diffraction of a plane wave from an interface with subwavelength dimensions is an extremely localized phenomenon.

The following chapters deal with systems in which the interaction with the evanescent field has a delocalized character. This means that the interaction with the evanescent field arises within a lateral domain larger than the square of the wavelength of the light used.

In Chap. 4 the characteristics of evanescent-field fiber-optic couplers will be described. A survey of the analysis of the coupling between two waveguides is also provided in this chapter. We also describe some applications of evanescent-field couplers. Chapter 5 is devoted to a different type of evanescent-field couplers, known as integrated-optical couplers.

Chapter 6 deals with the utilization of the sensitivity of the evanescent field to physical parameters for the production of fiber-optic and integrated-optical sensors. These sensors are based on the perturbation exerted by a given parameter on the evanescent field of guided modes. Some applications of these sensors are also presented. A related topic is internal-reflection spectroscopy, which is treated in Chap. 7. Here again, extensions of this technique will be presented.

Chapter 8 is rather different in character, and addresses the applications of the properties of the evanescent field in the area of atom optics. In particular, we describe experiments on the reflection and deflection of atoms using evanescent fields. The very recent schemes of atom guiding are also presented.

In Chap. 9 is a description of microscopy systems based on the use of properties of evanescent waves. We present two different instruments, the dark-field microscope developed in the nineteenth century and the photon tunneling microscope. The fundamental difference between these microscopes and the systems described in the three last chapters of the book is that the latter use a subwavelength probe for the detection of the evanescent field. A further difference between these systems is that the resolution of the microscopes described in Chap. 9 is still within the Rayleigh limit.

4. Evanescent-Field Optical-Fiber Couplers

In classical optics, the use of a beam splitter is required to be able to sum and compare data transported by distinct light waves. Without such an element, a large range of optical functions cannot be achieved. In guided optics, beam splitters are replaced by optical-fiber couplers.

The idea of data transfer necessarily implies the possibility not only of summing signals with distinct origins, but also of selecting a part of the propagating signal. Couplers are precisely those devices which have been developed in order to fulfill this function. We first present evanescent-field couplers realized from optical fibers. Integrated-optical evanescent-field couplers will be treated in the next chapter.

4.1 Types of Couplers

In the chapters devoted to the description of the evanescent field, we described the modes of optical fibers. The field of these modes has an evanescent part, which is generally located within the cladding of the fiber [Adams 1981]. If two optical fibers are sufficiently close to each other, a coupling between the two fibers arises. In other words, the light energy travels from one of the fibers into the other, and conversely. The coupling between two fibers can be expressed in a simple form as a function of the overlap integral of the fields of the fibers.

The couplers based on these principles are referred to as evanescent-field couplers. A large range of components derive from this type of couplers: some of them will be presented hereafter. Evanescent-field coupling is also used in planar optics. The applications in these fields are treated in the next chapter. The fundamental principle of evanescent-field coupling is schematized in Fig. 4.1.

Before turning to evanescent-field coupling, it may first be recalled that evanescent-field couplers are not the only type of optical-fiber couplers. Indeed, the simplest method for summing two information signals coming from different sources consists in joining two waveguides. Figure 4.2 represents such an arrangement, known under the name of a Y-branch coupler. The name refers to the form of the coupler, which consists of two waveguides converging at a branch into a single waveguide. This device can be used either

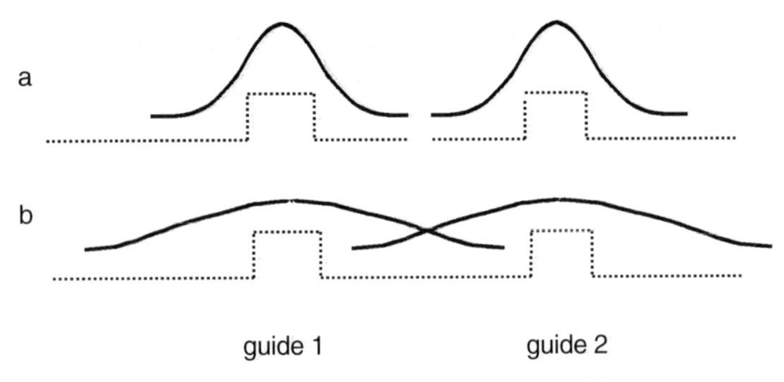

Fig. 4.1. Principle of an evanescent-field coupler (**a**) if the distance between the two waveguides is very large, the evanescent fields do not overlap, and the waveguides are not coupled, (**b**) if the waveguides are close enough, the fields overlap: a part of the light energy present in one of the waveguides can transfer into the other waveguide

as a multiplexer or as a demultiplexer. The Y-branch coupler represented in Fig. 4.2 in fact works as a divider.

These couplers can be fabricated only from fibers with large diameter. Further, the distribution of the field of the two fibers is not identical with the repartition of the field of the fiber where the signals are multiplexed, and this induces nonnegligible losses. For these reasons, evanescent-field couplers are generally preferred.

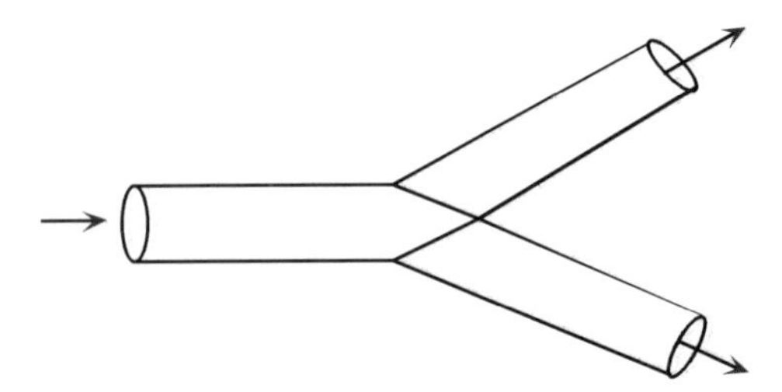

Fig. 4.2. Y-branch coupler used as a divider. The three fibers are designed in such a way that the core of the first fiber is in contact with the cores of the other two: the light beam is divided with nonnegligible loss

4.2 Fabrication Techniques of Evanescent-Field Fiber-Optic Couplers

4.2.1 Twist-Etched Fiber Couplers

Early evanescent-field couplers were produced by chemically etching two fibers at the same time in order to reduce the distance between the cores of the fibers. This technique was described by Sheem in 1979. Two twisted optical fibers are immersed in a bottle filled with hydrofluoric acid (Fig. 4.3). When the desired coupling ratio is reached, the hydrofluoric acid is replaced by water in order to stop the chemical etching of the silica and rinse the fibers. Finally, the bottle is filled with a resist which, when solidifying, reduces the fragility of the coupler [Sheem 1979, Liao 1980, Sheem 1981]. This technique has been presently abandoned, for the devices thus produced were lacking stability and durability.

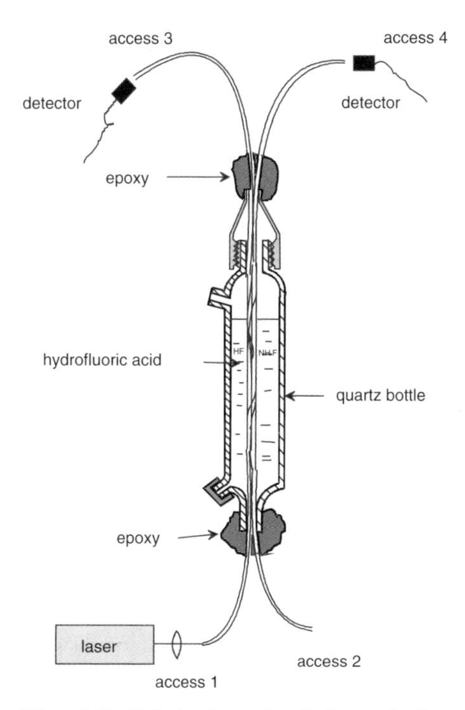

Fig. 4.3. Fabrication of a twist-etched coupler [Sheem 1979, with the permission of the Optical Society of America]

4.2.2 Mechanically Polished Fiber Couplers

The lack of stability and durability of twist-etched fibers that results from the etching of the fiber by hydrofluoric acid has led to the proposal of other fabrication techniques. The method that we describe has been developed by Bergh. The distance between the cores is reduced by moving closer together two locally polished fibers [Bergh 1980].

As shown in Fig. 4.4, a slot is first cut into two quartz blocks, and a fiber is bonded in the slot of each of these blocks. The blocks are then mechanically

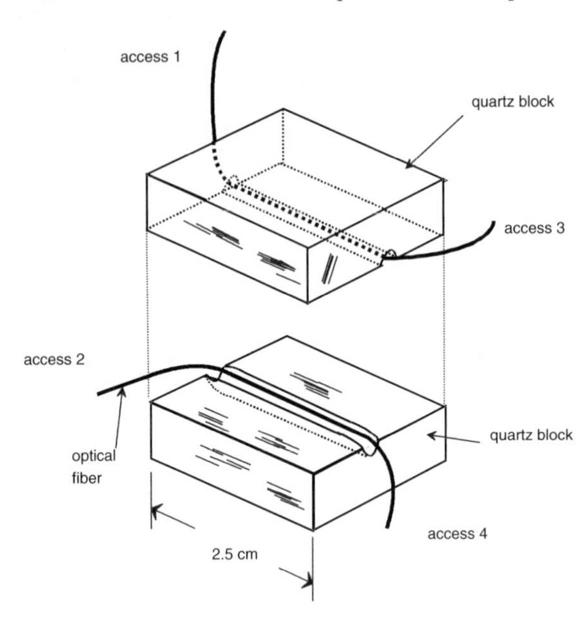

Fig. 4.4. Fabrication of a polished-fiber coupler [Bergh 1980, with the permission of the IEE]

polished and moved closer to each other. By adjusting the distance between the fibers, any value of the coupling ratio can be achieved. Refractive-index matching oil can be inserted between the fibers [Bergh 1980]. The couplers realized with this technique present low losses, high directivity, and more generally good characteristics, and accordingly they have become widely used.

4.2.3 Fused-Tapered Fiber Couplers

The fabrication technique of fused-tapered fiber couplers is schematically represented in Fig. 4.5. Two optical fibers are twisted together and mechanically pulled either under the flame of a burner or inside a furnace [Kawasaki 1981, Ragdale 1983]. Light is then injected into one of the fibers, which is pulled until the desired coupling ratio is reached. The implementation of this technique is simple, but requires good control of the pulling process if reproducible coupling ratios are wanted.

Fused-tapered fiber couplers differ from the two previously described types of couplers by the fact that the overlap of the fields is achieved by reducing simultaneously the distance between the cores and their diameter, thus causing a spreading of the evanescent field inside the cladding.

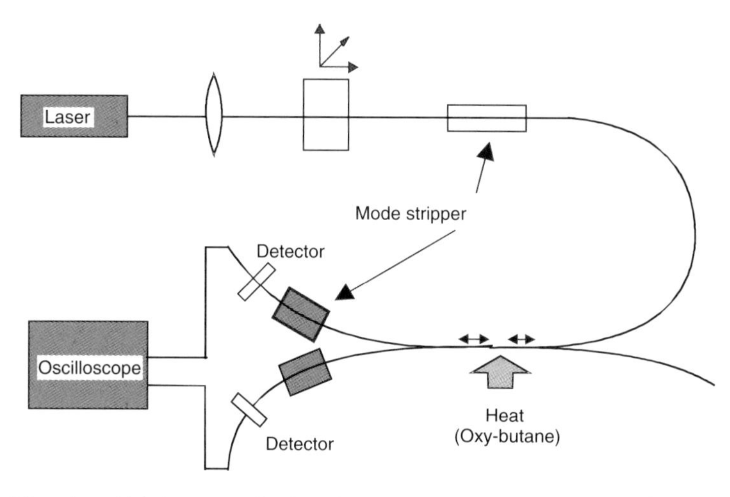

Fig. 4.5. Fabrication of a fused-tapered optical-fiber coupler [Ragdale 1983]

4.2.4 Comparison Between the Different Types of Couplers

A coupler, for example a (2×2) coupler, can be schematically illustrated in a simple form as a black box with two inputs and two outputs, as represented in Fig. 4.6. An $(n \times n)$ coupler has n inputs and n outputs.

The characteristics of a coupler with a single input at I_1 are determined by the following set of parameters.

- Excess loss: $-10 \log \left(\dfrac{O_1 + O_2}{I_1} \right)$.
- Coupling ratio: in %, $\dfrac{O_2}{O_1 + O_2}$; in dB, $-10 \log \left(\dfrac{O_2}{O_1 + O_2} \right)$.
- Insertion loss: excess loss + coupling ratio.
- Directivity: $10 \log \left(\dfrac{I_2}{I_1} \right)$.

Here I_1 and I_2 are, as illustrated in Fig. 4.6, the values of the intensity at each input of the coupler, while O_1 and O_2 are the values of the intensity at each output.

We end this rapid survey of the techniques of production of couplers with a comparison of the characteristics of fused-tapered and polished couplers.

Fig. 4.6. Schematic representation of the parameters in a coupler

Table 4.1 has been set from data indicated by different manufacturers, such as Gould [1992]. The characteristics indicated in this table correspond to a standard coupler, and single-mode fiber couplers with lower losses than the limits reported can be designed. The coupling ratios of the couplers have not been indicated, for the value of the losses remains almost unaffected by variations of this parameter.

Table 4.1. Comparison of fused-tapered couplers and mechanically polished couplers.

Parameters of the coupler	Excess loss (dB)	Directivity (dB)
Fused-tapered single-mode fiber couplers	< 0.2	> 55
Polished single-mode fiber couplers	< 0.2	> 55
Fused-tapered multimode fiber couplers	< 1.5	> 40

The utilization of evanescent-field coupling enables the production of single-mode fiber couplers with very low losses. In contrast, multimode fiber couplers still present residual losses. The theoretical analysis of the coupling presented in the next section is designed to provide the materials necessary for understanding the reasons for the difference between the behaviors of these two couplers.

4.3 Analysis of the Coupling

An evanescent-field coupler consists at least of two waveguides placed in conditions such that the evanescent parts of the fields of their modes partially overlap. The system thus produced is a coupled system. If the fields slightly overlap, the system is defined by the two waveguides where the eigenmodes of the fields propagate. The first studies on this subject were carried out in the field of microwaves [Marcuse 1973].

The coupling between the two waveguides is expressed by the value of the rate of energy transfer from one of the waveguides into the other, which is approximately in proportion to the overlap integral of the modes of the two waveguides. This is true only if the presence of each waveguide does not much perturb the mode propagating inside the other. Otherwise, the ensemble system of the two waveguides has to be described. This means that the eigenmodes of the system thus formed have to be taken into account. We shall limit ourselves here to the case of evanescent-field couplers, which corresponds to low coupling.

The analysis presented hereafter resumes works by Marcuse for microwave frequencies [Marcuse 1973]. These calculations are based on the assumption that each waveguide is not much perturbed by the presence of the other waveguide. In other words, the resulting field can be expressed in terms of the modes of the two waveguides, as in the equations

$$\begin{aligned} \mathbf{E}(r,\theta,z) &= A_1(z)\mathbf{E_1}(r,\theta,z) + A_2(z)\mathbf{E_2}(r,\theta,z), \\ \mathbf{H}(r,\theta,z) &= A_1(z)\mathbf{H_1}(r,\theta,z) + A_2(z)\mathbf{H_2}(r,\theta,z). \end{aligned} \tag{4.1}$$

The functions $A_1(z)$ and $A_2(z)$ yield the relative evolution of the fields in each waveguide. The resulting fields are interrelated by Maxwell's equations

$$\begin{aligned} \mathbf{rotH}(r,\theta,z) &= \mathrm{j}\omega\epsilon_0 n^2\mathbf{E}(r,\theta,z), \\ \mathbf{rotE}(r,\theta,z) &= -\mathrm{j}\omega\mu_0\mathbf{H}(r,\theta,z), \end{aligned} \tag{4.2}$$

where the refractive index n is defined in terms of the indices of each waveguide

$$n^2 = (n_1^2 - n_3^2) + n_3^2, \tag{4.3}$$

where n_1 and n_2 are the indices of the two fibers. Each of these indices depends on r and θ. n_3 is the index of the medium in which the fibers have been placed. We consider the case where the value of n_3 is constant.

The functions $A_1(z)$ and $A_2(z)$, which express the evolution of the field, are obtained by solving the equation deduced from the previous relations

$$\begin{aligned} &\frac{\mathrm{d}A_1(z)}{\mathrm{d}z}\int_0^\infty\int_0^{2\pi}[\mathbf{E}_\mathbf{1}^-(\mathbf{e_z}\wedge\mathbf{H_1}) + \mathbf{H}_\mathbf{1}^-(\mathbf{e_z}\wedge\mathbf{E_1})]\mathrm{d}r\mathrm{d}\theta \\ &= A_2(z)\mathrm{j}\omega\epsilon_0\int_0^\infty\int_0^{2\pi}(n_1^2 - n_3^2)\mathbf{E}_\mathbf{1}^-\mathbf{E_2}\mathrm{d}r\mathrm{d}\theta, \end{aligned} \tag{4.4}$$

where

$$\mathbf{E_{1,2}} = \mathbf{E_{1,2}}\exp[\mathrm{j}(\omega t - \beta_{1,2}z)] \text{ and } \mathbf{E}_\mathbf{1,2}^- = \mathbf{E}_\mathbf{1,2}^-\exp[-\mathrm{j}(\omega t - \beta_{1,2}z)]. \tag{4.5}$$

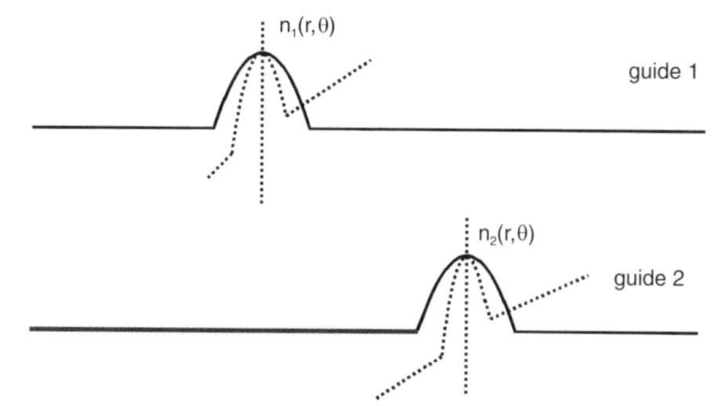

Fig. 4.7. Index profiles of the two waveguides

The subscripts in the equation indicate the waveguide to be considered. In the case of a fiber with diameter gradually and adiabetically extending with z, the eigenmodes of the waveguide can always be locally determined.

The previous equation can also be written in the form

$$A_2(z)c_1(z) = \mathrm{j}(\mathrm{d}A_1(z)/\mathrm{d}z)\exp[-\mathrm{j}(\beta_1 - \beta_2)], \tag{4.6}$$

where $c_1(z)$ is given by the equation

$$c_1(z) = -\omega\epsilon_0 \frac{\displaystyle\int_0^\infty \int_0^{2\pi} (n_1^2(r) - n_3^2)\mathbf{E}_1^*\mathbf{E}_2 r\mathrm{d}r\mathrm{d}\theta}{\displaystyle\int_0^\infty \int_0^{2\pi} \mathbf{e_z}(\mathbf{E}_1^* \wedge \mathbf{H}_1 - \mathbf{E}_1 \wedge \mathbf{H}_1^*) r\mathrm{d}r\mathrm{d}\theta}, \tag{4.7}$$

r being defined with respect to the axis of the first fiber. The equation for $c_2(z)$ is exactly the same as (4.7) except that the subscript 1 is replaced by the subscript 2. The integration is then carried over the second fiber instead of the first.

These coefficients characterize the coupling between the modes of two fibers. In order to determine the coupling between the modes, it is necessary to measure the transmitted power. We first examine the ideal case of two parallel uniform fibers.

4.3.1 Coupled Power Between Two Parallel Uniform Fibers

If the two fibers are perfectly uniform and parallel, the value of the coupling coefficient is constant. In the case of two fibers with the same characteristics, it is readily seen that functions $A_1(z)$ and $A_2(z)$ become

$$\begin{aligned} A_1(z) &= [A_1(0)\cos(cz) + A_2(0)\sin(cz)]\exp\mathrm{j}\beta z, \\ A_2(z) &= [A_1(0)\sin(cz) + A_2(0)\cos(cz)]\exp\mathrm{j}\beta z. \end{aligned} \tag{4.8}$$

Assuming that the light power was injected in the first waveguide at $z = 0$, the values of the power transferred in the two waveguides are respectively

$$\begin{aligned} P_1(z) &= P_0\cos^2(cz), \\ P_2(z) &= P_0\sin^2(cz), \end{aligned} \tag{4.9}$$

where P_0 represents the initial power at $z = 0$.

It is essential to understand the physical meaning of the expression of the coupling coefficient. The coupling coefficient is in proportion to the overlap integral of the fields of the two fibers. In the current conditions of fabrication of couplers, the only terms expressing the coupling are the products between the evanescent part of the field and the propagative part guided in the core, or the cross-products of the evanescent parts of the fields. The coupling can be seen as arising via the evanescent part of the field of the guided mode. For this reason, these couplers are in general referred to under the name of 'evanescent-field couplers'.

4.3.2 Step-Index Fibers

The analytical expression of the coupling ratio of two step-index fibers is [Snyder 1972]

$$C = \frac{\sqrt{(n_1^2 - n_2^2)}u^2 K_0(wD/a)}{aV^3 K_1^2(w)}, \tag{4.10}$$

where n_1 is the index of the cladding of the fibers, and K_0 and K_1 are the modified Bessel functions of the second kind of order 0 and 1 respectively. u and w are variables characterizing the propagation of the mode, and V is the normalized frequency of the fibers. The expression of the normalized frequency is

$$V = (2\pi/\lambda)a\sqrt{n_1^2 - n_2^2} = \sqrt{u^2 + w^2}. \tag{4.11}$$

Here u and w represent the solutions to the equation

$$uK_0(w)J_1(u) = wK_1(w)J_0(u), \tag{4.12}$$

where J_N is the Bessel function of the first kind of order N. Figure 4.8 represents the coupling coefficient C expressed as a function of the normalized frequency V [Bures 1983].

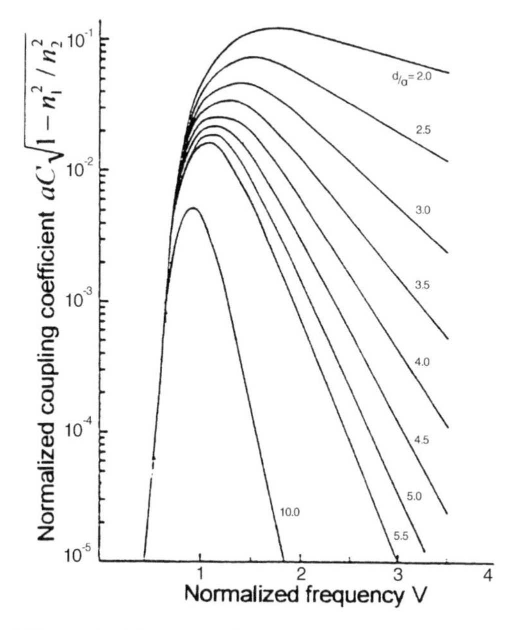

Fig. 4.8. Theoretical curves of the normalized coupling as a function of the normalized frequency V, with D/a as parameter. D is the distance between the centers of the two fibers and a is the radius of the core [Bures 1983, with the permission of the Optical Society of America]

From observation of these curves, it is apparent that the coupling arises only at values of the normalized frequency contained in a certain interval. The coupling is non-existent at very high values of V, for the evanescent fields do not in this case overlap. As the normalized frequency decreases, the evanescent field extends farther from the core, thereby inducing an excitation of the modes of the other waveguide (Fig. 4.9). At very low values of the normalized frequency, the modes of the two waveguides merge, and therefore the waveguides become undifferentiated. In this case, the model of the evanescent-field coupling fails to apply.

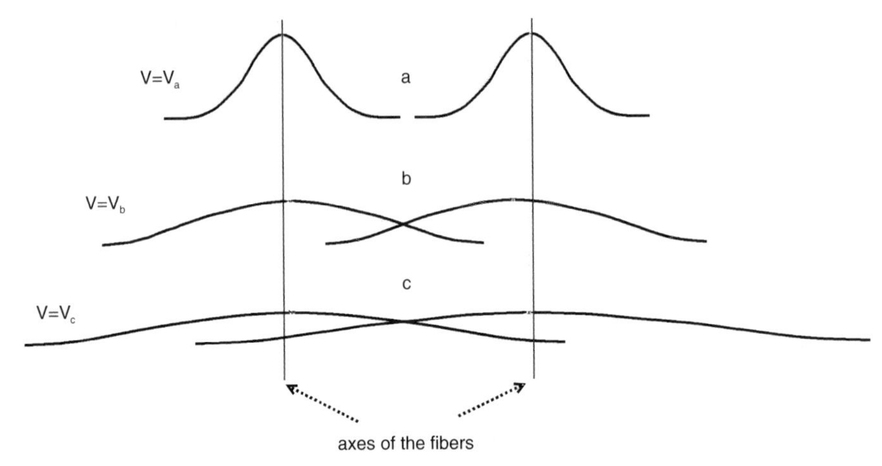

Fig. 4.9. Spreading of the fields as a function of the normalized frequency $V_c < V_b < V_a$. The amplitudes of the two fields were shifted in order to distinguish them

4.3.3 Inner-Cladding Fibers

Step-index fibers are not the only type of single-mode fibers. In fact, enhancement of single-mode fibers as regards to their guiding qualities can be achieved by creating an intermediate area with variable refractive index between the core and the optical cladding. The index profile of such a fiber is schematized in Fig. 4.10. The coupling between fibers of this type, referred to as 'inner-cladding fibers', will be compared with the coupling between step-index fibers.

The complete analysis of the coupling between these fibers is quite cumbersome and we shall hereafter content ourselves with presenting the curves representing the dependence of the coupling on the various characteristics of the fibers (Figs. 4.11, 4.12 and 4.13). The coupling between the modes of two fibers is dependent on the parameters of each fiber, as well as on intrinsic characteristics of the couplers, for example the value of the refractive index in the intermediate area or the distance between the two cores.

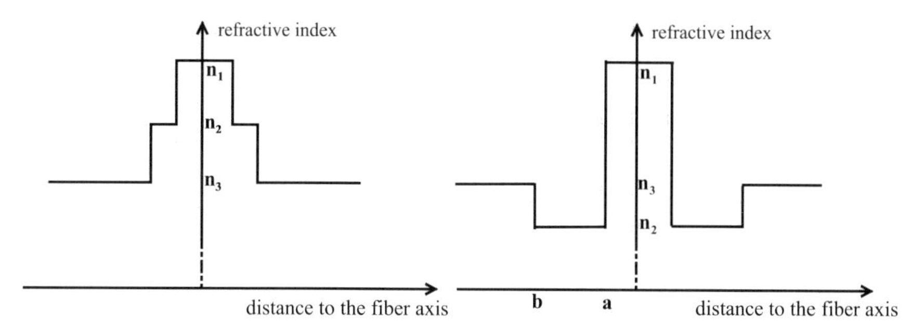

Fig. 4.10. Index profiles of raised and depressed inner-cladding fibers

The dependence of the coupling on the index of the intermediate area is illustrated in Fig. 4.11. From this curve, it can be seen that, depending on the value of the refractive index, the intermediate area behaves like an extension either of the core or of the optical cladding. In the first case, it can be noted that the fundamental mode of this structure has a nonzero cutoff frequency. This characteristic has to be taken into account in the production of couplers.

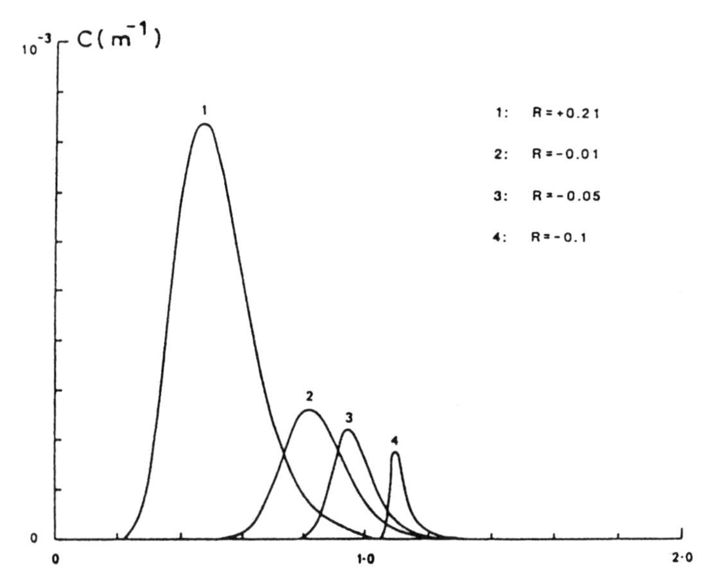

Fig. 4.11. The coupling as a function of the normalized frequency V, for different values of the parameter R, $S = 5$

If the refractive index of the intermediary cladding is very low, the field is highly confined inside the core. The confinement of the field results in this case from the great difference between the refractive indices of the core and

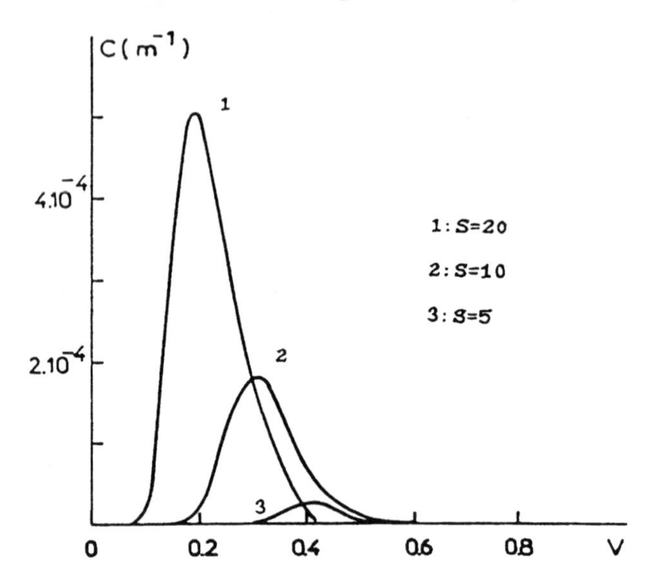

Fig. 4.12. The coupling ratio of two parallel uniform fibers as a function of the width of the intermediate area, for different values of the parameter S, $R = 0.1$

of the cladding. Positive values of R correspond to the case of less confined fields, where the coupling between the modes can arise more easily. The same remarks can be made concerning Figure 4.12, where curves obtained at different values of the distance between the cores of the fibers are represented.

These variations of the coupling ratio reveal the importance of the characteristics of the waveguide in the coupling. Nevertheless, it remains difficult to grasp the physical meaning of the value of the coupling ratio, and it is therefore preferable to characterize a coupler through the value of the coupling length. The coupling length of a coupler is defined as the distance at which a transfer of a given power P arises from one waveguide into the other. The coupling length L is related to the power P in the second waveguide by the equation

$$L = (1/C) \arcsin \sqrt{P/P_0}. \tag{4.13}$$

The first complete transfer of the power ($P = P_0$) takes place at the distance L_T, which is equal to $\pi C/2$. At values of the coupling length equal to odd multiples of L_T, the power is totally transferred into the second waveguide. As can be seen from Fig. 4.13, the value of L_T expands from a few millimeters to some centimeters depending on the characteristics of the waveguide and on the geometrical parameters of the coupler.

From observation of these curves, it may be noted that the values of the coupling length of depressed inner-cladding fibers are higher than the values of the coupling length of raised cladding fibers. This difference is related to

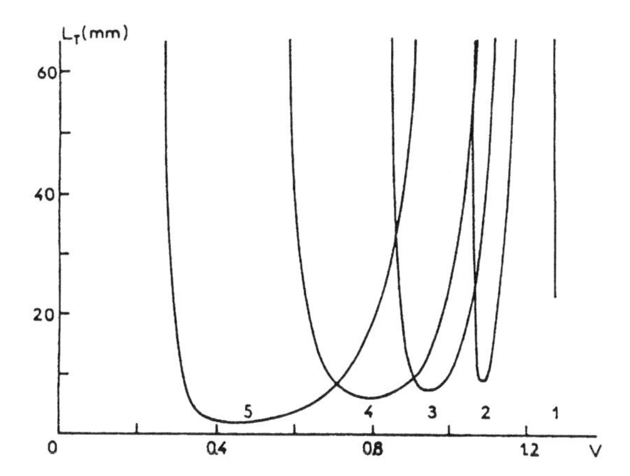

Fig. 4.13. The coupling length L_T as a function of the normalized frequency of the depressed inner cladding: $S = 5$, $D = 4b$ (**a**), $R = -0.2$, (**b**) $R = -0.1$, (**c**) $R = -0.05$, (**d**) $R = -0.001$, (**e**) $R = 0.21$

the greater confinement of the field in the core that results from the index depression.

4.3.4 Variable-Diameter Couplers

The preceding results were obtained in the ideal case of two perfectly uniform fibers. Therefore, they apply only approximately to actual couplers where, irrespective of the technique used for their fabrication, there always remains an area where the coupling is not uniform.

If the two fibers have characteristics varying with the distance z, the coupling coefficient at $z = L$ is defined by the equation

$$C(L) = \frac{1}{L}\int_0^L c(z)\mathrm{d}z. \tag{4.14}$$

The normalized frequency V is a parameter which frequently appears in guided optics. In the case of step-index fibers, the equation for V is

$$V = \frac{2\pi r_\mathrm{c}}{\lambda_0}\sqrt{n_\mathrm{c}^2 - n_3^2}, \tag{4.15}$$

where λ_0 is the wavelength of light in free space, r_c is the radius of the core of the fiber, n_c is the refractive index of the core of the fiber, and n_3 is the refractive index of the optical cladding.

The coupling coefficient can be rewritten as

$$C(L) = \frac{1}{L}\int_0^{V(L)} c(z(V))\frac{\mathrm{d}z}{\mathrm{d}V}\mathrm{d}V. \tag{4.16}$$

The local coupling ratio $c(z(V))$ will be referred to simply as $C(V)$. This simplification is justified in so far as we specialize to the case where the function c is monotonic.

In the case of a coupler where the normalized frequency depends linearly on the distance, we obtain the equation

$$C(V_\mathrm{f}) = \frac{1}{V_\mathrm{i} - V_\mathrm{f}} \int_{V_\mathrm{i}}^{V_\mathrm{f}} C(V)\mathrm{d}V, \tag{4.17}$$

where V_i and V_f are the values of the normalized frequency at each end of the coupler.

In the case of an exponential dependence on distance, the last equation becomes

$$C(V_\mathrm{f}) = -\log\frac{V_\mathrm{i}}{V_\mathrm{f}} \int_{V_\mathrm{i}}^{V_\mathrm{f}} \frac{C(V)}{V}\mathrm{d}V. \tag{4.18}$$

We shall examine more realistically the two most employed types of couplers: mechanically polished couplers and fused-tapered couplers [Digonnet 1983].

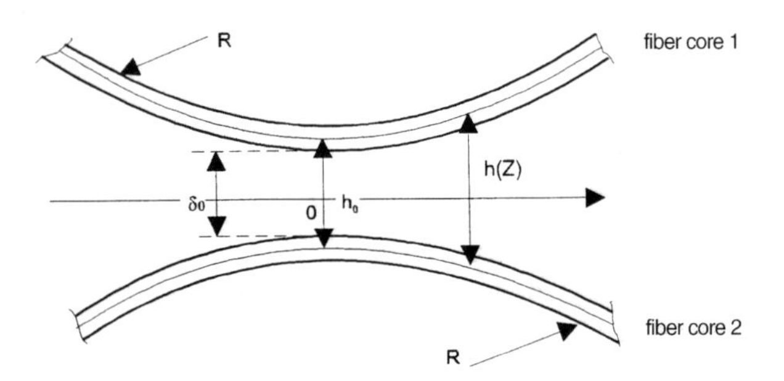

Fig. 4.14. Relative positions of the cores in a polished coupler. A shift normal to z, which has not been indicated in the figure, might also arise [Digonnet 1983, with the permission of the Optical Society of America]

A coupler realized by mechanical polishing is schematically represented in Fig. 4.14. The two fibers are bent with a radius of curvature R. The minimum distance between the cores of the two fibers will be denoted by h_0. By using a parabolic approximation, Digonnet *et al.* have demonstrated that the distance between the cores of the fibers is determined by the equation

$$h(z) = \sqrt{\left(h_0 + \frac{z^2}{r}\right)^2 + y^2}, \tag{4.19}$$

where y is the lateral shift between the fibers. The determination of the value of the complete coupled power from this equation is, as can be seen in Fig. 4.15, in agreement with the experimental measurements. The experimental curve was obtained by varying the lateral shift between the two fibers.

The parameters involved in the fabrication of a fused-tapered coupler are in general specific to each coupler (Fig. 4.16). Therefore, the general equation for the form of these couplers cannot be determined as readily as in the case of mechanically polished couplers. Measurements carried out with isolated fibers fused and pulled in the flame of a blowlamp have shown that their diameter varies according to the relation

$$R(l) = R_0 \exp\left[-\left(\frac{l-l_0}{2l_0}\right)\right], \tag{4.20}$$

where l_0 is the width of the burner used for fusing the fiber [de Fornel 1984a].

The fact that the fibers were twisted before being fused obviously modifies the shape of the tapered area of the coupler. In spite of this, the previous equation can be used as a first approximation.

Further, the shape of the coupler depends also on the temperature of the heating system and on the system used for lengthening the fiber (by gravitation, or using a linear motor, for example). For this reason, other expressions for the shape of these couplers have been suggested. As an example, Bures has advanced the following equation [Bures 1983]

$$R(z) = R_0(1+\gamma z^2) \quad \text{where } \gamma = 4\,\frac{\exp[(L-l_0)/2l_0 - 1]}{L^2}. \tag{4.21}$$

The coupling obviously depends on the shape of the coupler. As an example, we compare a coupler where the diameter depends linearly on the length of the thinned area with a coupler where the diameter varies exponentially. The dependence of the coupling ratio on the shape of the thinned area is represented in Fig. 4.17 for each type of coupler. In this example, it can be seen that a coupler with an exponential variation of its diameter presents a coupling ratio higher than a coupler with linearly varying diameter.

In the case of step-index fibers, the above equation for the coupling ratio is valid provided that the medium surrounding the fibers has the same refractive index as the cladding of the fibers. If the field at the interface between

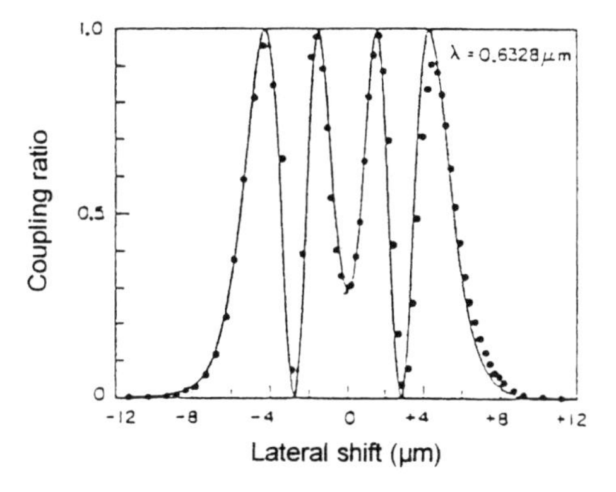

Fig. 4.15. Comparison between the experimental curve obtained with a source of wavelength $\lambda = 0.6328$ μm ($R = 100$ cm) and the theoretical curve, $h_0 = 4.40$ μm [Digonnet 1983, with the permission of the Optical Society of America]

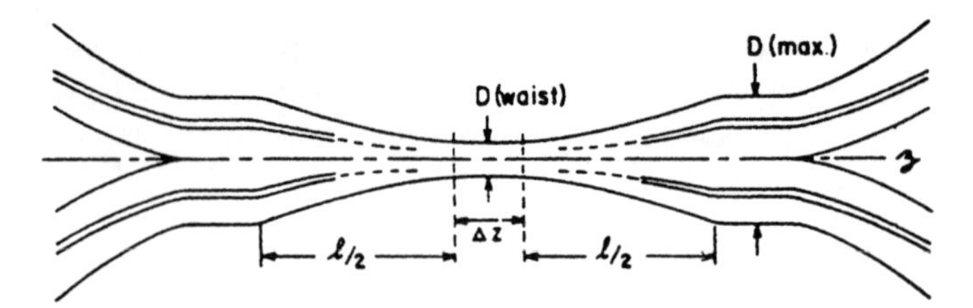

Fig. 4.16. Longitudinal view of a fused-tapered coupler. $D_{\max}$ is the initial diameter of the fused structure before pulling, D is the diameter of the waist w, l_0 is *the length of the fused area* and L is the total lengthening of the fiber [Bures 1983, with the permission of the Optical Society of America]

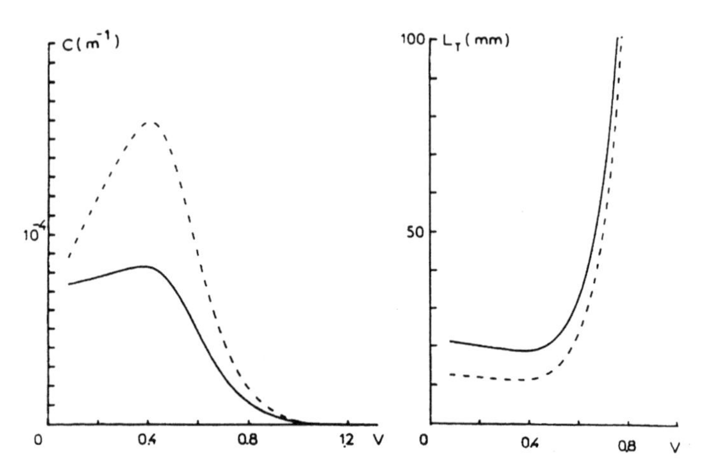

Fig. 4.17. Comparison between the coupling ratio and the length of complete transfer of couplers with different shapes: exponential (*dotted lines*) and linear (*solid lines*), as a function of the normalized frequency $R = 0.1$, $S = 5$, $D = 20a$

the cladding and the external medium is zero, this expression remains valid, irrespective of the value of the refractive index of the medium surrounding the fibers. The latter case can be seen as equivalent to that of an infinite optical cladding. If the field is nonzero outside the fiber, it it imperative not to neglect the effects of the medium in which the fibers are contained. In the calculations presented in Fig. 4.18, the presence of this medium has been taken into account [Pagnoux 1987].

These curves show that the model of an infinite cladding is adapted to high values of the normalized frequency, because the mode is not sensitive to the external medium in this case. If the value of the normalized frequency is very small, the field spreads inside the cladding and in the external medium, and so the core does not play any part in the guiding. Hence, the model of a negligible core can be used for low values of the normalized frequency.

Between these two extrema, the amplitude of the field as well as the coupling coefficient have to be determined from the complete model [Pagnoux 1987].

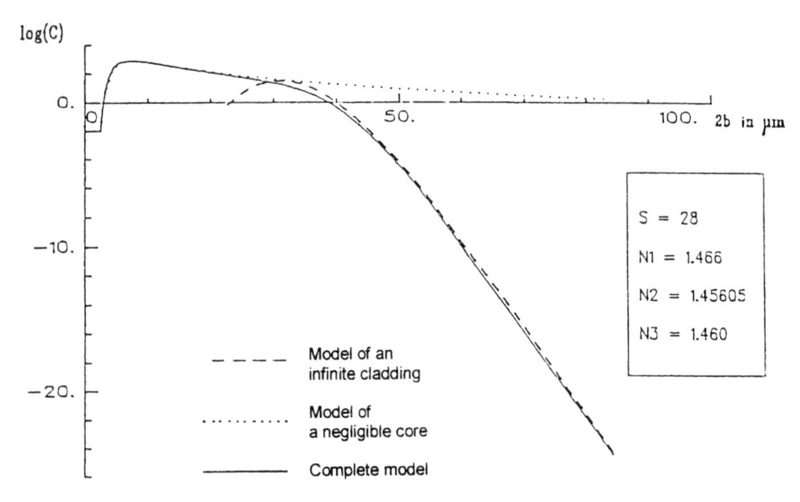

Fig. 4.18. The coupling ratio as a function of the normalized frequency. Comparison with the models of an infinite cladding and of a negligible core [Pagnoux 1987]

The results presented in Fig. 4.18 indicate that the guiding conditions are modified when the diameter of the fiber is reduced. The field is sensitive to the characteristics of the external medium. This phenomenon can be used for the production of sensors, for example temperature or concentration sensors.

Evanescent-field couplers present a very high level of sensitivity to polarization, wavelength and modal order. This high sensitivity would be a disadvantage in the case of a device like a beam splitter, but presents great interest for the production of optical components like polarizers or filters.

4.4 Spectral Filters and Spectral Multiplexers

As seen earlier, the coupling is highly sensitive to the normalized frequency, the latter being inversely proportional to the wavelength. For a given coupler, the intensity transferred in each branch depends on the wavelength used. In order to examine this dependence, we shall here review results presented by Digonnet *et al.* These results are represented in Fig. 4.19.

For a 2 µm vertical shift, the signal with wavelength 0.6328 µm is transferred into the second branch, while the signal with wavelength 1.064 µm is guided into the other output. The agreement between the numerical simulations and the experimental measurements can be noted.

A multiplexer and a demultiplexer are schematically represented in Fig. 4.20. The insertion losses are indicated for each wavelength. The isolation between the exits is also specified.

An application of multiplexers lies in the possibility of using them for increasing the transmission capabilities by multiplying the number of channels,

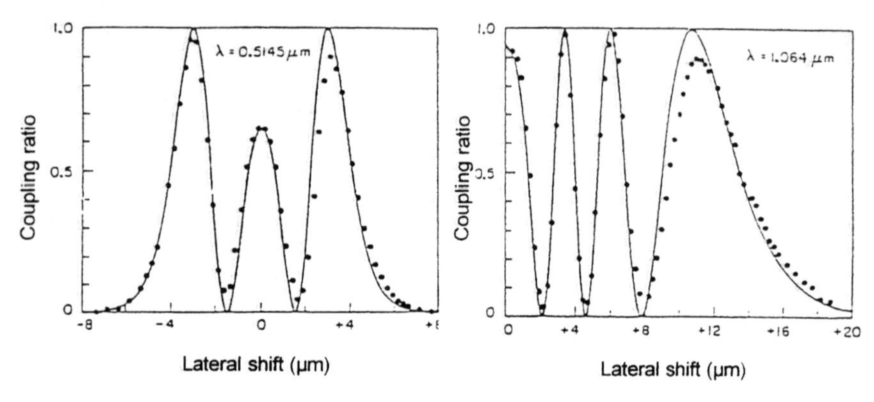

Fig. 4.19. Values of the coupling ratio calculated and measured for the same coupler (R = 100 cm) and for two sources with wavelength λ = 0.5145 μm and λ = 1.064 μm respectively (R = 100 cm, h_0 = 4.3 μm) with respect to the vertical shift [Digonnet 1983, with the permission of the Optical Society of America]

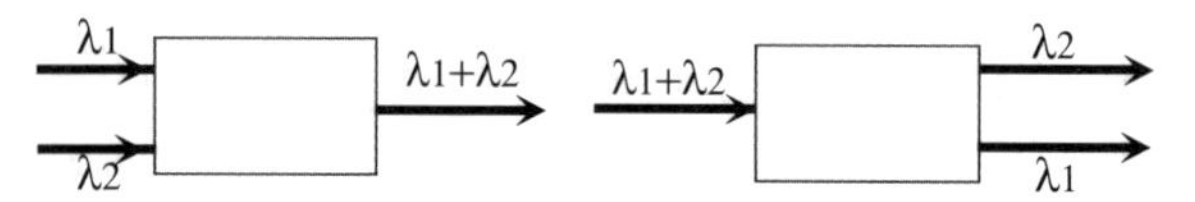

Fig. 4.20. (**a**) Multiplexer, (**b**) demultiplexer

each channel corresponding to a wavelength. Another application of multiplexers can be found in the production of fiber lasers. These devices will be described near the end of this chapter.

4.5 Polarization Splitters

As seen before, the coupling ratio is dependent on the propagation constant of the modes. A specific feature of polarization-preserving fibers is that polarized modes with different propagation constants can propagate inside them. As the coupling ratio of an evanescent-field coupler depends on the polarization of the mode, polarization splitters have been produced from couplers of this type.

The fabrication of fused-tapered couplers present some deficiencies. In particular, the axes of the twisted fibers can hardly be perfectly aligned. Further, in the case where inner-cladding fibers, for example bow-tie fibers, are being used, the reduction of the diameter of the fiber during the pulling process might bring the modes near their cutoff frequency, thereby inducing losses.

An advantage of couplers produced from mechanically polished fibers is that they can be used as polarization splitters. Fig. 4.21 represents a coupler working as a polarization splitter [Lefèvre 1984].

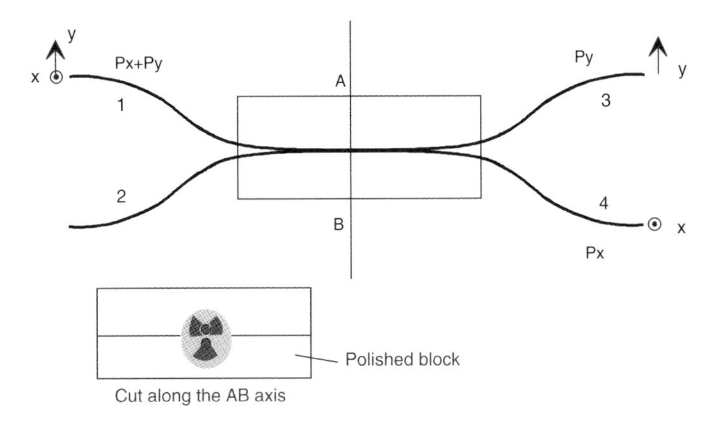

Fig. 4.21. Coupler used as a polarization splitter

Table 4.2. Parameters of a polished coupler used as a polarization splitter [Photonetics 1997a].

Polarization selectivity P3x/P4x or P4x/P3x (dB)	Cross polarization coupling P3x/P3y or P4y/P4x (dB)	Losses (P4 + P3)/P1 (dB)	Directivity (dB)	Fiber type
–10 to –13	–16 to –13	< 0.2	< –60	Bow-tie fiber

As an example, we present in Table 4.2 a few characteristics of a polarization splitter.

4.6 Production of Modal Filters

We previously specialized our discussion to the determination of the coupling between the fundamental modes of two single-mode fibers. In this section, we extend this description to the modes of higher order of weakly single-mode step-index fibers. The fibers are assumed to lie within a medium of the same refractive index as the optical cladding. Further, it is assumed that the modes are far from their cutoff frequency. Under these conditions, the approximation of linearly polarized modes can be used [Snyder 1969, Adams 1981].

The equation for the transverse component of the field is thus

$$\mathbf{E_0}(r,\theta) = -\mathrm{j}\Psi(r)\cos n\theta \mathbf{e}_x , \tag{4.22}$$

for the mode polarized along $\mathbf{e}_x$, and

$$\mathbf{E_0}(r,\theta) = \Psi(r)\cos n\theta \mathbf{e}_y, \tag{4.23}$$

for the mode polarized along $\mathbf{e}_y$.

Here

$$\begin{aligned} \Psi(r) &= J_N(ur/a) \qquad &\text{if } r < a, \\ \Psi(r) &= A_N K_N(wr/a) \qquad &\text{if } r > a, \end{aligned} \tag{4.24}$$

where $A_N = J_N(u)/K_N(w)$. Under these conditions, the coupling ratio can be expressed in the following form

$$C \simeq \frac{\omega\epsilon_0(n_0^2 - n_1^2)}{4P_0}\frac{\pi a}{2Dw} 2\pi I A_N \exp(-Dw/a), \tag{4.25}$$

where

$$I = \frac{a^2}{u^2 + w^2}[wJ_N(u)I_{N+1}(w) + uI_N(w)J_{N+1}(u)]. \tag{4.26}$$

Figures 4.22 and 4.23 represent the dependence of the coupling on the coupling length.

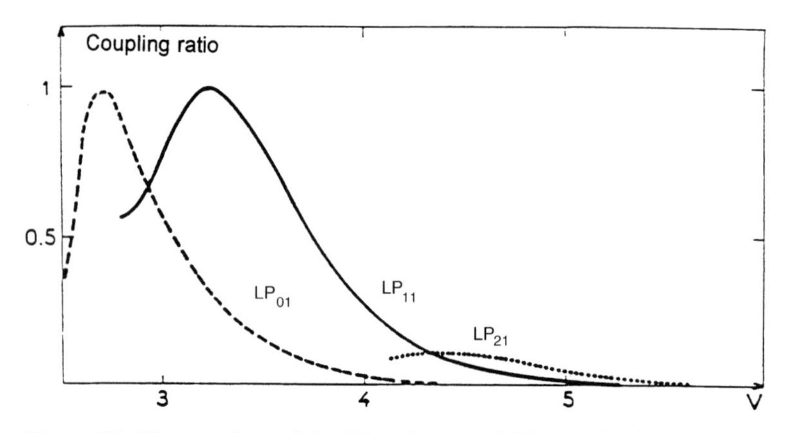

Fig. 4.22. The coupling of the LP_{01}, LP_{11} and LP_{21} modes between two step-index fibers as a function of the normalized frequency V, after a coupling length equal to 0.019 m, $D = 4R_c$, $\lambda = 0.633$ µm [Bennamane 1985]

These calculations, where the assumption was made that the diameter of the fibers remains constant, apply to the case of polished couplers. They indicate that the modes of step-index fibers can be multiplexed as well as selected [Bennamane 1985].

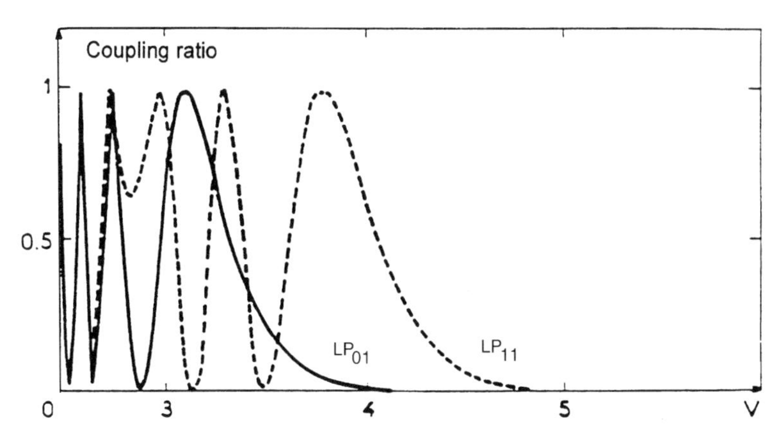

Fig. 4.23. The coupling of the LP_{01} and LP_{11} modes between two step-index-fibers as a function of the normalized frequency V, after a coupling length equal to 0.063 m, $D = 4R_c$, $\lambda = 0.633$ µm

4.7 Devices Produced from Evanescent-Field Couplers

We shall end this chapter by rapidly describing a few systems which are based on applications of evanescent-field coupling.

4.7.1 Optical-Fiber Gyroscope

An optical-fiber gyroscope is schematically represented in Fig. 4.24.

The light emitted by a laser is separated into two beams by a coupler [Lefèvre 1984]. The two beams propagate in inverse directions before recombining as they return through the coupler. If the system rotates about an

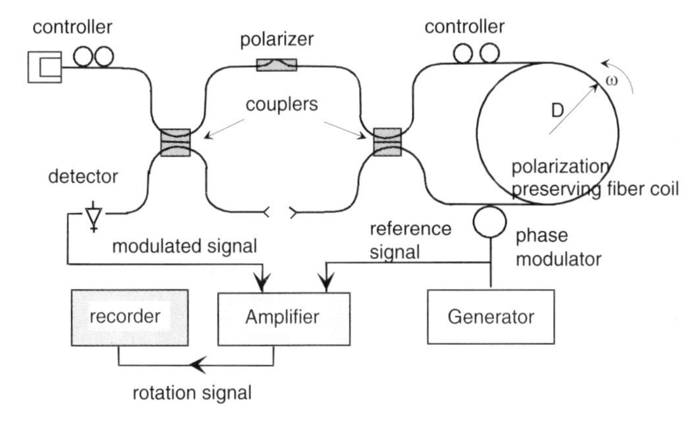

Fig. 4.24. Optical-fiber gyroscope

axis D, the Sagnac effect induces a phase difference between the inversely propagating beams. The phase difference between the two beams is expressed by the equation

$$\Delta\Phi = N\pi\omega r^2, \tag{4.27}$$

where N is the number of turns of the Sagnac loop, and r the radius of the loop. The speed of rotation of the system can be accurately determined from the phase difference thus measured.

4.7.2 Fiber Lasers

Thus far, we have not treated the case of active components, like gas lasers or solid-state lasers. We shall however make an exception with the description of fiber lasers. Figure 4.25 represents schematically an arrangement for such a device.

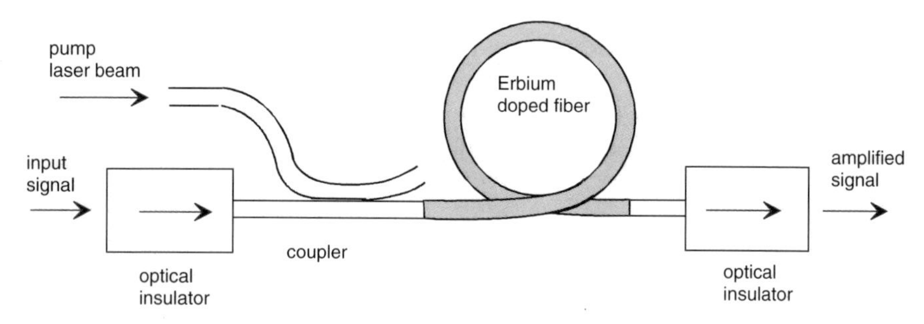

Fig. 4.25. Schematic illustration of a fiber laser

A fiber laser includes an amplifying medium generally formed from silica doped with rare earth elements, like erbium or neodymium. The production of such a device requires that the wavelength of the pump signal injected into the amplifying medium is different from the wavelength of the emitted signal. Table 4.3 summarizes the characteristics of a mechanically polished fiber-coupler produced with this application in view. The coupler was used in this arrangement for conveying the pump beam.

It must be borne in mind that the couplers used in such an arrangement have to resist light powers of the order of several hundred mW without having their characteristics changed.

These devices lie within the tendency towards the production of all-optical components. With a system of this type, a signal can be amplified without needing to perform successively first an optical-electric conversion and then an electric-optical conversion.

Table 4.3. Characteristics of a coupler designed for the production of fiber lasers [Photonetics 1997b].

	Spectral band (nm)	Insertion loss (dB)	Polarization sensitivity (dB)	Isolation (dB)
Pump	Between 1540 and 1560	< 0.1	< 0.05	> 15
Signal	980	< 0.2	< 0.005	

4.8 Conclusion

A first application of the properties of the evanescent field associated with the modes of an optical fiber has been described in this chapter. Evanescent-field coupling enables the realization of optical-fiber couplers with negligible losses for all values of the coupling ratio. In planar optics, an optical-fiber coupler fulfills the same functions as a separating plate in classical optics. For these reasons, evanescent-field coupling optical-fiber couplers have become widely used.

Beside the function of beam splitters, we have seen that these couplers could be used for example as mode or polarization splitters. A further application of these couplers is their use as spectral demultiplexers.

These applications have extensions in the production of more complex components, like electro-optical switches or amplifiers. It is likely that the present development of all-optical devices will lead to a more extensive use of evanescent-field couplers.

5. Integrated-Optical Evanescent-Field Couplers

In recent years, the trend towards the production of all-optical devices has given integrated optics a greater importance than ever before. The term 'integrated optics' assembles the different technologies involved in the production of active and passive couplers. These technologies are very different from the fabrication techniques of optical-fiber devices. Indeed, while optical-fiber devices are produced from already developed fibers, the waveguides used in integrated-optical components have to be designed at the same time as the components themselves.

This chapter begins with a description of the fabrication techniques of integrated-optical couplers. Before examining the applications of these couplers, an analysis of their properties is provided. Examples of applications of these couplers are then presented. As integrated-optical waveguides are often coupled with optical fibers, this type of coupling is treated here. Lastly, the coupling between an integrated-optical waveguide and a photodiode is presented.

5.1 Description of Integrated-Optical Couplers

An integrated-optical coupler includes at least two waveguides fabricated on the same substrate (Fig. 5.1). Unlike polished couplers, where any coupling ratio can be achieved depending on the distance separating the fibers, the parameters of an integrated-optical coupler comprising only two waveguides are in general definitively fixed.

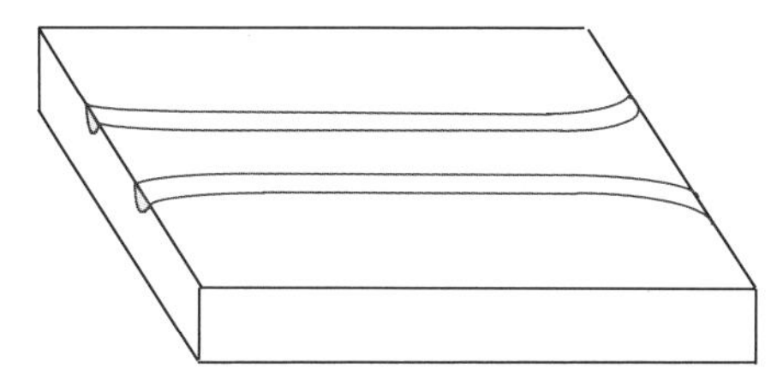

Fig. 5.1. Schematic representation of an integrated-optical coupler

The production of strip-loaded waveguides has been described in Chap. 3. The fabrication techniques of integrated-optical couplers are basically the same as the techniques used for fabricating these waveguides. Figure 5.2 illustrates the fabrication process of an ion-implanted coupler. Strip-loaded couplers are realized according to the method described in Chap. 4.

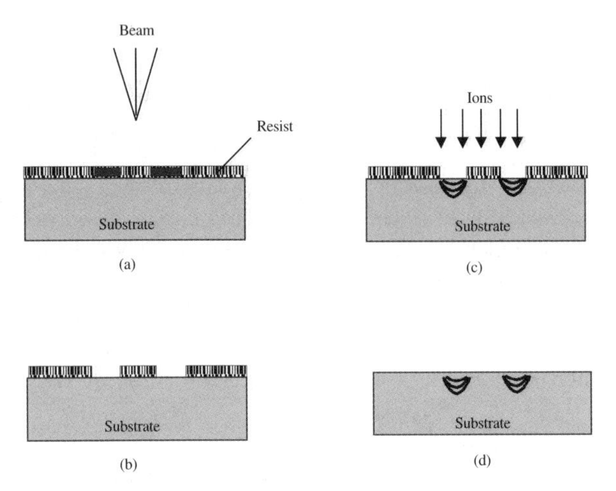

Fig. 5.2. Fabrication of a coupler by localized implantation (**a**) deposit of an electroresist or a photoresist and appropriate irradiation, (**b**) developing of the resist, (**c**) etching of the unexposed areas for producing the mask, (**d**) diffusion or implantation through the mask

5.2 Analysis of the Coupling Between Two Waveguides

The coupling between the modes of an integrated-optical waveguide can be analyzed using the same formalism as in the analysis of the coupling between optical fibers. Since these calculations have been extensively detailed in the preceding chapter, it does not seem necessary to repeat them. In what follows, emphasis will be placed instead on a few distinctive characteristics of these couplers.

5.3 Active Couplers

One of the reasons behind the interest in integrated optics is the possibility of realizing active components. As early as 1968, researches inquiring into the possibility of fabricating active integrated-optical devices were carried out at Bell Telephone Laboratories [Miller 1969]. The first active components, on lithium niobate ($LiNbO_3$) and on gallium arsenide (GaAs), were produced about 1975. Before examining these two components in technical detail, their fundamental principle has to be described.

The coupling ratio between the modes of two waveguides depends on the propagation constants of each of the two modes. If the modes have the same propagation constant and if the length of the coupler satisfies certain conditions, i.e. if it is a multiple of L_c, the transfer can be total. In contrast, if the modes have different propagation constants, the coupling between them remains always partial (Fig. 5.3).

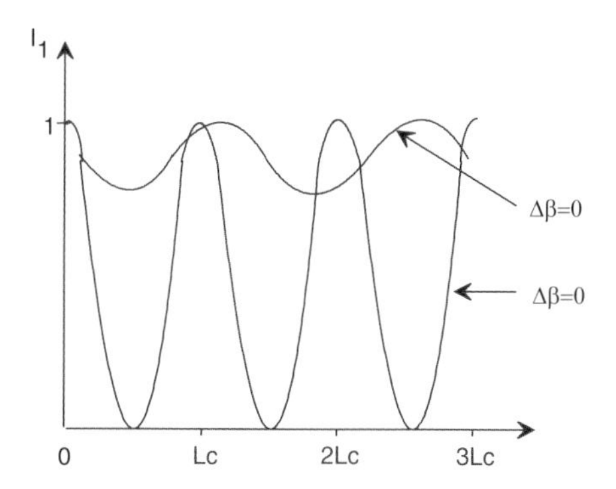

Fig. 5.3. Coupled intensity as a function of the propagation constant of the modes of each waveguide

The propagation constant of a mode depends on the index profile of the waveguide. Electro-optical devices are characterized by the fact that the refractive index can vary with respect to the electromagnetic field applied to them. The variation of the refractive index Δn is in proportion to the value of the electric field for the Pockels effect, and to the square of the value of the field for the Kerr effect. The latter effect will not be examined here.

In the case of the simple system represented in Fig. 5.4, the dependence of the variation of the refractive index on the crystal length traveled by the beam is given by the equation

$$\Delta n = \frac{n^3 r V L}{2d}, \tag{5.1}$$

where V is the voltage applied to the crystal, d is the thickness of the crystal and L is the distance that the light beam travels inside it. r and n are respectively the electro-optical coefficient and the refractive index of the material.

As a consequence of the geometry of the electrodes and of the waveguide, the equation for the variation of the refractive index in an integrated-optical coupler is not as simple as in the example above.

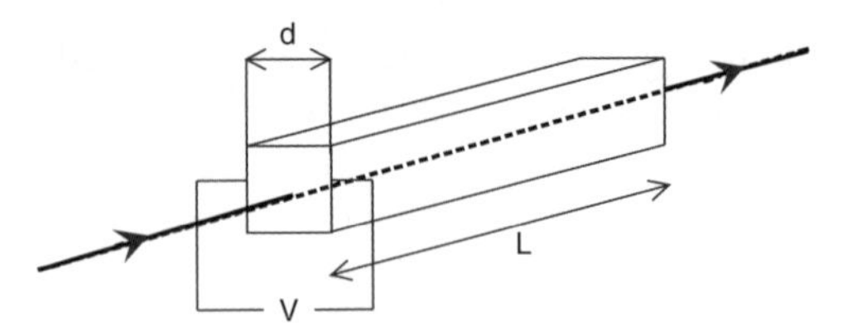

Fig. 5.4. Polarization of an electro-optical crystal

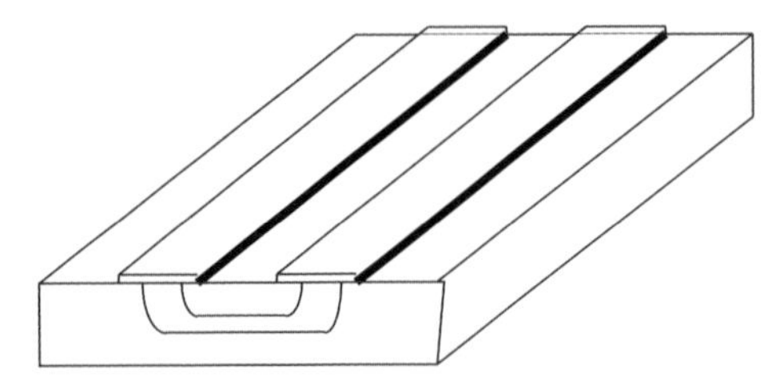

Fig. 5.5. Form of the field lines for ribbon electrodes

Let us consider here the simple case where the electrodes are just placed over each waveguide. As can be seen in Fig. 5.5, the field lines have a form more complex than previously.

Different schemes of the structure of the electrodes in a directive coupler can be designed. Two examples are represented in Figs. 5.6 and 5.7 [Carenco 1983].

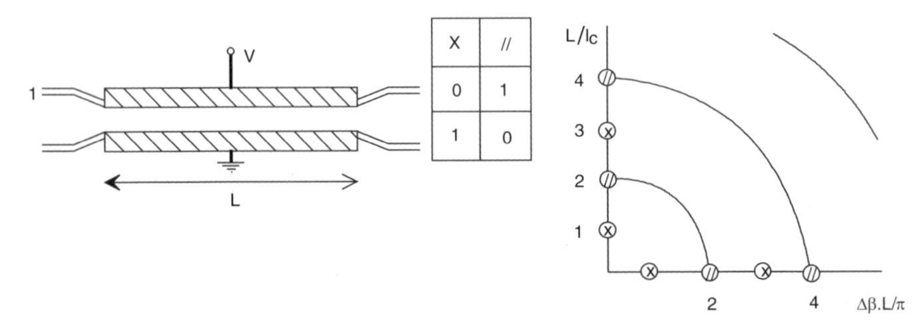

Fig. 5.6. Directive coupler with a structure of uniform electrodes and diagram of electric switching. L_c is the transfer length and $\Delta\beta$ the induced phase difference. The latter is in proportion to V

The phase difference is proportional to the voltage applied. In particular, it can be seen from Fig. 5.6 that a signal can be transferred from one of the waveguides into the other by simply reversing the polarization of the electrodes. For returning to the initial conditions, i.e. to a $\|$ state, the voltage has to be removed. This difficulty can be prevented by fabricating several

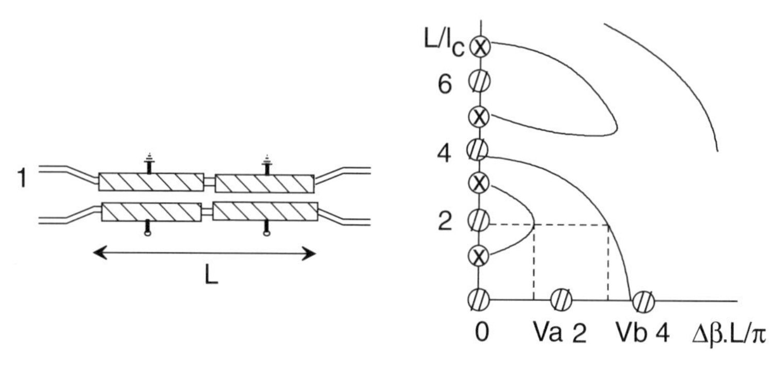

Fig. 5.7. Directive coupler with a structure of alternate electrodes and diagram of electric switching

successive electrodes. The transition from a $\|$ state to a $\perp$ state can then be controlled for example by modulating the voltage from Va to Vb.

5.4 Coupling from a Fiber to a Planar Waveguide

Recent works [Marcuse 1989, Panajotov 1994] have demonstrated the possibility of coupling a single-mode fiber to a planar waveguide. Transverse and longitudinal views of this fiber-to-planar waveguide coupler are shown in Fig. 5.8.

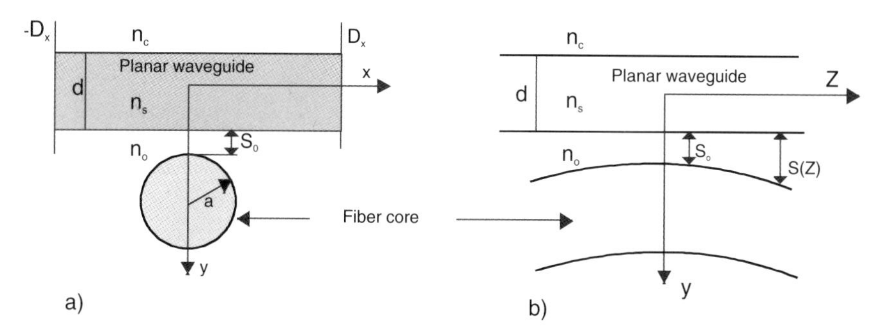

Fig. 5.8. Transverse (**a**) and longitudinal (**b**) views of a fiber-to-planar waveguide

As in the case of optical-fiber couplers, the field Ψ of this coupler can be expressed in terms of the sum of the field F of the fiber and of the fields S_ν of the planar waveguide

$$\Psi = a(z)F\exp(-\mathrm{j}\beta_{\mathrm{f}}z) + \sum_{\nu=0}^{N_{\mathrm{s}}} b_\nu(z)S_\nu\exp(-\mathrm{j}\beta_{\mathrm{s}\nu}z). \tag{5.2}$$

Assuming that the two waveguides do not perturb each other much, the equation of the field of the coupler is deduced from Maxwell's equations. This leads to the following coupled-mode equations

$$\frac{\mathrm{d}a}{\mathrm{d}z} = -\mathrm{j}(Q_{\mathrm{fs}} + Q_{\mathrm{fc}})a - \mathrm{j}\sum_{\mu=0}^{N_s} K_{\mathrm{fs}\mu} b_\mu \exp[\mathrm{j}(\beta_\mathrm{f} - \beta_{\mathrm{s}\mu})z], \tag{5.3}$$

$$\frac{\mathrm{d}b_\nu}{\mathrm{d}z} = -\mathrm{j}(K_{\mathrm{s}\nu} + K_{\mathrm{c}\nu})a \exp[\mathrm{j}(\beta_{\mathrm{s}\nu} - \beta_\mathrm{f})z] \\ -\mathrm{j}\sum_{\mu=0}^{N_s} Q_{\mathrm{s}\nu\mu} b_\mu \exp[\mathrm{j}(\beta_{\mathrm{s}\nu} - \beta_{\mathrm{s}\mu})z], \tag{5.4}$$

where

$$Q_{\mathrm{f}x} = \frac{k^2}{2\beta_\mathrm{f}} \int\int (n_x^2 - n_0^2) F^2 \mathrm{d}x\mathrm{d}y, \tag{5.5}$$

and

$$K_{x\nu} = \frac{k^2}{2\beta_{\mathrm{s}\nu}} \int\int (n_x^2 - n_0^2) F S_\nu \mathrm{d}x\mathrm{d}y, \tag{5.6}$$

where x stands either for s or c.
Here

$$Q_{\mathrm{s}\nu\mu} = \frac{k^2}{2\beta_{\mathrm{s}\nu}} \int\int (n_\mathrm{f}^2 - n_0^2) S_\nu S_\mu \mathrm{d}x\mathrm{d}y, \tag{5.7}$$

$$K_{\mathrm{fs}\nu} = \frac{k^2}{2\beta_{\mathrm{f}\nu}} \int\int (n_\mathrm{f}^2 - n_0^2) F S_\nu \mathrm{d}x\mathrm{d}y. \tag{5.8}$$

From these equations, Panajotov has obtained the numerical results represented in Fig. 5.9.

The high sensitivity of the coupling to the value of the refractive index of the waveguide can be emphasized. These curves indicate the potential of this type of coupling. Indeed, if the planar waveguide is realized from an electro-optical material, a system of this type can be used as a polarizer, where either of the two polarizations can be selected from the polarization applied to the system. Applications to different types of sensors might also be developed.

5.5 Integration of a Waveguide and a Photodiode

Besides the miniaturization of components, the interest in optics is linked with the possibility of enhancing the capacities of data transfer and data processing. The utilization of the evanescent-field coupling from a photodiode to a waveguide already permits us to increase the bandpass of these components [Umbach 1995].

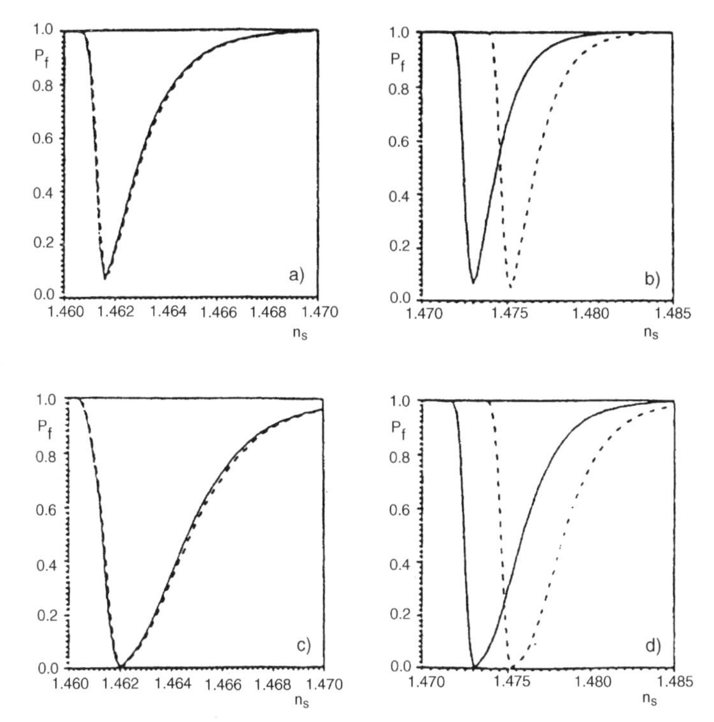

Fig. 5.9. Power P_f calculated at the end of the fiber for TE (*solid lines*) and TM (*dashed lines*) polarization modes, as a function of the index n_s of the planar waveguide. The value of the coupling length is 2500 μm (**a**) $s_0 = 3$ μm, $n_c = 1.453$, (**b**) $s_0 = 3$ μm, $n_c = 1.0$, (**c**) $s_0 = 2$ μm, $n_c = 1.453$, (**d**) $s_0 = 2$ μm, $n_c = 1.0$

Umbach *et al.* have integrated a GaInAsP waveguide and a GaInAs photodiode on InP. The waveguide used was made of a GaInAsP:Fe layer with a 600 μm thickness ($\lambda_G = 1.06$ μm) and of a 40 nm GaInAsP:Fe tape ($\lambda_G = 1.3$ μm). The structure of this component is detailed in Fig. 5.10.

Figure 5.11 represents the time response of the photodiode to a 2 ps impulse.

5.6 Conclusion

Evanescent-field integrated-optical couplers present a number of advantages over optical-fiber couplers. By using masking techniques, couplers with any shape can be realized. This allows to reduce the losses arising at the curvature of the waveguide. Further, the fact that couplers can also be realized from active electro-optical materials leads to the possibility of using them for a number of optical functions, for example as multiplexers or as switches.

The excitation of the modes of a waveguide generally induces some losses. By directly exciting the modes from the evanescent field of a fiber, these

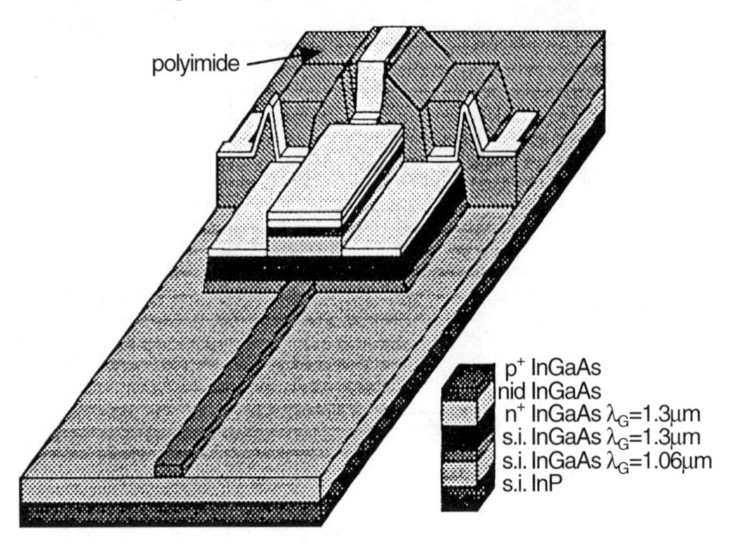

Fig. 5.10. Integrated waveguide and photodiode

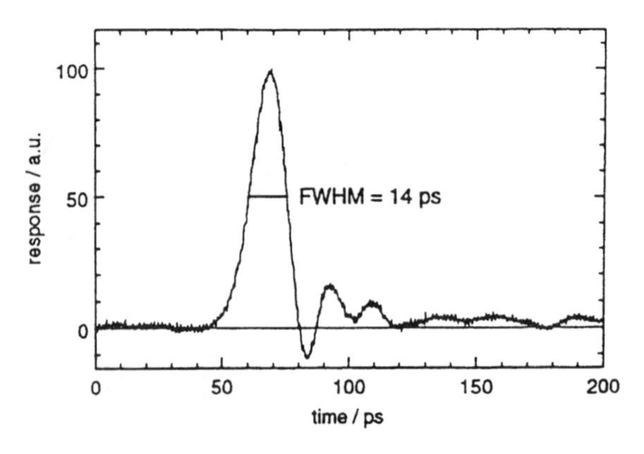

Fig. 5.11. Response to an impulse of an integrated photodiode with length 10 μm and thickness 6 μm [Umbach 1995]

losses can be reduced. Further, since the coupling arises along a relatively large length of the waveguide, the adjustment of the fiber can be achieved in conditions less critical than with a classical excitation.

In the same way, an evanescent field can be used for directly coupling a source to a waveguide. With such a coupling, the rapidity of the response of the system has been shown to be greatly enhanced in comparison with current systems.

The few examples presented in this chapter show the ever increasing part of evanescent-field coupling in the field of integrated optics.

6. Evanescent-Field Waveguide Sensors

The phenomenon of total internal reflection, whether it be generated by a prism or by a waveguide, is highly sensitive to all the parameters involved, namely, to the wavelength, to the refractive indices of the different media, to the roughness, etc. This sensitivity has given rise to the production of several different types of sensors.

The first sensors developed were designed for spectroscopic measurements and used the total internal reflection of light inside a prism. An extensive literature is available on these spectroscopes, which have been developed from the 1960s. Besides, their widespread use in research as well as in industry indicates that these devices have reached maturity. These are the reasons for which a separate chapter has been devoted to internal-reflection spectroscopy (Chap. 7).

In this chapter, we shall describe sensors where the properties of guided modes in fibers or in integrated-optical waveguides are utilized. In these sensors, the evanescent field of the guided modes is perturbed by the phenomenon to be detected. When affected by this perturbation, these modes transfer the information thus detected. This behavior is evidently of great interest for measurements in distant, dangerous or hard to reach environments.

The propagation of a mode can be extremely affected by local modifications of the conditions of the guiding. Accordingly, the optical properties of the cladding in which the evanescent field lies can easily be modified. This characteristic is present in most evanescent-field sensors. The properties of the evanescent field are also exploited in sensors designed for the detection of microbending.

6.1 General Points on Sensors

The different types of sensors can be grouped into two classes: intrinsic sensors and extrinsic sensors [Dakin 1988, Ferdinand 1992, Udd 1995]. With an intrinsic sensor, the value of the physical parameter to be measured is determined from the effect of this parameter on some intrinsic properties of the waveguide. In contrast, with a sensor of the second type, the measurement of the physical parameter is based on extrinsic properties of the waveguide, that is, on characteristics of the waveguide resulting from some modifications this waveguide has been submitted to. The sensors that will be described in this chapter belong in general to the class of extrinsic sensors.

The characteristics of a guided mode depend on the parameters of the waveguide. If the waveguide is being locally perturbed, the perturbation induced might cause a modification in the distribution of the mode, a change in the propagation constant of the mode, a delay, and even losses.

Different methods can be used for perturbing a mode. The perturbation can for example be achieved either:

- by curving the waveguide or by submitting it to microbending, or
- by modifying the geometrical parameters of the waveguide (thinning of the waveguide by fusion and pulling, chemical etching of the cladding), or
- by modifying the refractive index of the cladding of the waveguide, either by diffusion or by gas adsorption.

The physical parameters that can be detected with a given waveguide depend on the perturbation induced on the waveguide. Since the main subject of this book is the evanescent field, we shall specialize to evanescent-wave sensing. Therefore, the sensors we describe are all based on the use of the sensitivity of the evanescent field to external perturbations exerted on the waveguide. We examine separately sensors realized from fibers and sensors realized from planar or integrated waveguides.

6.2 Fiber-Optic Sensors

The fibers used in telecommunication systems are produced in such a way that the field is zero at the interface between the external medium (e.g. air) and the surface of the fiber. To this end, these fibers are fabricated from materials that do not depend much on the physical conditions, for example on the ambient temperature, on the index of the external medium or on the bending of the fiber. A consequence of these characteristics of the raw materials of fibers is that a fiber cannot be sensitized without first having been modified.

A technique currently used consists in treating the fiber in such a way that the evanescent part of the guided modes is nonzero in the medium tested. Another method is by using fibers with such characteristics that for certain parameters, for example the temperature or the wavelength, the evanescent field has a nonzero value at the interface between the core and the cladding.

The perturbation of the evanescent field is generally accompanied by an increase of the losses. If the losses are distributed along the fiber or at least along a great length of the fiber, the measurement can be performed using an optical-fiber reflectometer. The basic principle of this device is represented in Fig. 6.1.

An impulse is emitted by the source. A small percentage of the injected energy is reflected at the input side of the fiber. A part of the light is

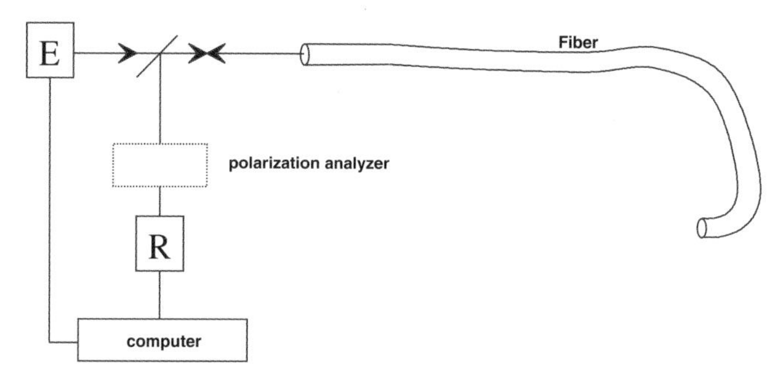

Fig. 6.1. Principle of a fiber-optic reflectometer; E is an impulse source, R is a detector. The latter can be preceded by a polarization analyzer

then backscattered along the entire fiber. This phenomenon results from the Rayleigh backscattering, from defects inside the fiber, from bending and from microbending. The shapes of the backscattering curves are schematized in Fig. 6.2.

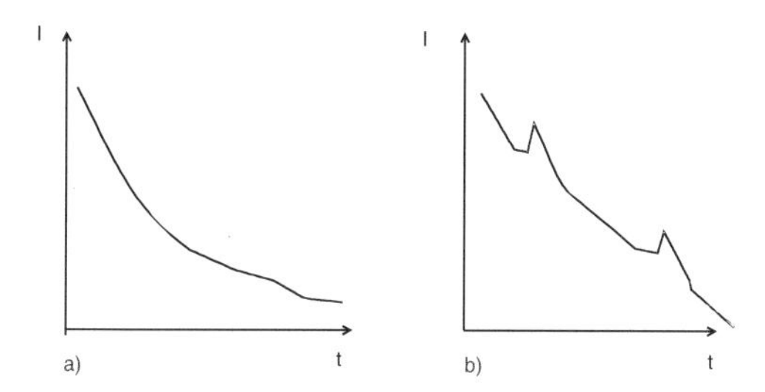

Fig. 6.2. Time-dependence of the intensity I of the backscattered signal (**a**) case of a fiber without defects, (**b**) case of several fibers fused together

The losses caused by some physical parameter result in a perturbation of the propagation of the mode. Accordingly, a physical parameter whose variations induce losses can be measured in reflectometry.

The phenomenon of Rayleigh backscattering can be analyzed on the basis on the model represented in Fig. 6.3. A part of the beams originating in the scattered field is guided in the fiber along the two directions of propagation. The losses of the fiber correspond to the beams not totally internally reflected at the interface between the core and the cladding.

The time dependence of the backscattered signal is expressed by the equation

$$P_{\mathrm{R}}(t) = 0.5 E_0 \nu_{\mathrm{g}} \alpha_{\mathrm{d}} B(t) \exp[-\alpha(t) \nu_{\mathrm{g}} t], \tag{6.1}$$

where E_0 is the initial injected energy, ν_{g} is the group velocity in the fiber, $B(t)$ is the ratio of the backscattered energy to the total energy scattered at the same point, and $\alpha = \alpha_{\mathrm{a}} + \alpha_{\mathrm{d}}$ is the loss coefficient of the fiber, where α_{a}

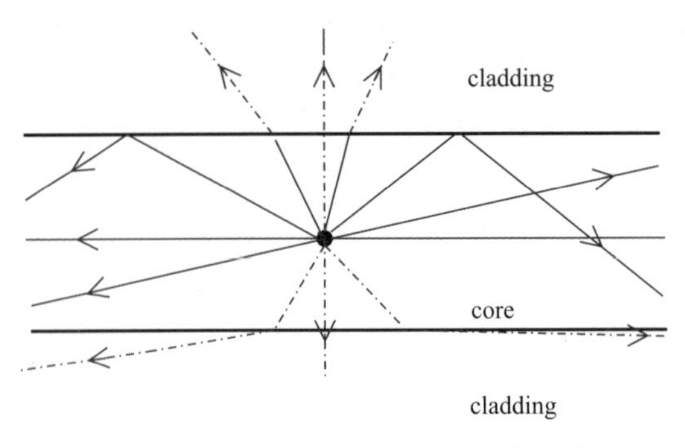

Fig. 6.3. Schematic representation of the losses caused by a scattering center. The beams represented in dashed lines correspond to the losses of the fiber

and α_d are respectively the adsorption coefficient and the diffusion coefficient of the fiber.

The sensitivity of the measurement of the scattered signal depends on the width of the impulse. Two typical values indicated by Ferdinand are reported in Table 6.1.

Table 6.1. Characteristics of reflectometric measurements for a single-mode fiber illuminated by a laser diode [Ferdinand 1992].

Width of the impulse (ns)	Spatial resolution (m)	Typical range	Backscattered level on initial level (dB)	Bandpass (MHz)
10	1	A few 100 m	− 60	> 100
100	10	Some km	− 50	> 10

The form of the backscattered signal depends on several parameters. As an example, we present here two cases reported by Ferdinand. The first of these concerns the determination of the level of a liquid from the backscattered signal. For these measurements, Yoshikawa used an eccentrically clad fiber.

An effect of the eccentricity of the core of the fiber is that the evanescent field spreads outside the core. The fiber was immersed in a liquid with high index, in order to prevent a part of the light from reflecting at the interface between the cladding and the external medium, which would have resulted in an increase of the value of the losses [Yoshikawa 1988]. Figure 6.4 represents the structure of the fiber, the arrangement used and finally the characteristic shape of the backscattering curves.

The presence of bending or microbending also causes a leak of the light beams outside the core of the fiber. This can be seen in the numerical simulation represented in Fig. 6.5.

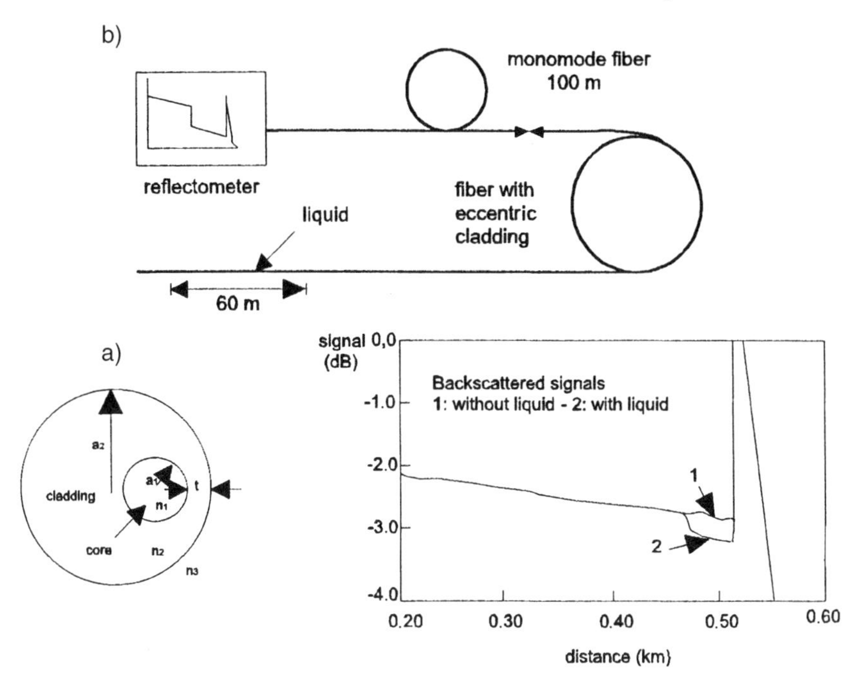

Fig. 6.4. Backscattering measurement of the level of a liquid

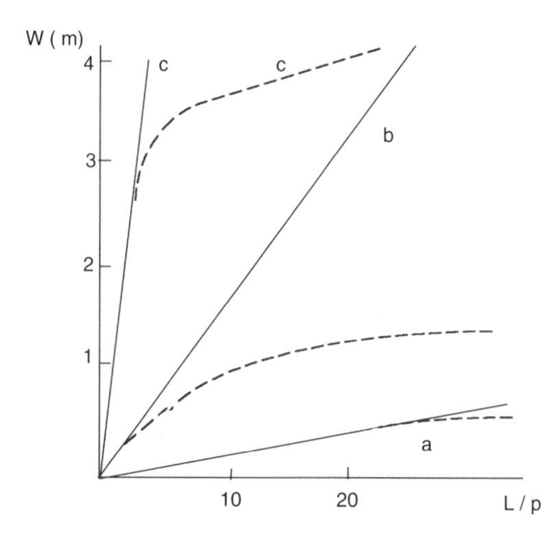

Fig. 6.5. Spreading of a helicoidal mode (w is in μm) propagating inside a fiber as a function of the length of the fiber. The fiber was subjected to periodic microbending with amplitude (**a**) 1.25 nm, (**b**) 2.5 nm, (**c**) 125 nm respectively. The fiber has either a parabolic index profile (*solid curve*) or a triangular index profile (*dashed curve*)

The backscattered signal also reflects the amount of the losses, and so a variation of the pressure exerted along the fiber can be recognized from the analysis of the backscattered signal (Fig. 6.6). As can be seen in the example represented in Fig. 6.7, this property can be used for the time-resolved monitoring of the evolution of a pressure exerted on a fiber [Ferdinand 1992].

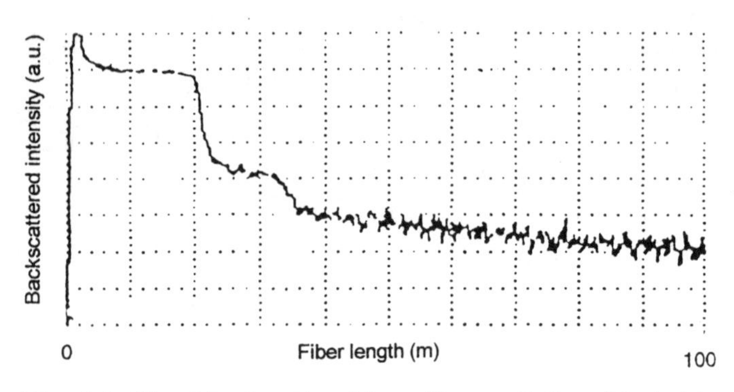

Fig. 6.6. Signal backscattered by a fiber partially subjected to microbending

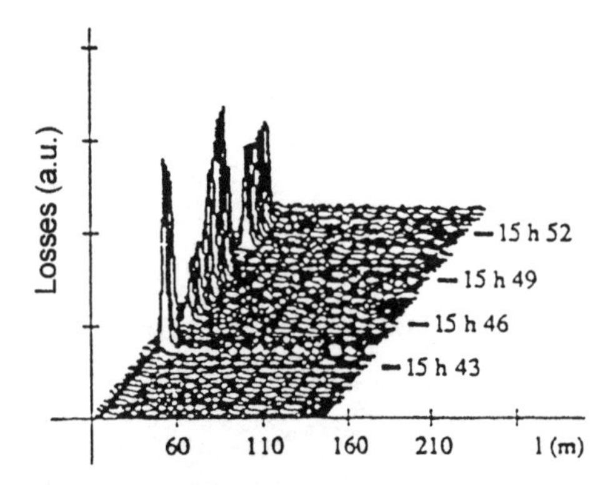

Fig. 6.7. Profile of the perturbations exerted on an optical fiber as a function of time [Ferdinand 1992]

Recently, a type of sensor based on a different principle has been suggested [Brunner 1995]. The technology used for the production of this sensor is descended from the liquid-core fibers which were developed during the 1970s [Hartog 1983]. The basic principle of these sensors is the dependence of the reflectometric response on the local temperature of the fiber.

The sensor devised by Brunner is based on the use of a capillary filled with a liquid with refractive index smaller than the index of the capillary. The light is therefore guided by a ring, as represented in Fig. 6.8.

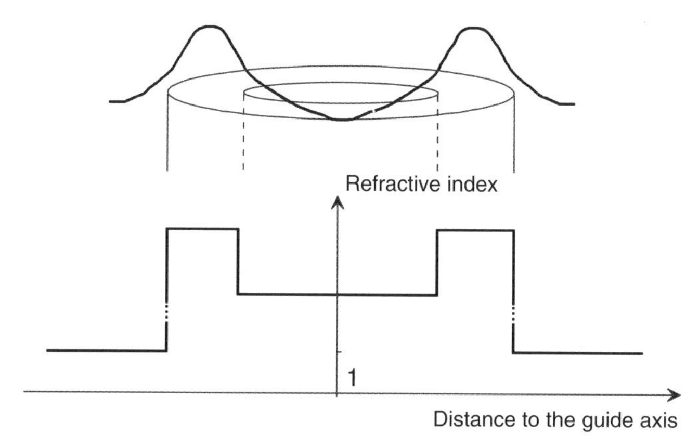

Fig. 6.8. Annular waveguiding

Only an evanescent part of the guided mode propagates inside the liquid. The parameters of the guiding of the mode(s) depend on the value of the refractive index of the liquid. By observing which modes are guided inside the fiber, an eventual modification of the liquid can be detected. The description of the modes that are sustained by such a structure can be found in the chapter devoted to evanescent-field atom optics.

The interest of these backscattering measurements is in the possibility of thereby measuring phenomena located at a great distance from the source. The advantage of such measurements is perceptible in the case of measurements practiced in polluted or dangerous environments (e.g. nuclear, mines). As can be seen from Table 6.1, the high sensitivity of a measurement is accompanied by a less accurate localization of the defect. In each particular case, a compromise between the level of sensitivity of the measurement and the accuracy of the localization has to be found.

In order to measure the effects of the variation of a given physical parameter, a part of the optical cladding of the fiber can be removed so that the field can 'see' the external medium. The same result can be achieved using a fiber thinned by fusion and pulling, as represented in Fig. 6.9 [de Fornel 1984b, Golden 1994, Egami 1996].

Fig. 6.10 represents the dependence of the sensitivity of a tapered fiber on the value of the refractive index of the external medium [de Fornel 1984b].

For understanding the physical meaning of such a curve, it only has to be recalled that the polarized HE_{11x} and HE_{11y} modes of a bow-tie fiber do not have the same cutoff frequency. Under these conditions, the evanescent fields of these modes extend more or less outside the fiber. Hence, the effect of a given variation of the refractive index of the external medium depends on the polarization of the modes. If the fiber has been tapered in such a way that one of the modes is near its cutoff frequency, even a very weak variation of the index of the external medium can significantly modify the value of

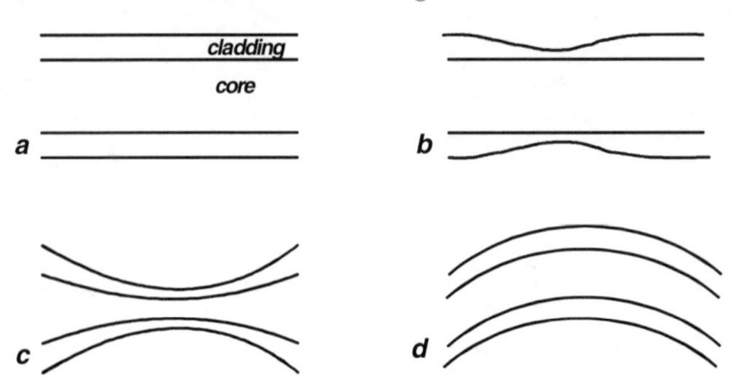

Fig. 6.9. Different techniques used for having a part of the evanescent field penetrates into the tested medium (**a**) non-perturbed fiber, (**b**) removal of a part of the cladding, (**c**) thinning of the fiber by fusion and pulling, (**d**) bending of the fiber

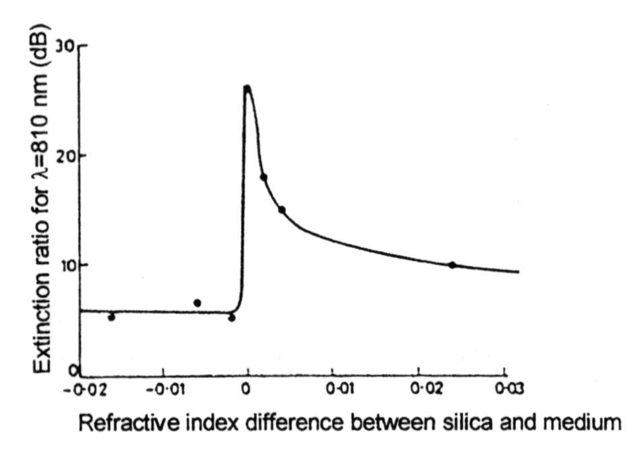

Fig. 6.10. Measurement of the extinction coefficient between the two polarized modes of a polarization-preserving fiber as a function of the difference between the refractive indices of the silica fiber and of the medium which contains the tapered fiber. Comparison with direct calculations of the reflectivity between the two media

the losses. Figure 6.11 summarizes the effects of three different values of the index difference.

The extreme sensitivity of these measurements can be noted. A variation of 25 dB was obtained for a 0.001 variation of the refractive index. An inherent disadvantage of this technique is the lack of durability and the fragility of the device thus produced.

6.2.1 Monitoring of a Chemical Reaction by Fluorescence Detection

Let us examine the case of an optical fiber where the evanescent field extends outside the fiber or of a fiber whose cladding contains molecules that can adsorb fluorescent markers. If the fiber is immersed in a medium where flu-

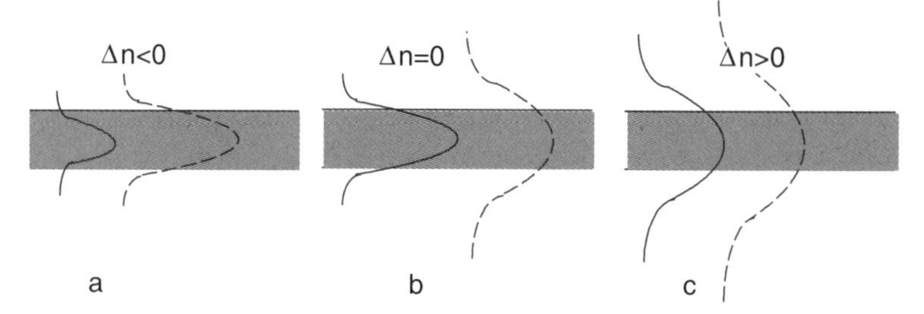

Fig. 6.11. The waveguide and the modes for three values of the index difference (**a**) the two modes are well-guided, (**b**) one of the modes is near its cutoff frequency, (**c**) the two modes are near their cutoff frequencies

orescent markers released in it have interacted with the molecules present in this medium, the fluorescent molecules can be excited. Different techniques have been developed for exciting the molecules and detecting the fluorescence emission. As represented in Fig. 6.12, the detection of the fluorescence emission can be used for the monitoring of a chemical reaction.

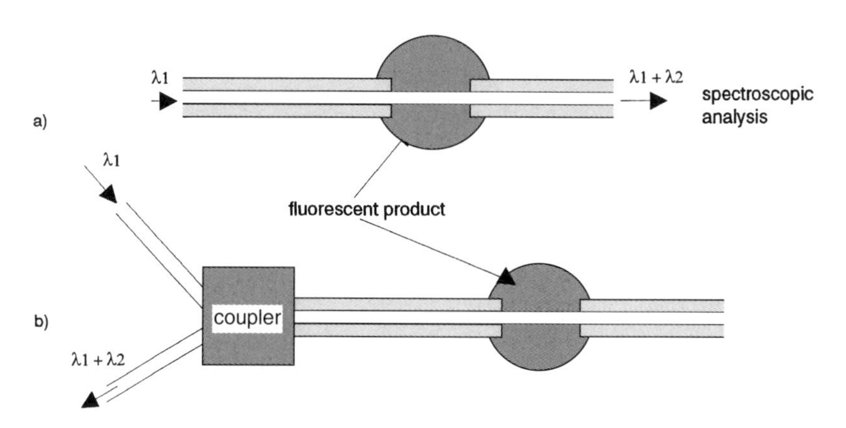

Fig. 6.12. Two methods of monitoring a chemical reaction by the measurement of fluorescence (**a**) in transmission, (**b**) in reflection

Figure 6.13 represents a typical arrangement used for measuring the antigen concentration in a liquid. The measurement is based on the detection of the fluorescence excited by the evanescent field using an optical fiber coated with antibodies [Bluestein 1990].

Under these conditions, the fiber is placed in the medium to be examined. As represented in Fig. 6.14, by varying the concentration of a given product, the optical response can be correlated with the concentration. The high level of sensitivity of these results is apparent in the curves. Similarly, chemical

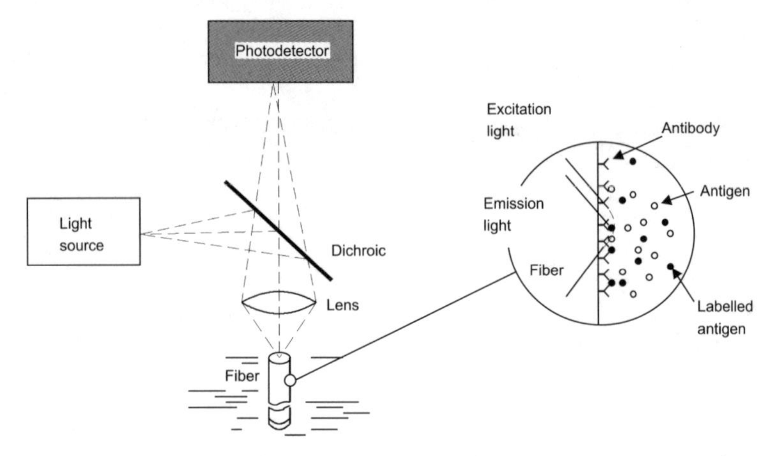

Fig. 6.13. Principle of the measurement of antigen concentration from the detection of the fluorescence reflected by a fiber coated with antibodies

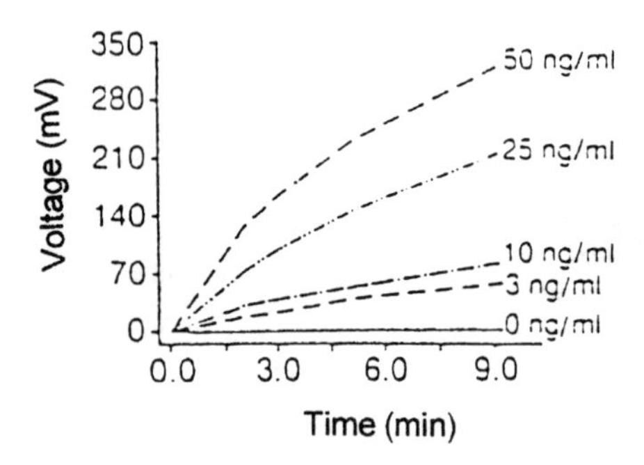

Fig. 6.14. Fluorescence detection applied to biology: detection of different antigen concentrations versus time; the response is determined from the voltage measured in mV [Bluestein 1990]

reactions can be monitored in real time using the arrangement shown in Fig. 6.13.

All evanescent-field optical-fiber sensor systems have not been commercially developed. This results from the fact that most of them were designed with specific and limited studies in view. The characteristics of some optical-fiber sensors are summarized in Table 6.2. Researches on optical-fiber sensors are presently still being carried out. Therefore, the characteristics presented here are given only for information.

As evanescent-field couplers were described in the preceding chapter, the case of sensors produced from couplers has not been addressed here. We have seen that evanescent-field optical-fiber sensors are based on the use of either the interaction between the evanescent-field of the modes of the fiber and the external medium or the variations in the conditions of total internal reflection

Table 6.2. Characteristics of optical-fiber sensors based on the use of the evanescent field.

Parameter detected	Perturbation type	Measurement
Temperature	Hollow-core fiber	Reflectometry
Temperature	Unclad fiber	Transmission
Refractive index	Unclad fiber, pulled fiber, thinned fiber	Transmission
Refractive index	Eccentrically clad fiber	Reflectometry
Pressure	Fiber submitted to pressure	Reflectometry, transmission
Chemical reaction	Unclad fiber, doped cladding	Transmission, reflection

associated with the perturbation. Integrated-optical sensors are based on the same type of interactions with respect to the specific constraints involved by their technology.

6.3 Integrated-Optical Sensors

6.3.1 Analysis of the Sensitivity of Integrated-Optical Sensors

The main field of applications of planar waveguides is, as for optical fibers, information transmission. Nevertheless, they might also be employed as sensors [Zhou 1991, Helmers 1995, Helmers 1996]. Planar waveguides generally sustain only a few modes and present a simple index profile, which directly depends on the technique used for their production (diffusion of silver salts, sol–gel solution, ion implantation, etc.). As in optical fibers, the essential part of the field is the evanescent part.

Enhancement of the sensitivity of the sensor can be obtained if the experimental conditions are such that the evanescent field reaches its maximal amplitude and confinement. Under these conditions, a modification arising at the surface of the waveguide will then have a maximal perturbing effect.

The simplest waveguides to be studied are asymmetrical index-profile waveguides, that is waveguides presenting only three refractive indices (Fig. 6.15). As demonstrated by Parriaux, for the different parameters which characterize these waveguides, the values corresponding to a maximal amplitude of the field at the interface between the waveguide and air can be analytically determined [Parriaux 1994a, Parriaux 1994b].

The square of the amplitude of the TE_0 mode of this waveguide is

$$A_{\mathrm{c}}^2 = \frac{4\omega\mu_0 P}{n_{\mathrm{e}}(\varepsilon_{\mathrm{g}} - \varepsilon_{\mathrm{c}})} \frac{\varepsilon_{\mathrm{g}} - \varepsilon_{\mathrm{c}}}{k_0 w_{\mathrm{e}}}, \tag{6.2}$$

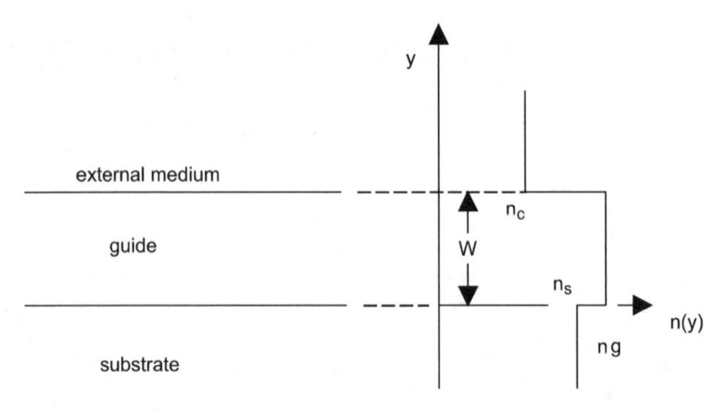

Fig. 6.15. Step-index planar waveguide. w is the width of the waveguide, n_c, n_g and n_c are the indices of the core, substrate and external medium respectively. The asymmetry parameter is $a = (n_g^2 - n_c^2)/(n_g^2 - n_s^2)$

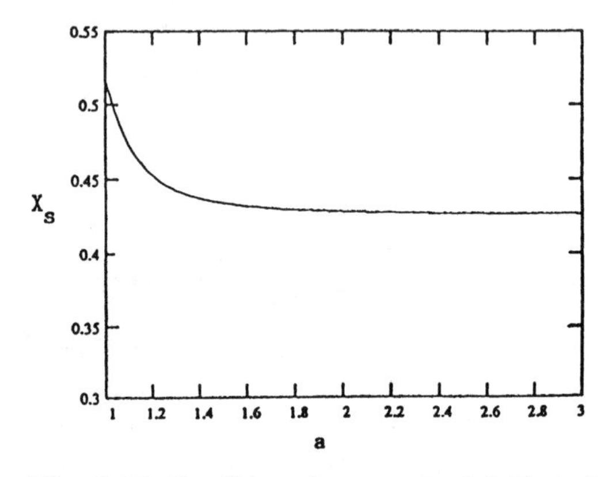

Fig. 6.16. Conditions for a maximal field at the interface between the waveguide and the external medium. $s = [(\varepsilon_e - \varepsilon_s)/(\varepsilon_g^2 - \varepsilon_e)]^{1/2} = [(n_c^2 - n_s^2)/(n_g^2 - n_c^2)]^{1/2}$, and a is as defined in the caption of Fig. 6.15

where P is the total power guided by the mode. Figure 6.16 indicates the conditions for a maximal value of the field at the interface between the waveguide and the external medium.

The maximal amplitude of the field is given by the equation

$$A_{cm}^2 = 4\omega\mu_0 P \left(\frac{1 - \varepsilon_s/\varepsilon_g}{x_{s0}^2 + \varepsilon_s/\varepsilon_g} \right)^{1/2} \times [(1 + x_{c0}^2)(\arctan x_{s0} + \arctan x_{c0} + m\pi + x_{c0}^{-1} + x_{s0}^{-1})]^{-1}. \tag{6.3}$$

From this equation, the maximal sensitivity of the evanescent-field sensor can be determined in the case of materials deposited in the form of thin layers at the interface between the waveguide and the external medium.

However, the results of the analysis presented here are valid only as a first approach in the determination of the sensitivity of one-parameter sensors of the type described above.

6.3.2 Creating the Sensing Region

An integrated-optical element has to fulfill different functions.

- Receive and guide the optical signal.
- Be sensitive to local perturbations.
- Transfer the signal onto the detector.

Different techniques can be used for creating the sensing region of an integrated-optical element (Fig.6.17). A standard waveguide can be locally polished, in order to be able to perturb more easily the evanescent field. Waveguides with variable depth may also be used [Clauss 1994]. Another method is by using a waveguide with convenient length, where the perturbation of the evanescent field from the tested medium will be much higher than the losses occurring in the guiding region.

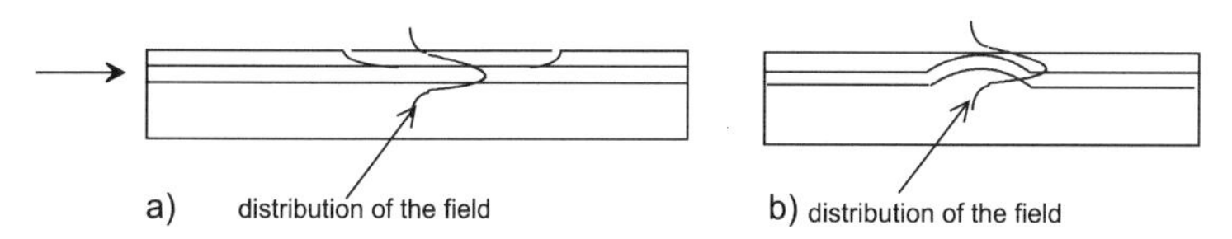

Fig. 6.17. Two different methods for accessing the evanescent field (**a**) local removal of the cladding, (**b**) waveguide with variable depth

6.3.3 Evanescent-Field Interferometric Sensors

The mode whose field extends within the external medium is sensitive to the parameters of this medium. Thus far, only the losses associated with the modes have been taken into account. There are however other parameters of the mode, in particular its velocity, which undergo significant changes.

The basic principle of the sensor developed by Gréco *et al.* is schematized in Fig. 6.18. Two types of interferometric sensors have been realized. A rectilinear waveguide (I) is divided into two arms at a Y-junction (II) where light is distributed equally between waveguides (III) and (IV). Branch III presents locally a depression where the waveguide is in contact with a material whose refractive index varies with the presence of a gas or of an antigen. The material may be either a polymer or antibodies grafted on glass [Gréco 1994].

The effective index of the mode in branch III can thus differ from the index of branch IV. For the same distance traveled inside the waveguide, the optical paths in each branch will be different. After the second junction at

Y(V), the two modes recombine. Depending on their phase difference, their summation has either a constructive or a destructive character. The second element diverges from the first one by the fact that the junction is followed by a widened waveguide. This arrangement permits the observation of the effect of the summation of the two modes, which is reflected by interference fringes.

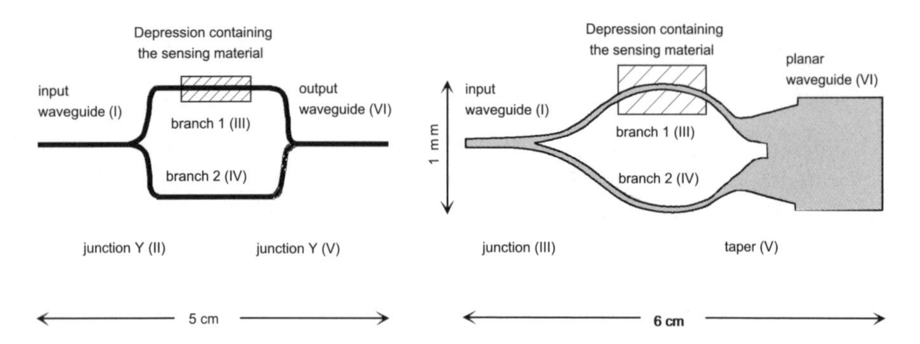

Fig. 6.18. Evanescent-field extended fields interferometer [Gréco 1994]

The device used comprises an area where the width of the waveguide increases adiabatically, in order to prevent losses as well as a coupling of the modes. The two waves interfere in this section of the device, thereby causing interference fringes like those represented in Fig. 6.19.

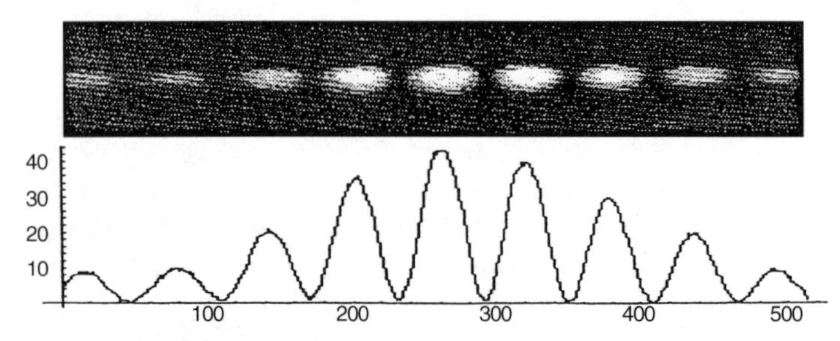

Fig. 6.19. Interference fringes observed at the end of the interferometer

The waveguides used in this study were realized from microscope glass plates using the method of ion exchange Na^+-K^+, for a 4 hr time of exchange at 450°. The results obtained suggest a high level of sensitivity. The curve represented in Fig. 6.20 represents the detection of hexane vapors for hexane concentrations comprised between 1.21% and 0.19%.

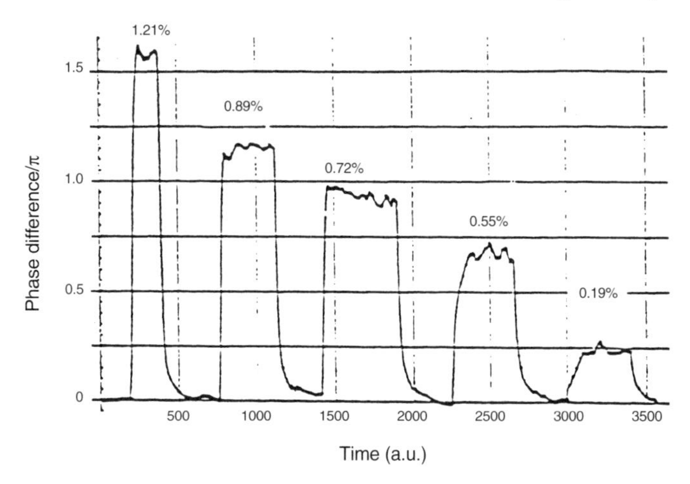

Fig. 6.20. The phase difference as a function of hexane concentration [Gréco 1994]

A different method of enhancing the sensitivity of a sensor of this type consists in amplifying the value of the evanescent field in contact with the substance to be detected.

6.3.4 Amplification of the Evanescent Field by a Multilayered System and Applications to Biosensors

The evanescent field can be amplified with a system consisting of a set of layers with high and low indices. A system of this type ensures a very high level of sensitivity for the measurements which are to be performed. Therefore, the application of this method for the measurement or the detection of biological parameters has led to the production of very efficient biosensors [Buckle 1993, Davies 1993, Georges 1995, Georges 1999]. The scheme drawn in Fig. 6.21 represents the technique developed by few companies.

The reader is referred to Chap. 8 for an analysis of the field present in such a structure. The values of the incidence angle and of the wavelength are determined in such a way that, for a given set of layers, the amplitude of the field at the interface between the last layer and the external medium is maximal. Further, the last layer has an intrinsic roughness of such a value that a part of the field is diffused, thereby causing a reduction of the amplitude of the reflected beam.

A change, due for example to the absorption of certain molecules, of the characteristics of the medium where the evanescent field is present will result

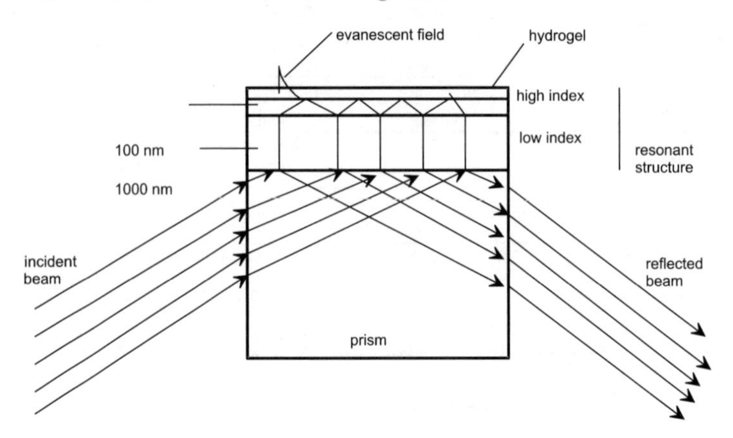

Fig. 6.21. Principle of an evanescent-field biosensor

in a decrease of the amplitude of the evanescent field. This in turn causes a relative increase of the amplitude of the reflected beam. As can be seen in Fig. 6.22, the experimental parameters of the system can be selected in order that the resonance corresponds to a particular stage of a chemical reaction.

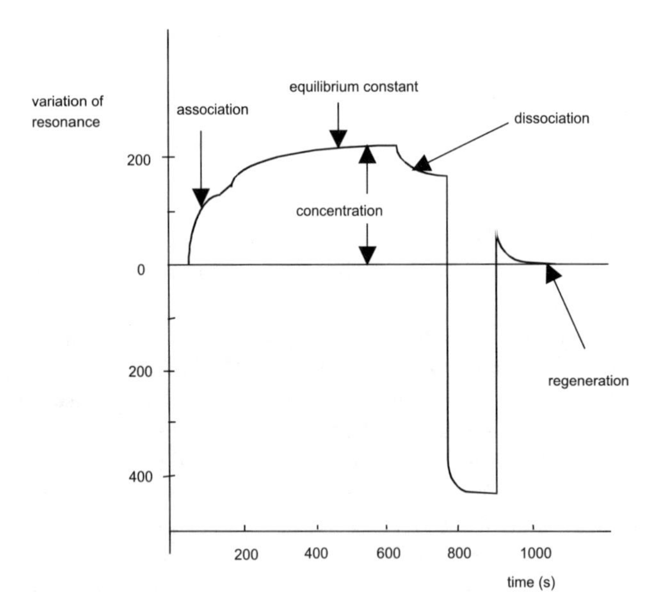

Fig. 6.22. Example of a variation of the resonance with respect to biological parameters

6.4 Conclusion

The light propagating inside a waveguide is highly sensitive to perturbations affecting the propagation of the modes. While a large variety of sensors are based on this interaction, we have restricted ourselves to sensors where the perturbation is exerted on the evanescent part of the modes. For amplifying the sensitivity of these sensors, we have seen that the waveguide needs to be placed under such conditions that the evanescent part of its field spreads out of the cladding into the medium to be investigated. This can be achieved either by bending the waveguide or by removing a part of the optical cladding. The few examples described in this chapter have shown that the sensors thus produced have a very high sensitivity. Researches are being carried out in order to determine the optimal production conditions of these sensors.

The advantage of optical-fiber sensors is that they can be used for measuring certain physical parameters in distant or dangerous environments. These sensors have applications in a variety of different areas, and therefore are presently being extensively studied. Nevertheless, a normalization of the different types of fiber-optic sensors is still necessary.

The spectral response of a material is one of the physical parameters from the measurement of which a material can be characterized. The techniques developed for detecting this parameter, which are referred to under the general term of 'internal-reflection spectroscopy', have not been described in this chapter. Indeed, the technique of internal-reflection spectroscopy has been rapidly expanding since the 1960s and is at present in a much more advanced state of development than the techniques presented herein. Therefore, it seemed to us justified to treat this topic in a separate chapter.

7. Internal-Reflection Spectroscopy

The refractive index and the absorption coefficient of homogenous objects larger than the wavelength can be readily determined. This does not hold when the materials to be measured are present in small quantities. In this case, it is necessary to use cautiously the very notion of refractive index. Researches have shown that the properties of total internal reflection could be efficiently used in the analysis of thin layers. The possibility of using total internal reflection to this end gave rise during the 1960s to the development of the technique known as internal-reflection spectroscopy (IRS) [Harrick 1979, Wilks 1967, Mirabella 1985]. This technique can be based on the use of either frustrated or non-frustrated total internal reflection.

The principles themselves on which internal-reflection spectroscopy is based were already known for a long time (Fig. 7.1). In fact, experiments in this direction had already been carried out much earlier [Taylor 1933a, 1933b, 1933c]. Nevertheless, the potential of internal-reflection spectroscopy really started to be exploited with the researches conducted by Harrick [1960].

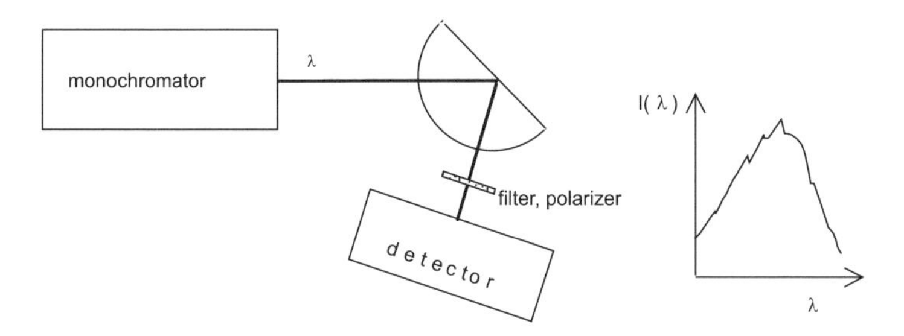

Fig. 7.1. Principle of a spectroscope using total internal reflection

Before examining the applications of internal-reflection spectroscopy, it is necessary to recall first that the phenomenon of total internal reflection is sensitive to the refractive indices, and in particular to the absorption coefficients, of the different elements. We then describe the most significant extensions of IRS. Lastly, we discuss the possibility of investigating the adsorption and desorption of molecules at the surface of a prism.

7.1 Effect of Index Variations on Total Internal Reflection

Internal-reflection spectroscopy, whether it employs frustrated or no-frustrated total internal reflection, is based on the fact that the quantity of light reflected at the interface between the prism and the material depends in either case on the absorption coefficient of the material where the evanescent waves have penetrated. As we have seen in the chapter devoted to total internal reflection, the components of the field inside the medium where the evanescent field is generated are

$$\mathbf{E}_{\mathrm{p}}(z) = E^{\mathrm{i}}_{\mathrm{p}} \frac{(2\cos\theta)\exp(-z/d_{\mathrm{p}})}{n^2\cos\theta + \mathrm{j}(\sin^2\theta - n^2)^{1/2}} \left[-\mathrm{j}(\sin^2\theta - n^2)^{1/2}\mathbf{e}_x + \sin\theta\mathbf{e}_z\right], \tag{7.1}$$

in p polarization. In s polarization, the equation for these components is

$$\mathbf{E}_{\mathrm{s}}(z) = E^{\mathrm{i}}_{\mathrm{s}} \frac{(2\cos\theta)\exp(-z/d_{\mathrm{p}})}{n^2\cos\theta + \mathrm{j}(\sin^2\theta - n^2)^{1/2}} \mathbf{e}_y, \tag{7.2}$$

where θ is the incidence angle of the plane wave inside the first medium. n is equal to n_2/n_1, n_1 is the index of the first medium, and n_2 is the index of the second medium, $n_1 > n_2$.

If the medium where the evanescent waves have been generated is absorbing, the field present in this medium has a more complex expression [Harrick 1979]. We address the problem of determining this field in the following subsection.

7.1.1 Effective Thickness

The field inside an absorbing medium can be determined using analytical methods. This field does not have a simple form, and it is usually necessary to proceed to a few calculations for determining it. It will be assumed here that the experimental conditions are such that the interaction of the evanescent field with the absorbing medium can easily be calculated. This can be achieved by selecting the values of the experimental parameters in order that the resulting absorption coefficient does not exceed an order of 10%.

Let us now return to the problem of determining the transmission coefficient of a material. Neglecting the losses arising at the reflection, the transmitted intensity decreases exponentially: $I/I_0 = \mathrm{e}^{-\alpha d}$. In the case of low absorption, that is, if αd is smaller than 1, we have: $I/I_0 \approx 1 - \alpha d$. In the case of an absorbing medium, the absorption coefficient α is related to the refractive index by the following equation

$$\alpha = 4\pi nk/\lambda. \tag{7.3}$$

Two extreme cases can be examined: a semiinfinite medium and a thin layer deposited at the surface of a prism [Mirabella 1985]. The losses inside a medium are measured in transmission. In the case of a material illuminated under total internal reflection, we determine the equivalent layer with thickness d_{e} of the same material, illuminated in transmission and undergoing the same losses.

In the case of a thin film with a thickness far less than the penetration depth of the field, we consider the effective thickness d_{e} of an equivalent medium with the same absorption coefficient. The effective thickness d_{e} of such a medium can be simply written in the form

$$d_{\mathrm{e}} = \frac{n_{21} d E_0^2}{\cos\theta}. \tag{7.4}$$

Replacing the expression of the value E_0 of the evanescent field at the interface between the first two media of indices n_1 and n_2, one obtains an expression of the effective thickness. For s and p polarizations of the incident wave respectively, the equations for the effective thickness are [Harrick 1966a, Zolotaryov 1970, Mirabella 1985]

$$d_{\mathrm{e}\perp} = \frac{4 n_{12}\cos\theta}{1 - n_{31}^2}, \tag{7.5}$$

$$d_{\mathrm{e}\parallel} = \frac{4 n_{21} d \cos\theta[(1 + n_{32}^4)\sin^2\theta - n_{31}^2]}{(1 - n_{31}^2)[(1 + n_{31}^2)\sin^2\theta - n_{31}^2]}, \tag{7.6}$$

where n_3 is the refractive index of the third medium.

These equations are of interest on several accounts. In particular, the spectral response of such a system can thus be correlated with the actual spectral response of the bulk material. Different authors, see for example Harrick and Mirabella, have stressed the fact that in the vicinity of the critical angle θ_c the measurements of the reflected beam yield a spectral response with only a few distortions caused by total internal reflection [Mirabella 1985, Harrick 1966a].

Let us return to a simpler two-media model. If total internal reflection arises within a second semiinfinite medium with low absorption, the losses are expressed by the equation

$$a = \frac{n_{21}\alpha E_0^2}{2\gamma\cos\theta}. \tag{7.7}$$

The equation for the effective thickness is then

$$d_{\mathrm{e}\perp} = \frac{n_{21}\lambda_1\cos\theta}{\pi(1 - n_{21}^2)(1 - n_{21}^2)(\sin^2\theta - n_{21}^2)^{1/2}} \tag{7.8}$$

in s polarization, and

$$d_{\mathrm{e}\parallel} = \frac{n_{21}\lambda_1(2\sin^2\theta - n_{21}^2)\cos\theta}{\pi(1 - n_{21}^2)(1 - n_{21}^2)[(1 + n_{21}^2)\sin^2\theta - n_{21}^2](\sin^2\theta - n_{21}^2)^{1/2}}, \tag{7.9}$$

in p polarization.

The effective thickness of a material is a very significant physical parameter. In particular, this notion is at the basis of any serious comparison between the values obtained under total internal reflection and the values measured in transmission. The penetration depth of the evanescent field can actually be quite different from the value of the effective thickness, as can be seen in the example given by Harrick (Fig. 7.2).

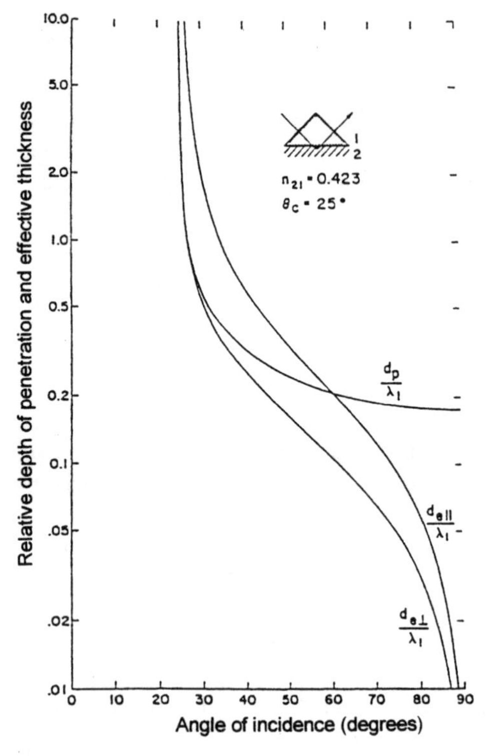

Fig. 7.2. Comparison between the penetration depth of the evanescent field in a medium of index $n_{21} = 0.423$ and the value of the effective thickness for each polarization [Harrick 1979]

In this example, it can be noted that in the vicinity of low-angled incidence, the value of the effective thickness goes far below the value of the penetration depth.

The value of the effective thickness depends on several parameters, such as the incidence angle, the wavelength or the refractive indices of the different media [Harrick 1979, Mirabella 1985]. When analyzing a spectrum, it is imperative to take into account the fact that the spectrum obtained depends on all these parameters.

As an example, a comparison between two spectra obtained under different experimental conditions is displayed in Fig. 7.3. Despite the difference in the experimental parameters, it can be seen that the two spectra represented present the same contrast.

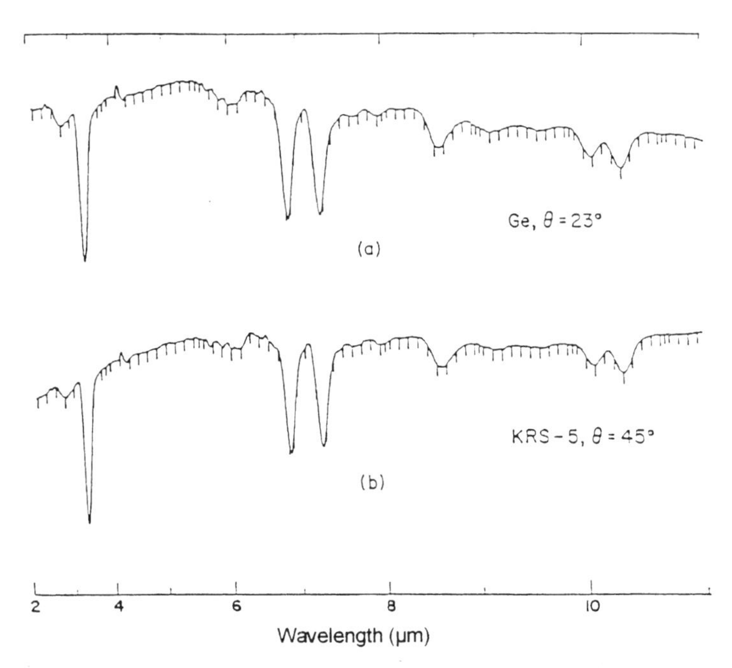

Fig. 7.3. Comparison between polypropylene spectra ($n_2 = 1.5$), in the case of a single reflection on a prism made of (**a**) germanium ($n_1 = 4$), for an incidence angle of 23°, (**b**) KRS-5 ($n_1 = 2.4$), for an incidence angle of 45° [Harrick 1979]

7.1.2 Measurement of the Dielectric Constants in an Arbitrary Medium

In the case of an arbitrary medium, the values of the reflected amplitude r in s and p polarizations respectively, for an incident field of amplitude 1, are given by the following equations [Mirabella 1985]

$$r_{\perp} = -\frac{\cos\theta - (\hat{n}_{21}^2 - \sin^2\theta)^{1/2}}{\cos\theta - (\hat{n}_{21}^2 + \sin^2\theta)^{1/2}}, \tag{7.10}$$

and

$$r_{\parallel} = \frac{(\hat{n}_{21}^2 - \sin^2\theta)^{1/2} - \hat{n}_{21}^2\cos\theta}{(\hat{n}_{21}^2 - \sin^2\theta)^{1/2} + \hat{n}_{21}^2\cos\theta}, \tag{7.11}$$

where $\hat{n}_{21} = n_{12}(1 + \mathrm{j}k_2)$.

From these relations, the refractive indices of any medium can theoretically be determined. The determination of the value of the indices can however be enhanced by varying parameters other than polarization. In fact, this usually turns out to be imperative when determining the imaginary component of the refractive index.

Before discussing the spectra obtained, it may be useful to describe in more detail some systems of internal-reflection spectroscopy. We begin with a description of the elements where total internal reflection arises.

7.2 Spectroscopy Devices Based on Total Internal Reflection

7.2.1 Description of Different Systems Generating Total Internal Reflection

Several configurations of spectroscopes have been developed. As an example, we may mention the spectroscope realized by Connecticut Instruments. Figure 7.4 displays the part of the spectroscope where total internal reflection is generated [Harrick 1966a, Harrick 1966b, Harrick 1979].

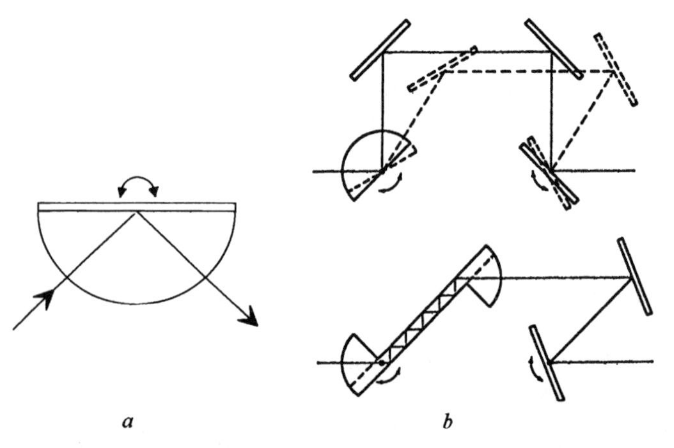

Fig. 7.4. Devices used for varying the incidence angle of the beam (**a**) single-reflection system, (**b**) multiple-reflection system [Harrick 1979]

As can be seen in Fig. 7.4b, the prism illuminated at total internal reflection (Fig. 7.4a) has been replaced by an optical device allowing multiple reflections [Harrick 1966b, Harrick 1966c, Harrick 1973]. An effect of this modification is the amplification of the absorption of the medium whose spectrum is desired. Fig. 7.5a represents a double-pass multiple internal reflection plate. The photograph in Fig. 7.5b shows the set of multiple reflections arising within the plate.

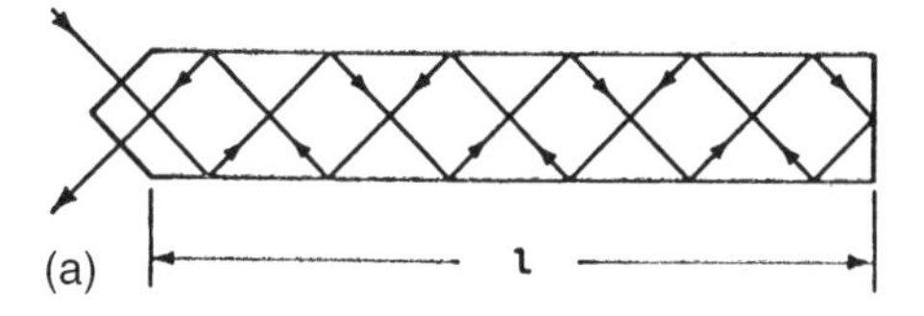

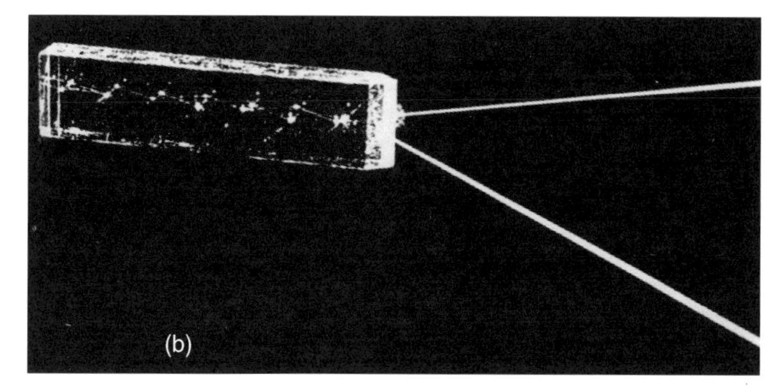

Fig. 7.5. (**a**) Double-pass multiple internal reflection plate, (**b**) photograph of the light path inside the plate [Harrick 1979]

It is sometimes necessary to separate slightly the directions of the reflected and incident beams. This can be obtained by modifying the form of the previous device (Figs. 7.6 and 7.7). As an example, Fig. 7.6 represents a device designed for separating the incident and reflected beams. If the incidence plane of the beam is not normal to the surface, the path of the rays can be a broken line. The number of total internal reflections is then multiplied even further.

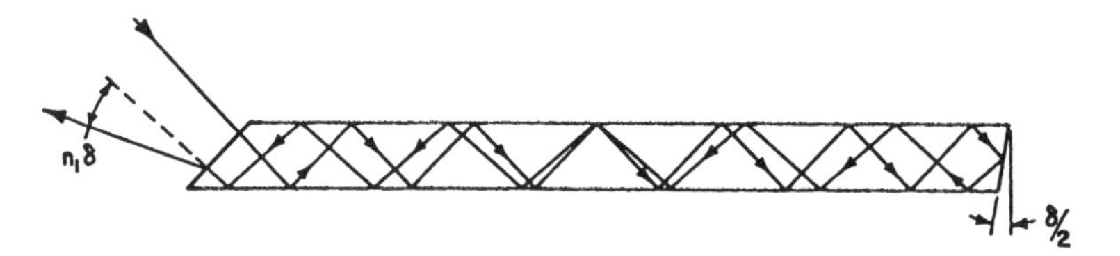

Fig. 7.6. Double-pass multiple internal reflection plate. The directions of the incident and reflected beams are almost identical [Harrick 1979]

The fabrication of devices of this type is not always easy. As seen in the preceding paragraphs, in order that reproducible absorption spectra can be obtained, the plate or the prism on which the sample is placed must be of a very high quality. As the development of optical fibers went on, it was recognized that fibers where the cladding had been either removed or

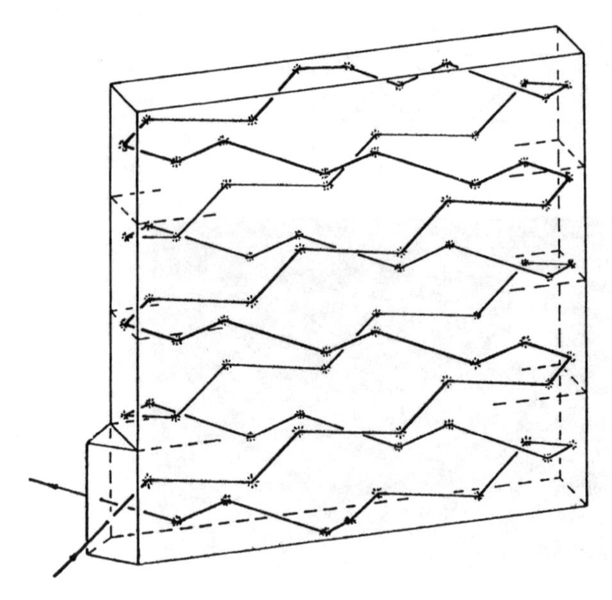

Fig. 7.7. Folded-path multiple-pass internal reflection plate [Harrick 1979]

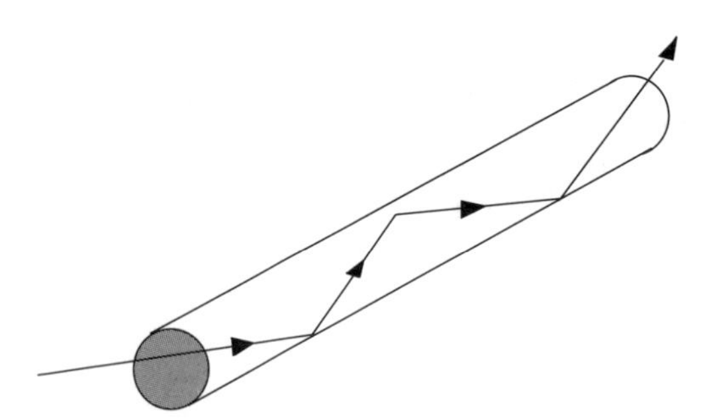

Fig. 7.8. Light path inside an optical fiber

reduced could fulfill the same functions as classical optical devices. Figure 7.8 represents the propagation of a ray in such a fiber.

The advantage of fibers is that their interface is in general of a very high quality. This is due to the fact that fibers are generally fabricated by fusion and pulling from a preform with larger diameter. Further, fibers are highly sensitive to the parameters of the external medium.

Fiber-optic evanescent-wave spectroscopy allows us to obtain spectra in the visible spectrum using silica fibers as well as infrared spectra using AgX silver halogenure fibers ($2\ \mu m < l < 20\ \mu m$) [Katz 1994]. As an example, unclad $AgCl_{0.3}Br_{0.7}$ fibers with a diameter 900 μm were used by Simhi [1996]. By constructing a cell for immersing the fiber, as represented in Fig. 7.9, reference spectra for a given medium can be obtained. If the physical param-

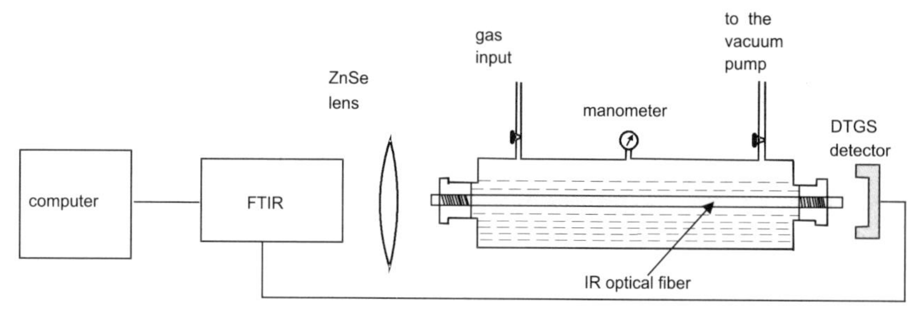

Fig. 7.9. Fiber spectroscope

eters of the medium are changed, the spectral response corresponds to this modification.

The advantage of using fibers for such purposes is that they permit us to perform measurements of very small quantities of materials [Chiacchiera 1992]. This can be seen in the spectrum represented in Fig. 7.10.

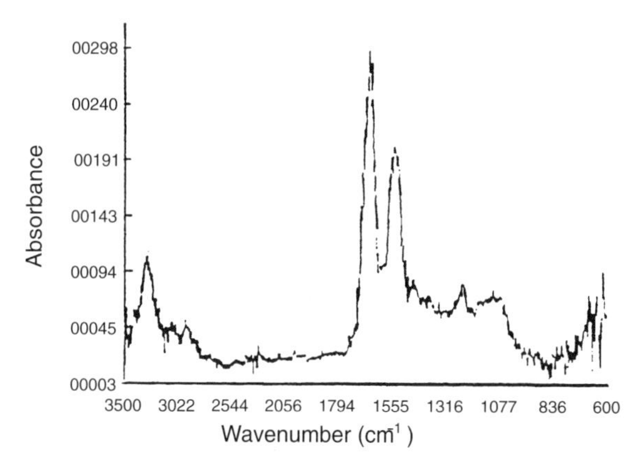

Fig. 7.10. Infrared spectrum of 10 μg trypsin on a silver halogenure fiber [Chiacchiera 1992]

The multiple internal reflection devices presented in Figs. 7.5, 7.6 and 7.8 can be used for measuring the spectrum of a liquid where the plate is partially immersed. This requires that a relatively great quantity of the product to be measured is available. In the case of a material in very small quantity or in the case of a powder, this disadvantage can be avoided by using an element in the form of a rosette. A device of this kind is represented in Fig. 7.11.

As described in Fig. 7.11, the light beam injected into the rosette is successively reflected several times at the same point [Harrick 1966c]. Spectra

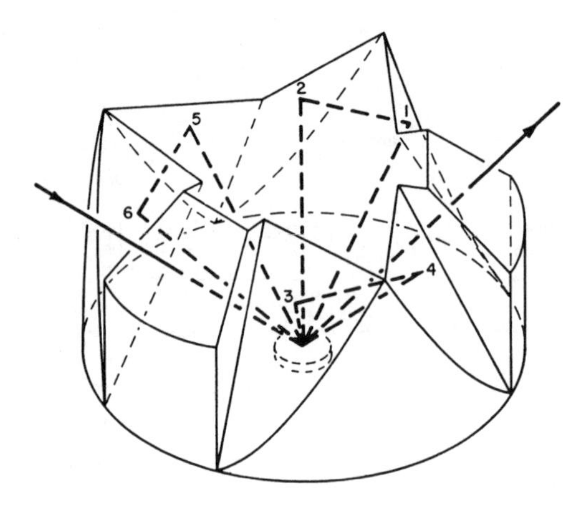

Fig. 7.11. Optical device in the form of a rosette for obtaining successive total internal reflections of a light beam at the same point [Harrick 1966c]

of materials of which only a very small quantity is available can thus be obtained.

7.2.2 Description of Internal-Reflection Spectroscopes

The element where the single or multiple total internal reflections arise is generally inserted inside a more complex system which consists of:

- a slot for injecting the light beams emitted from a monochromator,
- an optical device for the illumination of the prism under the required conditions,
- the system where total internal reflection arises,
- an optical system, comprising a lens, filters and polarizers in order to bring the light onto the detector,
- a system for detecting the signal after total internal reflection, and
- a data processing system for the detected information.

For carrying out measurements more accuratly, a double-prism spectroscope can be employed (Fig. 7.12). An advantage of using such a device is that the measurements actually performed can be compared to a reference spectrum.

The data processing of the spectrum consists in recording the spectrum and in analyzing it. The emerging light signal reflects the characteristics of the ensemble arrangement, and not only of the sample under study. In general, a reference spectrum is first recorded in the absence of the sample. The absorption spectrum of the sample is determined from the difference between

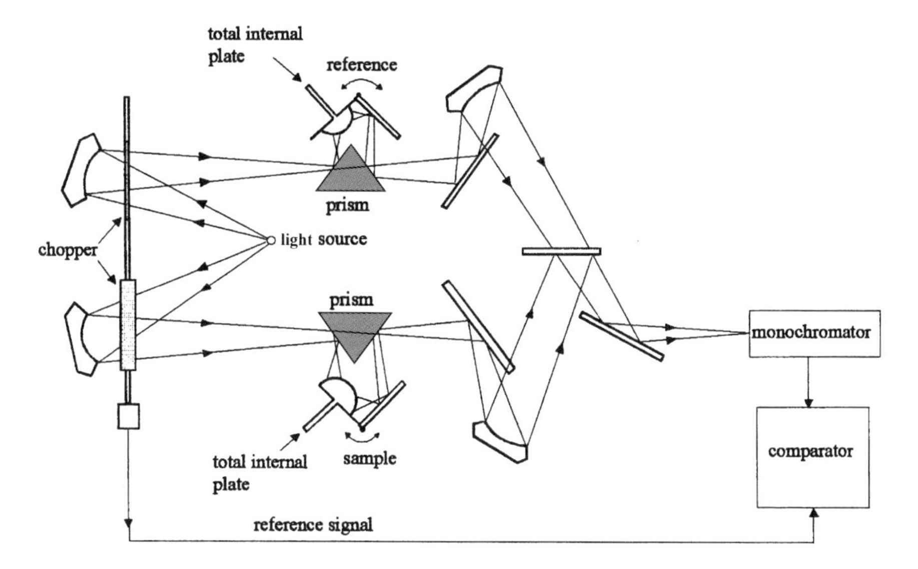

Fig. 7.12. Double-prism spectroscope

the spectrum of the sample and this reference spectrum. This procedure is illustrated in Fig. 7.13.

If the component which is studied needs to be diluted in a specific medium, in water or in alcohol for example, the reference spectrum has to be obtained in the presence of the liquid in its pure state. As an example, if the effect of a given physical parameter (e.g. temperature, pressure) on a given material is investigated, the spectrum of the material has to be recorded earlier in a reference state.

In recent years, the use of filtering techniques based on Fourier transforms has led to an enhancement of the signal-to-noise characteristics of internal-reflection spectroscopes in comparison with the current filtering methods.

7.2.3 Quality of the Reflective Element

The fabrication of the reflective element must satisfy certain conditions. The surface must be as even as possible in order that the material to be measured be in close contact with the surface of the prism. Not all materials can be prepared in such a way: for infrared measurements, glasses, silica or germanium are examples of materials presenting appropriate characteristics. The roughness of the internal reflection elements is generally of the order of a few nanometers.

On the other hand, the element where the total internal reflection arises must be easily cleaned. Hence, it must be insensitive to a number of solvents,

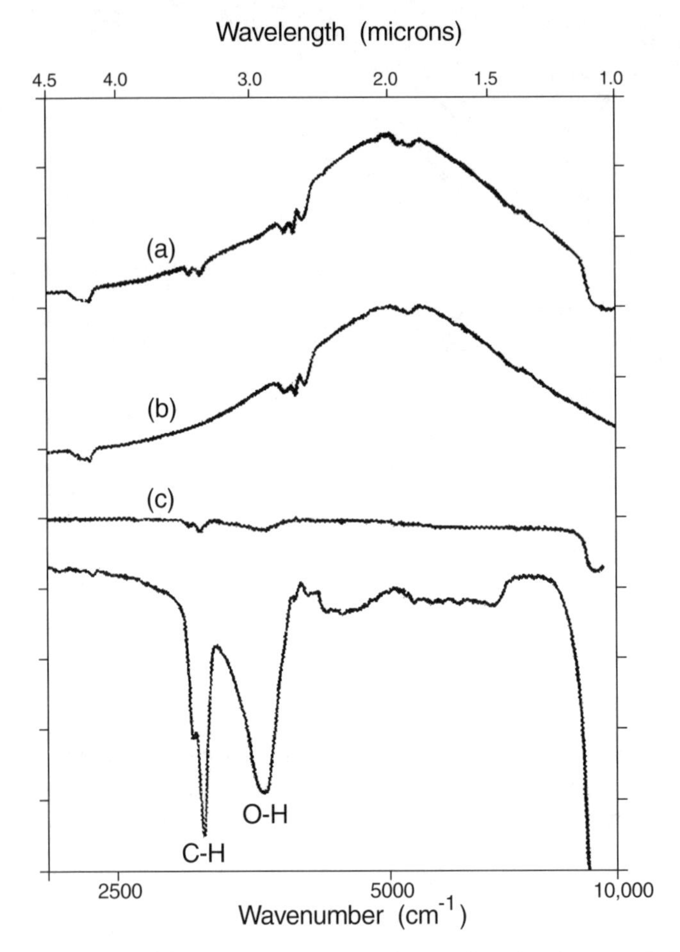

Fig. 7.13. (**a**) Reference intensity detected in the absence of the sample, (**b**) intensity detected in the presence of the sample, (**c**) difference between the two curves displaying the absorption spectrum of the sample, (**d**) the same absorption spectrum amplified with a factor 25

and especially, since internal-reflection spectroscopy is extensively used in biology, to organic solvents.

Fibers generally present a good state of surface and their use as internal-reflection elements has permitted us to extend the spectrum available and to enhance the sensitivity of the measurements.

7.2.4 Constraints in the Preparation of the Samples

The samples whose absorption spectrum is desired have to be deposited at the surface of the reflective element in the form either of a film or of a powder.

The refractive indices of the prism and of the medium to be studied are determined so that an evanescent wave can be generated at the interface between the sample and the prism. The choice of a given reflective element depends on the type of the sample to be analyzed.

7.3 Atom Spectroscopy in the Vicinity of Interfaces

Illumination under total internal reflection has also been employed for spectroscopic measurements different from those previously described [Oria 1989, Ducloy 1997, Deutschmann 1993]. As an example, measurements based on illumination in the conditions of total internal reflection of the interface between a gas and a prism have been carried out. Unlike volume measurements, these techniques can be used for obtaining information on specific effects related to the presence of the surface, such as, for instance:

- van der Waals interactions affecting the dipolar momentum of the atom, thereby inducing a displacement of the atomic resonance. While the range of these forces is of the order of 10^{-10} m, evanescent waves extend only between 10^{-6} m and 10^{-7} m. Accordingly, the detection of van der Waals interactions is quite difficult [Landragin 1996].

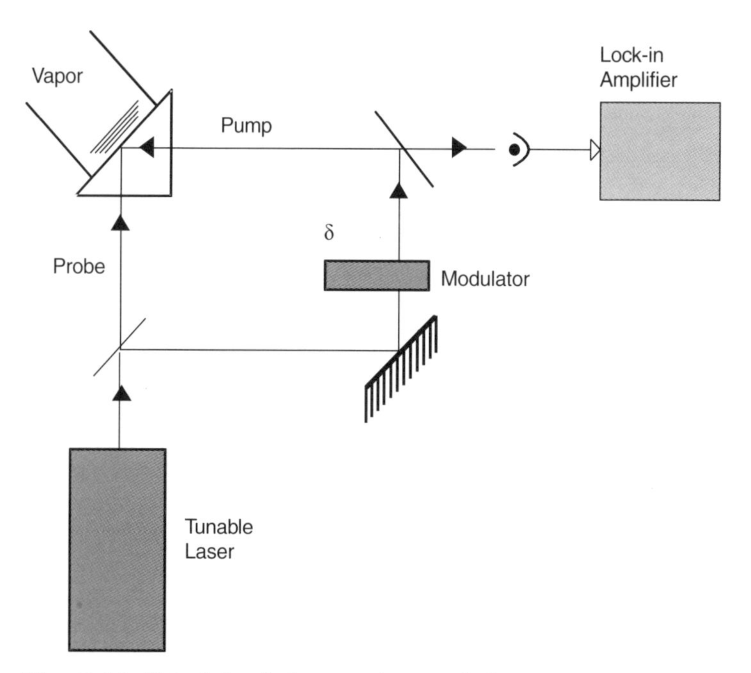

Fig. 7.14. Principle of the experiment of absorption saturated with evanescent waves

- modifications of the spontaneous radiation caused by the presence of the partially reflective interface: inhibition or amplification of the spontaneous radiation.
- collision effects with the wall. Experiments on Cs vapor (transition at 6S1/2-6P3/2, λ = 852 nm) have gained evidence for the redistribution of the velocities after the collision. These measurements have permitted the determination of the time of 'sticking' of the molecules with the surface.

Figure 7.14 illustrates the experimental arrangement used for carrying out these measurements. Two beams, referred to as 'pump' and 'probe', generated by the same laser, propagate along opposite directions until they reach a prism illuminated under total internal reflection. The pump beam is modulated with a frequency d, while the probe beam is used for detecting the modulations induced by the presence of the pump beam.

As an example, Fig. 7.15 represents the attenuation of the total internal reflection coefficient during an experiment on absorption saturated with evanescent waves on the Cs vapor around the transitions F=4 $\longrightarrow$ F' = 2,3,4,5 of the ray at 852 nm [Oria 1989]. The modulation frequency was equal to 10 kHz. The peaks corresponding to the transitions appear when the detection is in phase with the modulation. If the detection is in quadrature with the modulation, only the effect of the collision response appears. In this case, the response is out of phase and amplified.

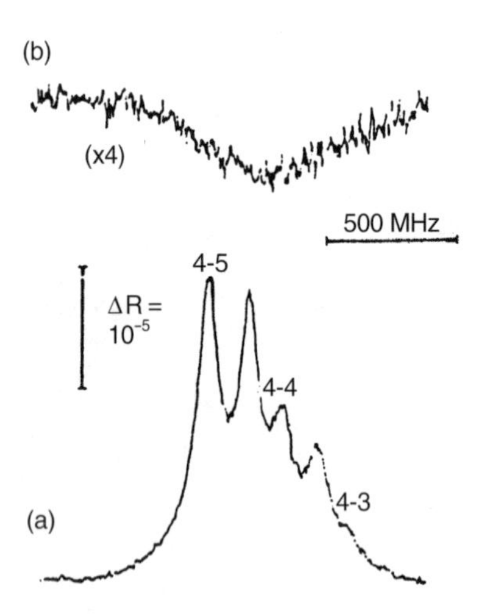

Fig. 7.15. Attenuation of the total internal reflection coefficient. Frequencies are represented decreasing from left to right (**a**) spectrum obtained with detection in phase with the modulation of the pump beam, (**b**) spectrum obtained with detection in quadrature with the modulation [Oria 1989]

7.4 Conclusion

The phenomenon of total internal reflection can be attenuated depending on the material deposited at the interface where it arises. This property is used for obtaining the absorption spectrum of materials. An extremely accurate determination of the refractive index of a material can be performed from these measures. Different systems based on the total internal reflection of light at a plane interface have been described in this chapter. An enhancement of the sensitivity of these spectroscopes has been obtained by the use of multiple internal reflections. Fibers can be used for obtaining information of the same type in a more direct and elegant way.

Internal-reflection spectroscopes present a high sensitivity which can be enhanced by an additional treatment of the information obtained by means of Fourier transforms. At present, these systems have apparently reached maturity, and should not undergo any significant evolution.

In contrast, the spectroscopy of atoms in the vicinity of interfaces is a relatively new technique. Advances in this area in the next few years are probable, this being related to the possibility of utilizing the evanescent field associated with plasmons or the evanescent field of guided modes of a fiber.

8. Evanescent-Wave Atom Optics

In recent years, extensive researches on the manipulation of atoms have been carried out. Besides the fundamental nature of these researches, a large number of applications should derive from them. As an example, an application of great interest would be the production of an atomic microscope. Indeed, since the de Broglie wavelengths associated with atoms have a very small extent, i.e. far below optical wavelengths, a resolution of the order of the angström could be reached. Applications of the manipulation of atoms also include high-precision atomic clocks [Aspect 1994] and the use of atom interferometry in researches on gravitation [Carnal 1992].

An important and far-reaching application of the properties of evanescent waves in the area of atom optics is the production of atom mirrors. Even if this subject lies outside the scope of this book, we shall begin this chapter by briefly presenting atom optics in general terms. This permits us to place the atom-optical applications of the evanescent field in context.

The advances which have arisen in this field would not have been possible if experiments dealing with atomic interferences or with the reflection of atoms had not been carried out before. These experiments will be briefly described at the beginning of this chapter. We first describe results presented by Mlynek on atomic interferences, and then describe in more detail experiments where the evanescent field is used for manipulating the trajectories of atoms. Different experiments performed in this field will be described. In particular, we shall pay attention to experiments dealing with atom reflections and atom deflections.

Near the end of the chapter we present recent results which lie within the prospect of fabricating atom waveguides. Since this is a still emerging field of research, it seemed to us interesting to include a section devoted to this subject.

8.1 Atomic Interferences

An atomic analog of Young's double-hole experiment can be carried outwith atomic de Broglie waves. Such an experiment indicates a similarity between atom optics and wave optics. The scheme of this experiment is represented in Fig. 8.1.

A beam of atoms is emitted from an aperture cut in the screen referred to as E. The atoms reaching the screen D have first traveled between the two apertures located in the screen DF. If the difference between the lengths of the two paths is a multiple of the wavelength associated with the atomic wave, a constructive effect arises. In this case, the intensity, which corresponds to the number of atoms detected during a given interval of time, reaches its maximal value.

In Fig. 8.2 values of the intensity are reported for different positions of a convenient grating placed at a distance L from each of the two apertures. The period of the interferences is equal to $x = L\lambda/d$, where λ is the wavelength of the metastable helium atoms used in the experiment. The de Broglie wavelength associated with these atoms is of the order of a few angströms.

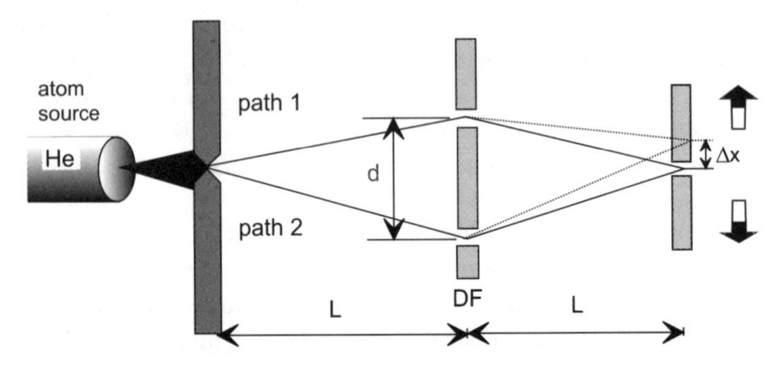

Fig. 8.1. Experimental arrangement used for the observation of atomic interferences [Carnal 1992]

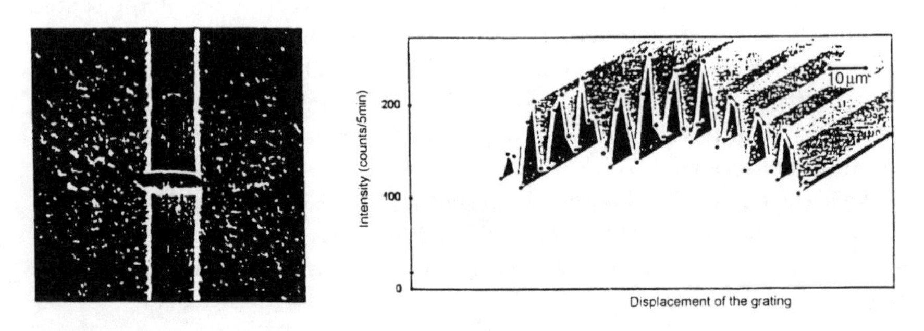

Fig. 8.2. Intensity detected in the screen D by displacing an appropriate grating. The periodic character of the intensity reflects the presence of interferences [Carnal 1992]

Experiments on atomic interferences are not the only type of experiments dealing with the wave character of atoms. It was known from the beginning of the twentieth century that atoms diffracted from a lattice exhibit a wave behavior. Further, in the same way that refractive indices are defined in classical optics, atom-optical indices for dilute media can be determined from certain physical parameters [Vigué 1995].

The radiation pressure exerted by light on atoms can also be used for manipulating atoms [Roosen 1972, Durrant 1995, Savage 1995]. Experiments based on this principle will be described in the following sections.

8.2 Reflection of Atoms

While atomic interferences reflect the wave character of atom optics, the ray-optical properties of atoms can be encountered in the manipulation of the trajectories of atoms based on the use of the evanescent field. Evanescent light fields have been used in particular for reflecting beams of atoms. The first experiments on the interactions between atoms and an evanescent field with a high intensity were carried out by Balykin [Balykin 1988, Aminoff 1993, Aspect 1994].

These experiments will be described at length in Sect. 8.3. We first examine experiments on atom rebounds. The reflection of atoms can be regarded as a special case of atom deflection, which will be examined in Sect. 8.3. The principle of these experiments is illustrated in Fig. 8.3.

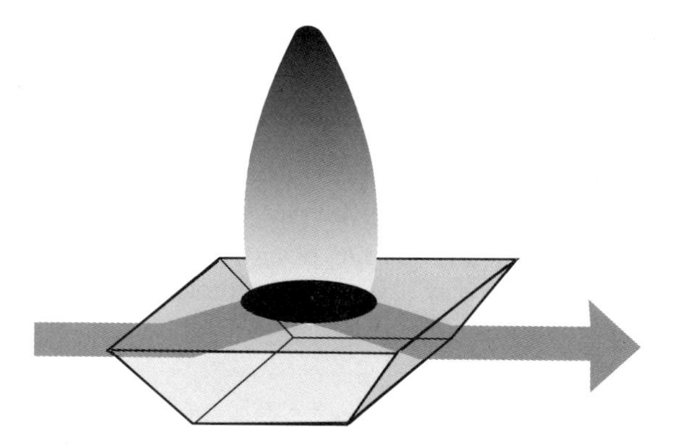

Fig. 8.3. Reflection of atoms at the surface of a prism illuminated in total internal reflection

Atoms cooled inside an 'atom trap' arrive near the surface with a reduced velocity. As they are subjected to the radiation pressure exerted by the evanescent field, they can be reflected. The evanescent field can be regarded here as acting like an 'atomic trampoline'.

If an atom with an electronic transition of frequency ω_{at} is released within an electric field, it will be subjected to a radiation pressure which can be expressed by a potential U

$$U = \hbar \frac{\omega_L - \omega_{at}}{2} \log \left(1 + \frac{I/I_{sat}}{1 + 4(\omega_L - \omega_{at})^2/\Gamma^2} \right), \tag{8.1}$$

where ω_L is the laser frequency, Γ is the width of the atomic transition, I_{sat} is its saturation intensity and I is the light intensity. The latter is related to the electric field by the following equation

$$I(r) = \frac{1}{2} \varepsilon_0 c \mathbf{E}^2(r) \tag{8.2}$$

at point r, where the electric field is expressed by the equation

$$\mathbf{E}(r,t) = \mathbf{E}(r) \cos[\omega_L + \phi(r)]. \tag{8.3}$$

The intensity of the evanescent wave $I(z)$ can be expressed by the function

$$I(z) = I_M \exp(-2kz), \tag{8.4}$$

where k and I_M are as defined earlier; k is the inverse of the penetration depth of the evanescent field in vacuum, while I_M is the light intensity in the $z = 0$ plane.

If the detuning is positive, i.e. if $d = \omega_L - \omega_{at} > 0$, an atom directed towards the prism–air interface will be subjected to a repulsive potential barrier. The height of this barrier is determined by the intensity of the field.

As described by Kaiser, laser intensities are generally too weak for producing a potential barrier higher than the thermal intensity of the atoms at ambient temperature. In the case of rubidium with a transition at 780 nm ($\Gamma/2\pi = 6$ MHz, $I_{sat} = 1.6$ mW/cm), a laser intensity of 1 mW/cm^2 gives a maximal barrier height too weak for overcoming the effect of thermal movement. It is therefore necessary to produce an amplification of the field.

The amplification of the field in the vicinity of a surface can be achieved by using a prism coated with several dielectric layers presenting such characteristics that the field will be highly amplified (Fig. 8.4) [Lévy 1985, Salomon 1991b, Seifert 1994]. This configuration has been described by Imbert and Lévy in 1975 [Imbert 1975]. Recently it was used again for obtaining local amplifications of the field higher than 300 [Kaiser 1994].

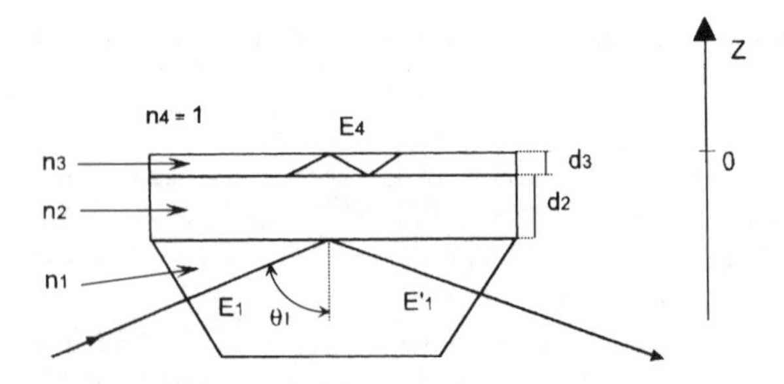

Fig. 8.4. Multilayered system used for the amplification of the evanescent field

The measurement of the value of the amplitude of the evanescent field is not always possible. An estimated value can be derived from the detection of the field which can be performed by frustrating the evanescent field from the tapered end of an optical fiber similar to those employed in near-field optical microscopy. The value of the amplitude of the evanescent field thus obtained remains approximate, because the fiber induces a perturbation on the evanescent field so that the interaction between the probe and the evanescent field is not entirely taken into account. Accordingly, the value of the field cannot be unequivocally determined using this method.

The distance between the tip and the surface is not an experimentally measurable value. The value of the evanescent field has therefore to be determined by other means. It is known that the field is amplified if the mode of the thin layer is excited, and that the excitation of this mode is highly sensitive. Hence, by measuring the reflectivity of the prism–layers system and by comparing it with a theoretical model, the amplification of the evanescent field can be estimated. As illustrated in Fig. 8.5, an amplification of a factor 300 can be determined for the layer under consideration, by comparing it with a numerical simulation [Kaiser 1994]

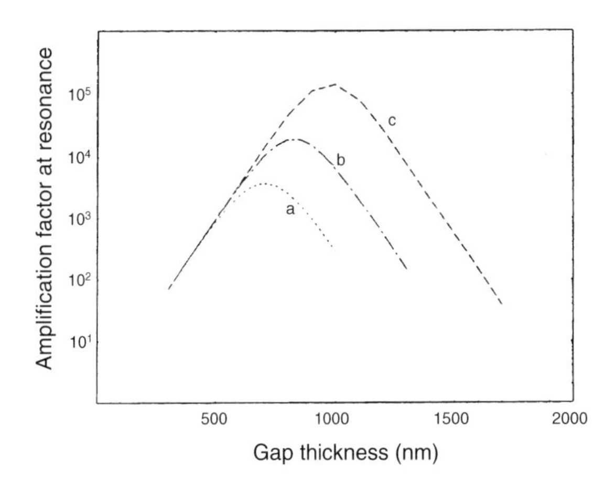

Fig. 8.5. Amplification of the field at the air–layer interface [Kaiser 1994]

It is apparent that the quality of the layer determines the maximal amplification of the field which can be reached [Landragin 1996]. The more the amplitude of the field has to be amplified, the higher must be the quality of the layer. In order to obtain an amplification of a factor 130, the layer was polished so that the residual roughness was of the order of 0.4 nm.

A different method for achieving the amplification of the field consists in coating the prism with a thin metallic layer and exciting a surface plasmon in the Kretschmann-geometry [Kretschmann 1968].

When the cloud of atoms arrives near the surface, the medium 'seen' by the incident field is somewhat different from the same medium in the absence of the atoms. A non-destructive measurement of the reflection of atoms on an evanescent wave has therefore been suggested recently [Vansteenkiste 1994, Aspect 1995]. The principle of this measurement is illustrated in Fig. 8.6.

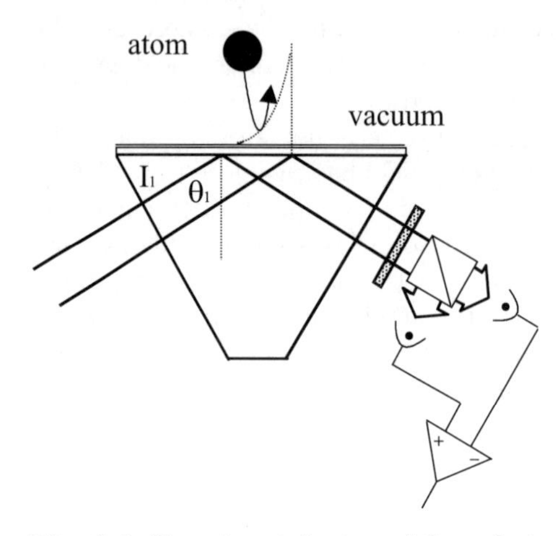

Fig. 8.6. Experimental setup of the polarization interferometer used for detecting the phase difference due to the reflection of the atoms on the evanescent wave [Aspect 1995]

The prism is illuminated by the sum of a TE wave and of a TM wave in such a way that a phenomenon of resonance arises for the TE beam. In these conditions, the phase difference induced by the total internal reflection is very important for the s-polarized beam, and negligible for the p-polarized beam. After the total internal reflection has arisen, the phase difference between the two polarizations can be measured by classical means. The authors have demonstrated that in the conditions of these experiments the phase difference can be measured.

8.3 Deflection of Atoms

Even if the reflection of atoms can be regarded as simply a particular case of the deflection of atoms, their respective applications are not exactly the same. In the deflection of atoms, the evanescent field is not used for causing the atoms to rebound, but only for deflecting them. The role of the evanescent field here can be regarded as similar to that of a mirror illuminated at an incidence angle not normal to the surface.

Two different types of experiments have been imagined for deflecting atoms: the first uses the configuration described earlier, where the evanescent field is simply generated by the total internal reflection of a plane wave. The second, which has not been experimentally implemented thus far, is based on the use of the evanescent field of the whispering-gallery modes of a dielectric microsphere.

8.3.1 Deflection Based on the Use of Evanescent Waves Generated at Total Internal Reflection

The first results concerning the deflection of atoms by an evanescent field were presented by Balykin *et al.* in 1988. The principles of this experiment are schematized in Fig. 8.7.

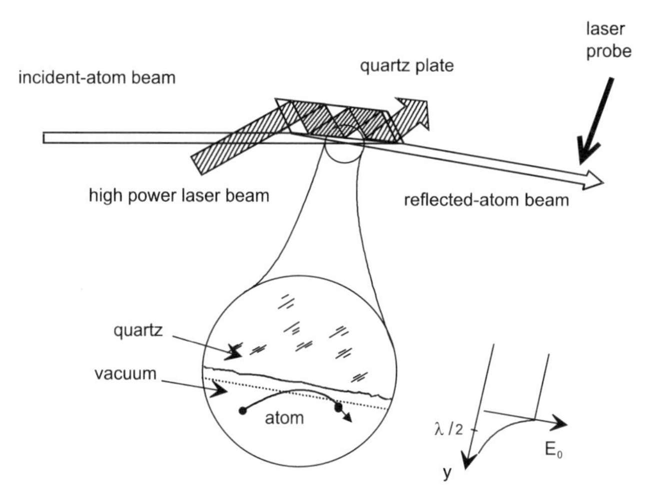

Fig. 8.7. Scheme of Balykin's experiment [Balykin 1988]

A prism is illuminated under total internal reflection. Atoms emerging inside the evanescent field are subjected to the pressure exerted by this field and are thus deflected. The atoms reflected by the evanescent field are then detected and localized from the fluorescence that they emit when illuminated with an appropriate source.

If the extent of the incident beam of atoms is large enough , the atoms divide into two distinct beams: while some of the atoms are deflected, the others continue along their initial direction. This arrangement acts therefore as an atom splitter, as can be seen in Fig. 8.8.

Experimental results presented by Sigel have demonstrated the dependence of the position of the deflected beam on the incidence angle. These results are represented in Fig. 8.9.

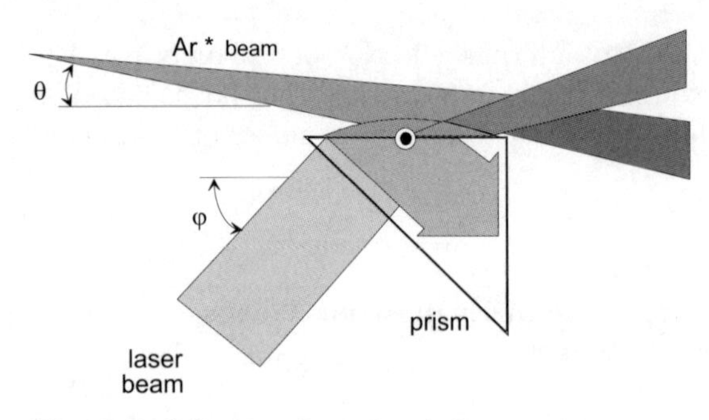

Fig. 8.8. Splitting of an atomic beam

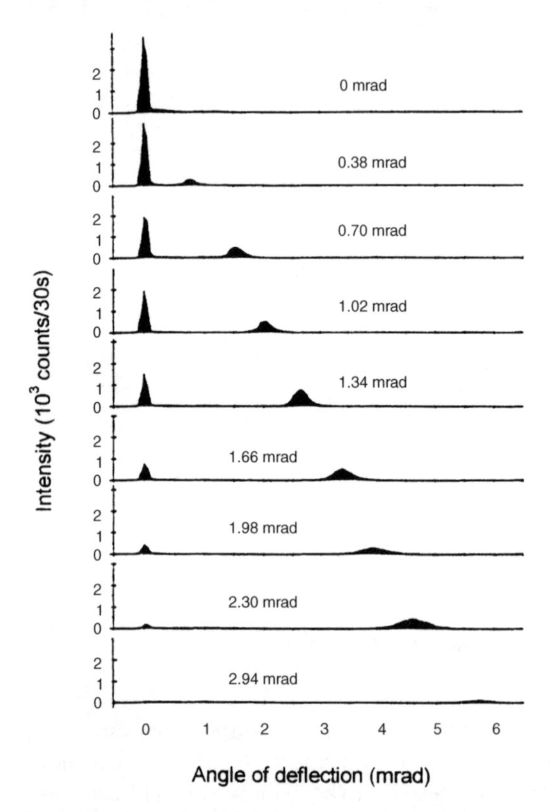

Fig. 8.9. Number of detected metastable argon atoms. The peak at 0 corresponds to the non-deflected atoms. The curves are given for different values of the angle of incidence [Sigel 1993]

An amplification of the evanescent field can be obtained by exciting a plasmon of a silver film placed upon a prism. Figure 8.10 displays the deflection of atoms obtained under these conditions. In their article, the authors stress the fact that the size of the intensity peak with respect to the angle depends on the variations of the thickness of the silver film [Sigel 1993].

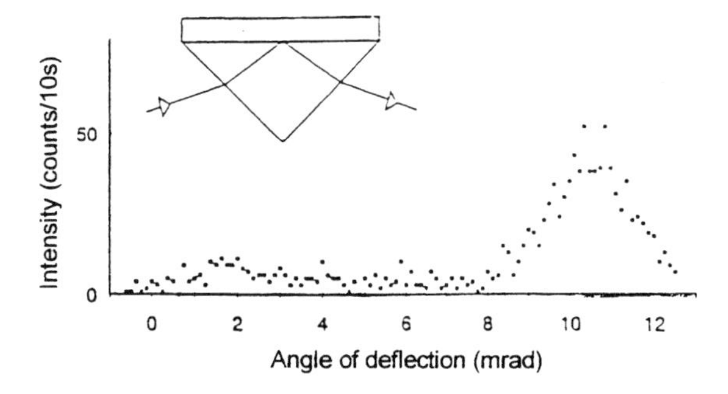

Fig. 8.10. Experimental results displaying the deflection of the atoms by the evanescent field of a surface plasmon of a silver film with thickness 52 nm [Sigel 1993]

8.3.2 Deflection Based on the Use of the Evanescent Field of Whispering-Gallery Modes of a Sphere

In this subsection, we shall describe a set of experiments which were primarily intended for demonstrating the relation between the quantization of the force exerted by an evanescent field on an atom and the quantization of the number of photons present within this field. Even if the aim of these experiments has not been presently attained, the fact that the evanescent field is involved in several respects in them has led us to incorporate in this chapter some of the results obtained with this end in view.

In particular, the preliminary experiments have resulted in a theoretical and experimental analysis of the whispering-gallery modes of a sphere. These modes have been mentioned near the end of Chap. 3.

Let us first describe the scheme of these experiments. The interaction force per photon is inversely proportional to the volume of the field, and therefore the quantization of the force exerted on atoms by an evanescent field can be measured only if extremely small cavities are used. The quality factor of these cavities must be high enough in order that the number of photons be kept constant. The experiment imagined at the Kastler–Brossel laboratory of the Ecole Normale Supérieure in Paris [Treussart 1994, Dubreuil 1995b] is schematized in Fig. 8.11.

From theoretical results, it follows that the quantization of the deviation of the atoms can be experimentally demonstrated to depend on the number of photons present within the whispering-gallery mode, as can be seen in Fig. 8.12 [Treussart 1994].

The performing of the experiments corresponding to these calculations requires the fabrication of microspheres with the required characteristics. The whispering-gallery modes of these spheres are then excited and characterized. The spheres are produced by fusing the extremity of a silica fiber or cylin-

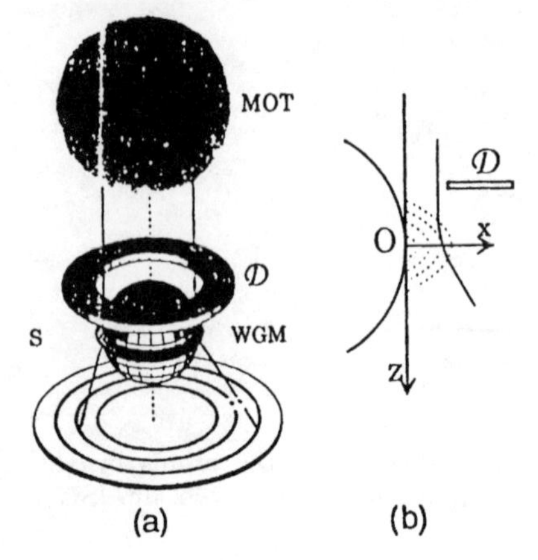

Fig. 8.11. Principle of the experiment. (**a**) Atoms released from a magneto-optical trap arrive between the sphere and a ring: they are deflected by the evanescent field of the whispering-gallery mode, (**b**) Path of an atom

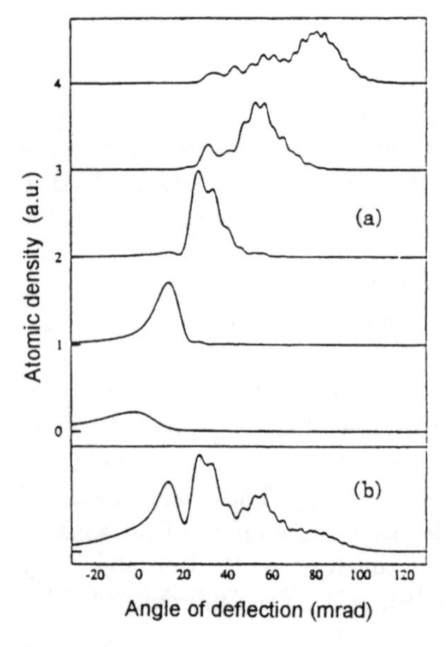

Fig. 8.12. Distribution of the deflections of atoms for different atomic densities

der. Using this fabrication method, spheres with a radius of a few tens of micrometers can be obtained (Fig. 8.13).

The modes are excited by coupling them to the evanescent field of a prism illuminated under total internal reflection or to the evanescent field of the modes of a polished optical fiber (Fig. 8.14).

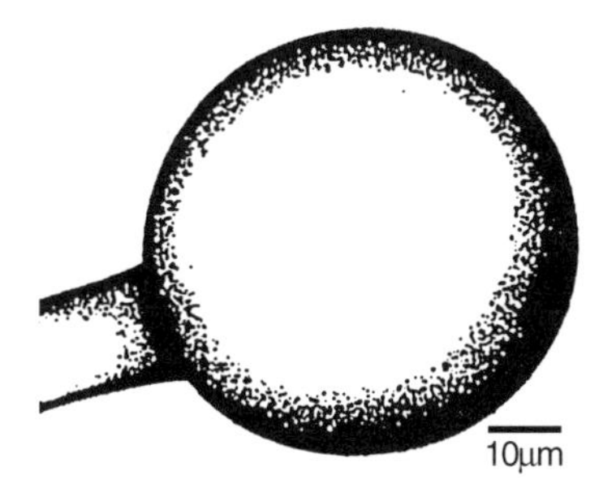

Fig. 8.13. Photograph of a microsphere [Treussart 1994]

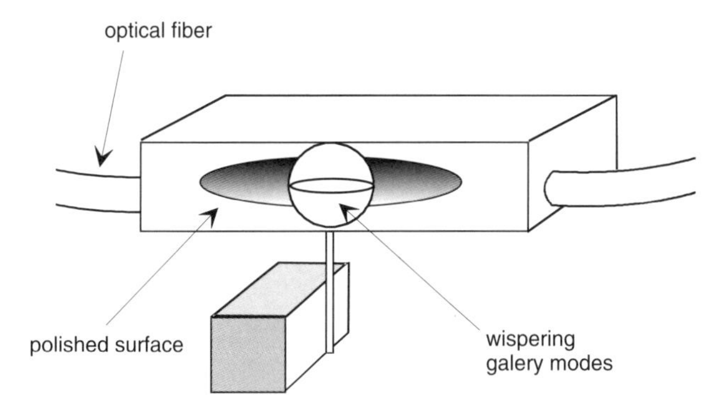

Fig. 8.14. Coupling from a sphere to a fiber

The resulting coupling ratio is of the order of 10%. For carrying out this experiment, both the sphere, the polishing of the fiber and the adjustment of the sphere with respect to the fiber had to be controlled. Before returning to the description of the excitation and coupling of whispering-gallery modes, it may be recalled here that, as mentioned in Chap. 1, similar components have been developed in the case of microwave frequencies. Since dielectric resonators possess a high quality factor, they can be utilized as multiplexers, as filters or as active components [Vedrenne 1982, Cros 1990a]. The basic principle of the excitation of whispering-gallery modes in such a resonator is schematically represented in Fig. 8.15.

Let us now return to the whispering-gallery modes of the microsphere. In order to characterize the modes excited in the sphere, a near-field detection was performed [Knight 1995]. To this end, a single-mode optical fiber with the end thinned by chemical etching was used as a probe and brought inside the near-field of the sphere (Fig. 8.16) [Lefèvre-Seguin 1995].

The end of the probe fiber is displaced with respect to a meridian. The angular dependence of the different whispering-gallery modes guided by the sphere is indirectly determined from the detected intensity which was found to be of the order of 0.1 μW. These modes are analytically expressed by the equation

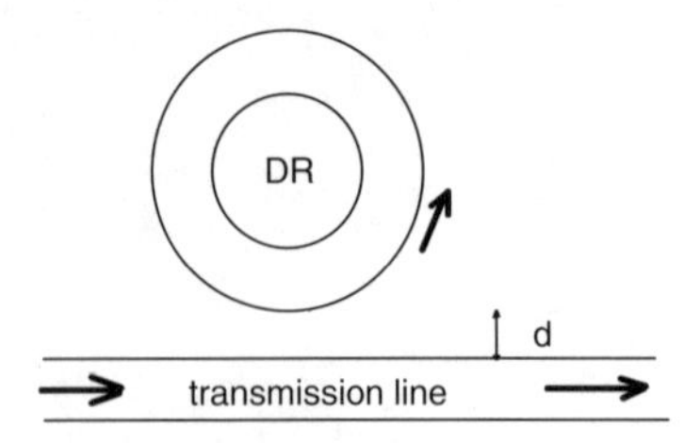

Fig. 8.15. Excitation of whispering-gallery modes inside a dielectric resonator (DR). d is the distance between the resonator and the excitation [Cros 1990]

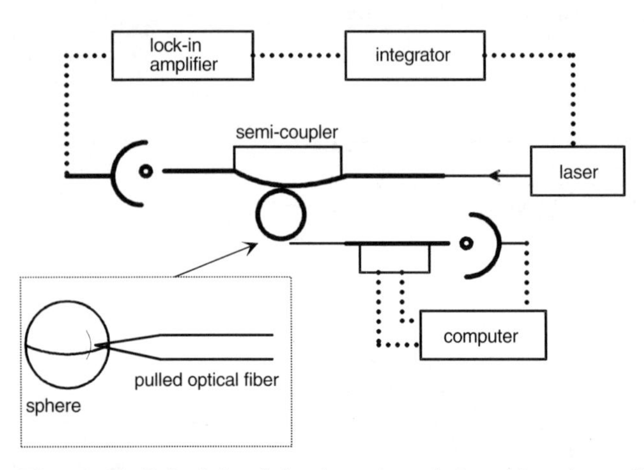

Fig. 8.16. Principle of the detection of the whispering-gallery modes of a sphere with use of a tapered fiber [Lefèvre-Seguin 1995]

$$I_{l,m}(\theta,\phi) \propto |H_{l-|m|}(I^{1/2}\cos\theta\sin^{|m|}\theta\exp(\mathrm{j}m\phi)|, \tag{8.5}$$

where $H_{l-|m|}$ is an Hermite polynomial, and θ and ϕ are respectively the polar angle and the angle in the equatorial plane of the sphere. As can be seen from Fig. 8.17, the experimental results are consistent with the theoretical values.

The presence of the detecting fiber slightly modifies the excitation frequencies of the whispering-gallery modes. Even if these first results do not concern directly the deflection of atoms, they prove the necessity of controlling both the fabrication and the characterization of every single element of the system.

8.4 Atom Guiding

The light guided inside an optical waveguide is totally reflected at the interface between the core and the cladding. This has led to the idea of producing atom waveguides where an evanescent field would be used for reflecting the atoms. The fundamental principle of such an atom waveguide,

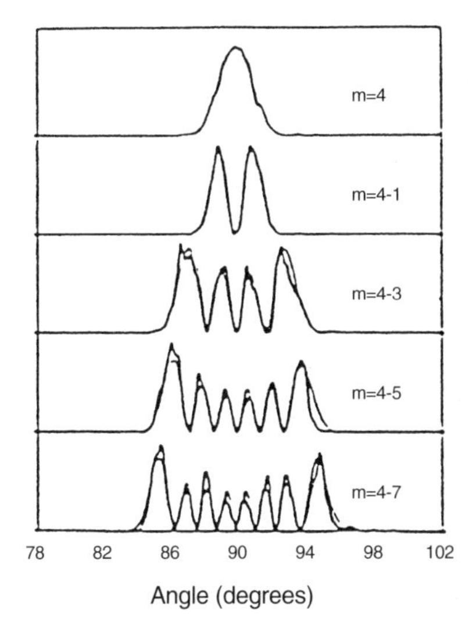

Fig. 8.17. Angular dependence of the whispering-gallery modes of a sphere [Lefèvre-Seguin 1995]

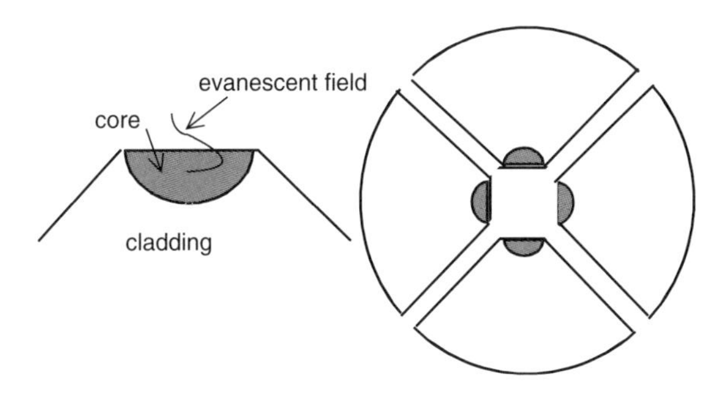

Fig. 8.18. Principle of an atom waveguide [Savage 1993]

as it has been imagined by different authors, is represented in Fig. 8.18 [Savage 1993, Jhe 1994, Kawata 1996].

The realization of an experimental arrangement of this type is prevented practically by a number of technical difficulties. Therefore, a different scheme based simply on the structure of the field of hollow waveguides has been proposed [Savage 1993, Ito 1995]. Figure 8.19 represents schematically the latter system.

The theoretical analysis of these waveguides had been developed during the 1980s, but, as these waveguides do not present any remarkable characteristics as regards to information transmission, their utilization remained at

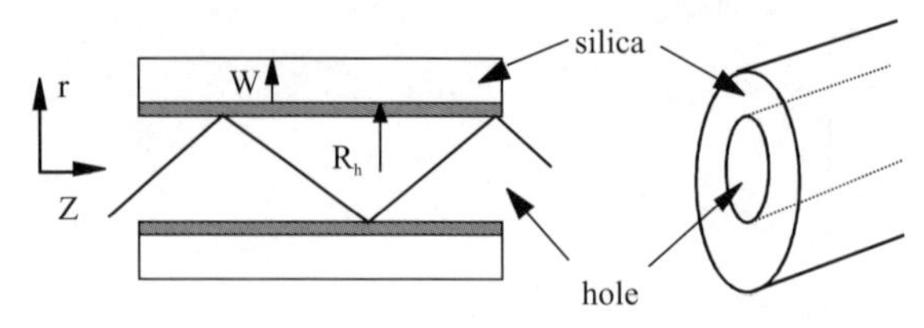

Fig. 8.19. Schematic of the waveguide used

the time limited. Later, as described in Chap. 6, it was recognized however that these fibers could be used for the production of sensors.

A distinctive feature of the eigenmodes of hollow waveguides is that the evanescent field they generate is present both in the center and outside the guiding region. Therefore, atoms propagating along the axis of the waveguide would be reflected by the evanescent field present near the cladding of the waveguide.

The equations for the modes of these waveguides have been given in Chap. 3. Since the expressions of the fields are quite cumbersome, we shall not return to them here and instead we present some related experimental results.

A prerequisite for the production of atom waveguides based on these principles is the control of the optical guiding of the light in hollow waveguides. The excitation by Ito *et al.* of the modes of a waveguide of this type with an efficiency of the order of 10% can be regarded as progress in this direction (Fig. 8.20). The very first results concerning the guiding of atoms have recently been published [Ito 1997].

The amplification of the guiding could be achieved with more complex waveguides sustaining only one mode, in which the atoms would be reflected evenly along the entire waveguide. Further, by using a waveguide presenting a convenient profile, a field with intensity higher than that of the waveguide described here could be produced.

8.5 Conclusion

The recent advances which have arisen in the manipulation of atomic trajectories have led to the development of the rapidly expanding field of atom optics. In classical optics, the behavior of light is modified by devices such as deflectors, mirrors and lenses. Researches in the field of atom optics are directed towards the production of atom-optical analogs of such devices. The respective roles of light and matter are reversed in comparison with classical optics: light fields are here being used for inducing displacements of atoms.

The evanescent field has been used for the production of mirrors for atoms, where it serves as an atom 'trampoline'. An important extension is the possibility of using the phenomenon of deflection of atoms for producing atom-optical analogs of beam splitters.

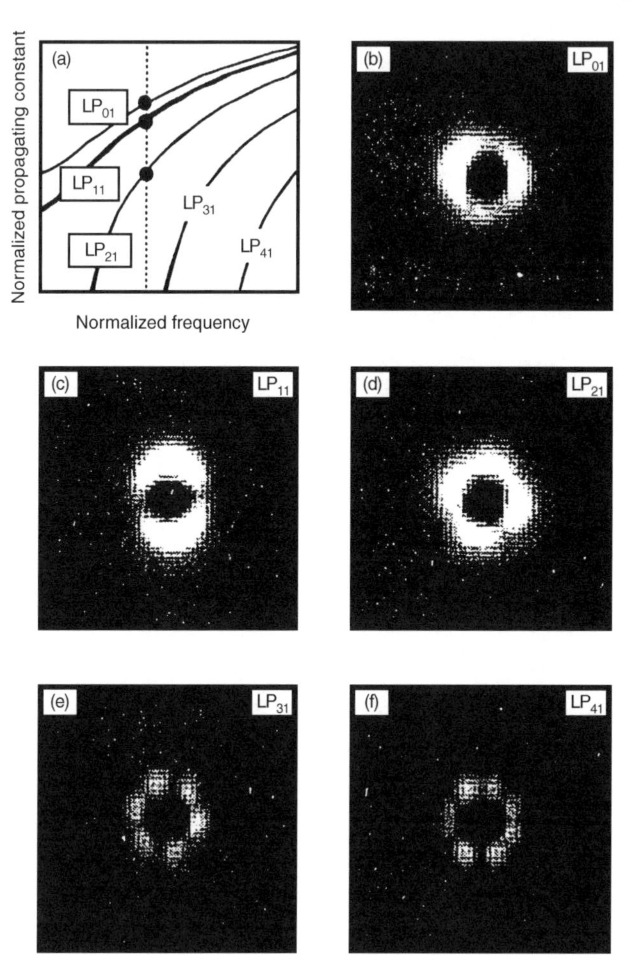

Fig. 8.20. Dispersion curve of annular waveguides and representation of a few modes excited in a waveguide of this type

Another application of the evanescent field in the area of atom optics is the production of atom waveguides. Presently, the researches in this direction are still at their very beginnings, but this might lead to the possibility of producing 'atomic lasers'. More generally, and as mentioned near the beginning of this chapter, atom optics have a large number of extremely promising applications such as atomic clocks and atomic interferometry.

9. Dark-Field Microscopy and Photon Tunneling Microscopy

From the early 1980s, the parallel advances which had arisen in the fields of piezoelectric microdisplacements, of feedback systems and of data processing led to a significant development of local-probe microscopies. The different microscopies assembled under this name are based on the use of the interactions arising between a local probe and the surface of a sample: these interactions can be either electric, as in the scanning tunneling microscope (STM), or resulting from van der Waals forces, as in the atomic force microscope (AFM). Local-probe microscopies have also turned out to be useful in the field of optics. The description of these microscopies will be addressed in the last part of this book.

In the present chapter, we examine two different types of microscopes based on the use of evanescent waves generated in the vicinity of the sample. We first describe the dark-field microscope developed by Nachet in 1847, and then the photon tunneling microscope devised by Guerra more than a century later in 1984.

9.1 Dark-Field Microscopy

9.1.1 Basic Principles

Let us first recall that a microscope consists of a generally incoherent light source, of a condenser equipped with devices like filters or polarizers, of the sample placed on the stage, and finally of the lens and the eyepiece for imaging the enlarged object. The function of the condenser is to focus light onto the sample under the conditions required by the type of observation, by the object to be observed and by the other elements in the system. These different elements are schematically represented in Fig. 9.1.

There exist two different ways of illuminating the sample to be observed [Boutry 1946, Wastiaux 1994]. With the bright-field illumination system, which is the more generally employed, the light beams detected by the lens are collinear with the beams illuminating the sample. In dark-field microscopy, the sample is illuminated in such a way that the beams transmitted onto the lens are not collinear with the beams illuminating the object. This is

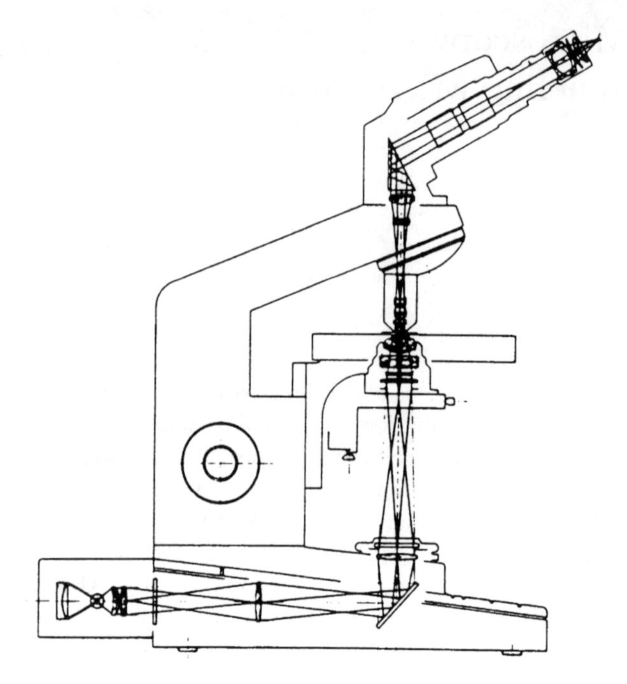

Fig. 9.1. Schematic representation of an optical microscope

achieved by illuminating the sample at a high incidence angle or even under total internal reflection.

Under these conditions, an object with extent larger than the half-wavelength diffuses light. The diffusion of light indirectly reveals the presence of the sample, which can thereby be imaged. In contrast, in the case of objects smaller than the half-wavelength, an effect of the diffraction phenomena that arise at their edges is that the apparent shape of a sample on the resulting images differs from its effective geometry. Unlike classical microscopy, where the samples appear upon a white background, the objects observed with a dark-field microscope are imaged upon a black background (Fig.9.2).

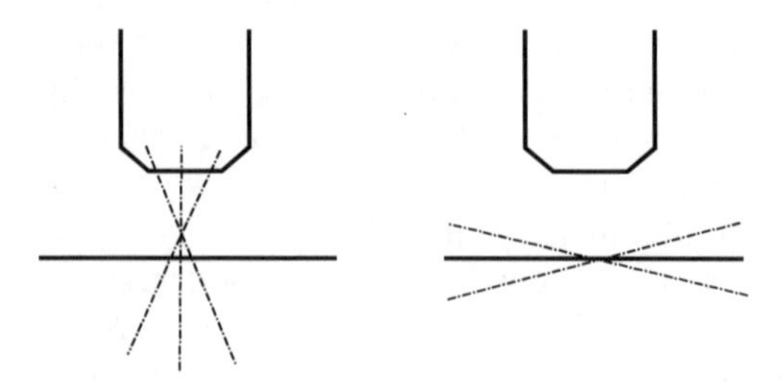

Fig. 9.2. Principles of the bright-field microscope and dark-field microscope respectively

9.1.2 Description of the Dark-Field Microscope

It follows from the very definition of a dark-field microscope that this instrument is a microscope where the sample is illuminated either at a high incidence angle or even in total internal reflection. Figures 9.3 and 9.4 are designed for a comparison between the two illumination systems: dark-field illumination and bright-field illumination.

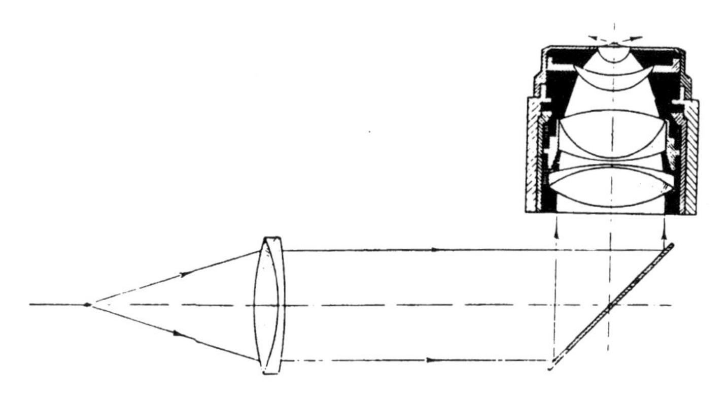

Fig. 9.3. Condenser used for bright-field illumination

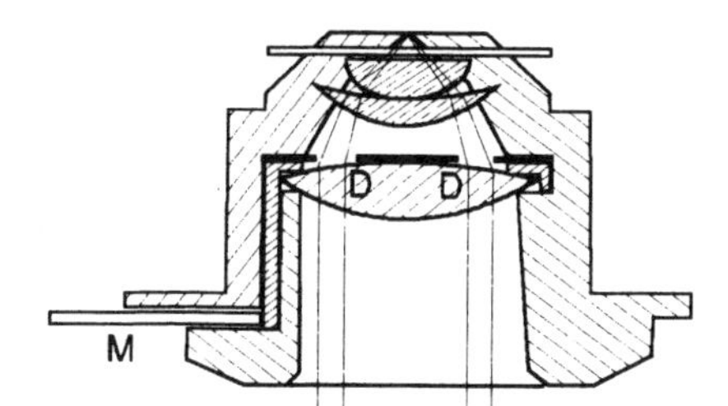

Fig. 9.4. Condenser used for dark-field illumination

With a microscope equipped with a condenser with a small numerical aperture, the sample appears upon a bright background. The resulting practical difficulties drastically restrict the possibilities of observing very small objects with this type of microscope. In contrast, if a dark-field condenser is used, the lens does not detect directly any light ray, but only the light diffused by the sample.

Historically, the dark-field microscope was the first microscopy system based on the use of properties of evanescent waves. The earliest dark-field illumination system was developed in 1847 by Nachet. As can be seen in Fig. 9.5, this device consisted of a glass cone for illuminating the sample at a high angle of incidence [Nachet 1847]

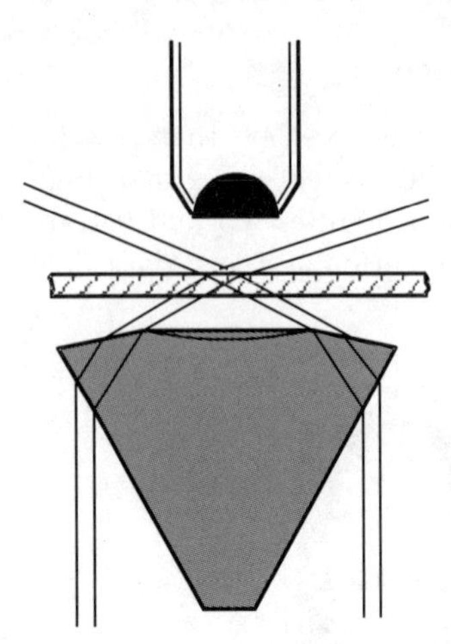

Fig. 9.5. Principle of the first dark-field microscope

The dark-field microscope could employ only objectives with a small numerical aperture. At the time, this type of microscope was not clearly understood, and the principles of dark-field microscopy were left behind until the researches conducted by A. Cotton and H. Mouton at the beginning of the twentieth century [Cotton 1906]. In the system developed by these authors, a light beam strikes a parallepipedal glass plate at such an incidence angle that the beam is focused under the conditions of total internal reflection onto the surface to be observed (Fig. 9.6).

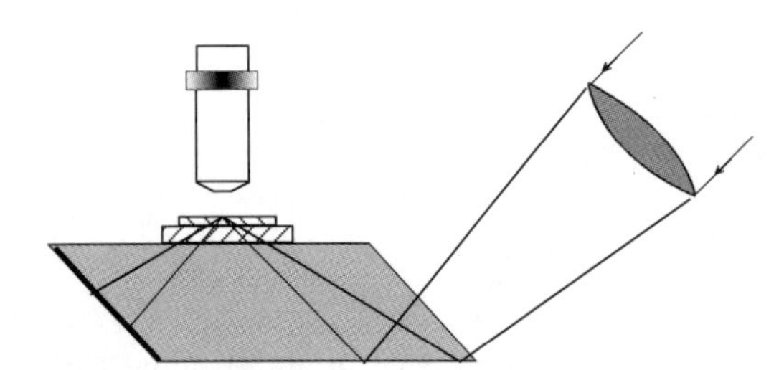

Fig. 9.6. Cotton–Mouton illumination

As these authors had suggested, an illumination system with a symmetry of revolution was later developed. This instrument, devised by Siedentopf, is represented in Fig. 9.7. The incident beam is reflected first on the convex surface, denoted by 1, and then on the concave surface, denoted by 2, each of the two surfaces being metallized. The reflected beams then reach the surface to be observed, either in total internal reflection or at such an incidence angle that the lens does not directly detect them (Fig. 9.7).

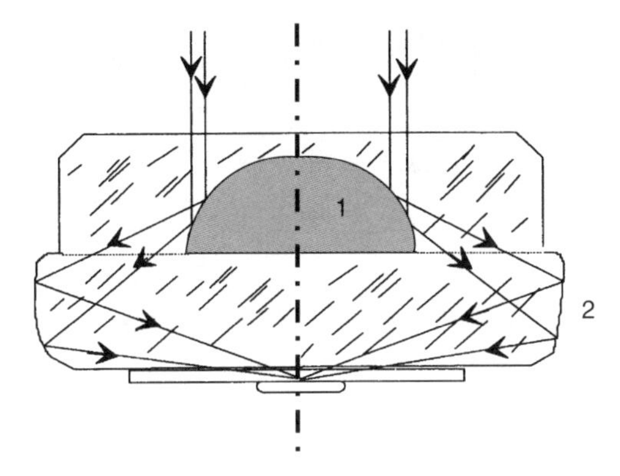

Fig. 9.7. Dark-field condenser devised by Siedentopf and referred to as 'cardioid condenser'

The utilization of the dark-field microscope includes in particular the observation of samples smaller than the wavelength of the light used for the illumination. Therefore, this technique has also been referred to under the name of ultramicroscopy [de Gramont 1945].

Although the history of dark-field microscopy dates back to the nineteenth century, it has recently been independently rediscovered and renamed under the name of total internal reflection microscopy (TIRM) [Nachet 1979, Temple 1981].

9.1.3 Comparison between Dark-Field and Bright-Field Images

As an example, the dark-field microscope can be used for determining the number of elements of subwavelength extent contained in a solution. The first experiments were carried out with solutions containing metallic particles [de Gramont 1945]. A direct way for comparing the features of images obtained with dark-field microscopy and bright-field microscopy respectively is to examine the bright-field and dark-field images of a given sample (Fig. 9.8).

It is apparent from these images that the contrast of the image obtained in dark-field is higher than in the bright-field image. However, the edges of the object are tainted with modulations which do not directly correspond to the effective surface of the object. In analyzing dark-field images, it is therefore necessary to take into account the fact that the intensity detected by the lens is related to the diffraction of light from the object.

As an example, the dark-field image in Fig. 9.8 exhibits thin lines caused by the presence of crystals and by the diffraction of light from the edges of these crystals. This phenomenon was not visible in the bright-field image of the sample. In the case of a total internal reflection illumination, the presence of the sample transforms the evanescent waves into propagative waves. As far

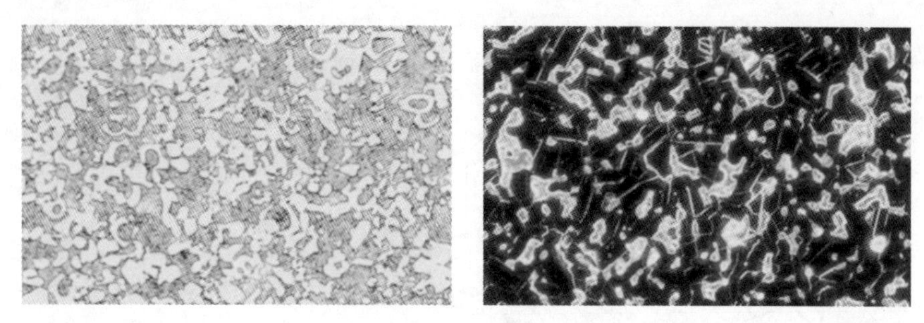

Fig. 9.8. Images of the same sample obtained using a bright-field condenser and a dark-field condenser respectively [Nachet 1998]

as we know, there does not exist any exhaustive theoretical analysis of the formation of images in dark-field microscopy.

Illumination systems where the sample is illuminated with two sources with markedly different wavelengths have also been proposed (Fig. 9.9). The use of a double-illumination system permits us to illuminate the object both in bright-field with one source and in dark-field with the other [Boutry 1946]. This method combines the advantages of the two illumination systems, namely the clearness of the images for objects larger than the wavelength, which characterizes bright-field illumination, and the possibility of imaging the presence of subwavelength structures or objects, as in dark-field microscopy. Using this system, the background upon which the samples observed appear depends on the first wavelength. For instance, if the dark-field illumination system uses blue light, while the bright-field illumination system uses red light, the sample will appear purple upon a red background.

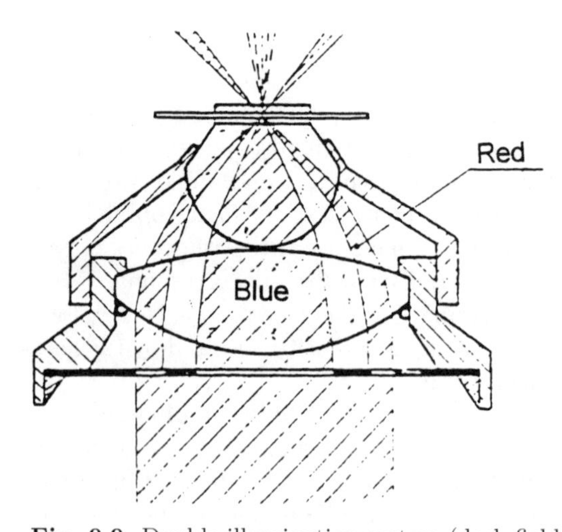

Fig. 9.9. Double-illumination system (dark-field and bright-field)

9.1.4 Dark-Field Microscopy and Fluorescence

An extension of the dark-field microscope is the illumination of a fluorescent sample in total internal reflection. This technique is known under the name of 'total internal reflection fluorescence' (TIRF). The first results on the selective excitation of a liquid–solid interface were obtained by Hirschfeld in 1965 [Hirschfeld 1965, Harrick 1979, Hirschfeld 1977, Lakowics 1992]. Different types of fluorescent materials can be used, for example adsorbed labels or grafted molecules. An advantage of this technique is that a section of a sample can be labeled with different types of markers fixed at different stages in the preparation of the sample. Further, depending on the value of the incidence angle, the section or the preparation to be observed can be illuminated along a variable depth.

The utilization of the dark-field microscope permits us to separate the illumination beam from the light emitted by the fluorescent materials. Since the excitation beam can be partially diffused by the object imaged, it may be useful to add a spectral filter. The arrangement used is represented in Fig. 9.10.

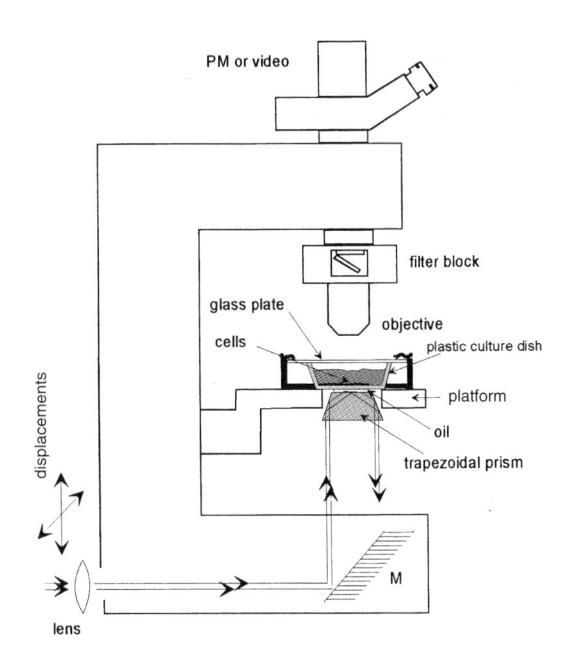

Fig. 9.10. Standard microscope equipped with a TIRF type microscope (enlarged in the drawing). For observations requiring measurements of a time-dependent parameter, the sample can be thermostated and subjected to a gas flux

The images obtained with such a system depend on the incidence angle, and therefore on the way the evanescent field penetrates into the preparation. If the angle of incidence has a high value, the penetration depth of the evanescent field is limited. In this case, the object is illuminated only along a thickness corresponding to the value of the penetration depth. By varying the incidence angle, a variable thickness of the sample can be illuminated. This can be seen in the two images presented in Fig. 9.11 [Axelrod 1992].

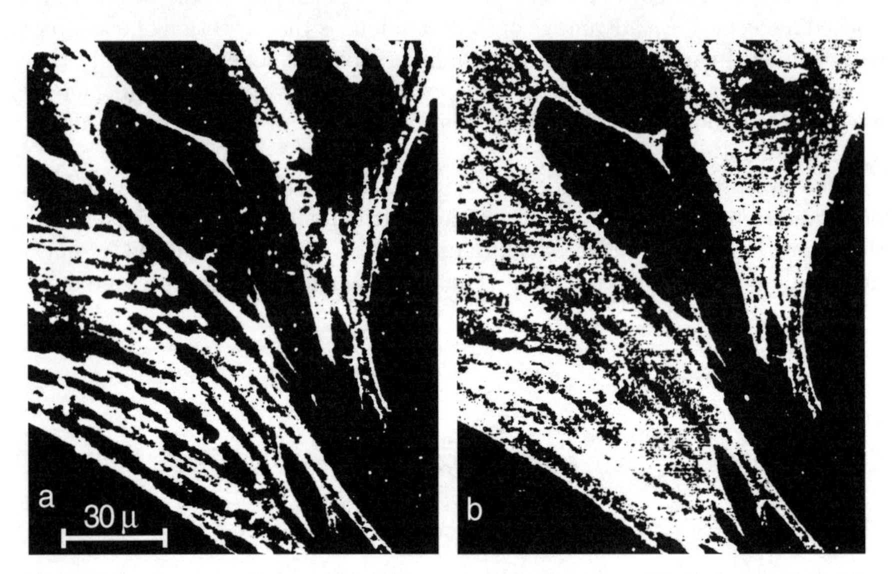

Fig. 9.11. Images of a sample of human skin fibroblast labeled with DiI. The sample was illuminated at two different values of the incidence angle, and therefore at two values of the penetration depth d (**a**) $\theta = 74.3°$, $d_p = 105$ nm and (**b**) $\theta = 67.9°$, $d_p = 406$ nm [Axelrod 1992]

The extension of the use of dark-field microscopy to researches on fluorescence indicates the interest of exploiting the properties of evanescent waves.

Evanescent waves can be transformed into propagative waves by diffraction or by fluorescence. The possibility of achieving the transformation of evanescent waves into propagative waves by frustration of a plane wave has been viewed in the theoretical analysis presented in Chap. 2. This characteristic of the evanescent waves is the basis of a recently developed microscope, the photon tunneling microscope.

9.2 Photon Tunneling Microscopy

The microscope that will be examined in this section uses the properties of the evanescent field associated with total internal reflection. In the case of the dark-field microscope, the presence of an object smaller than half the wavelength was detected from the diffusion of light caused by the sample [de Gramont 1945]. In the microscope described hereinafter, the measurement of very low height variations is based on the optical tunneling effect [Guerra 1990, Dyer 1993].

The optical tunneling effect refers basically to the same phenomenon as the frustration of an evanescent wave in classical optics. The term 'optical tunneling effect' was coined in reference to the electronic tunneling effect [Binning 1982]. Indeed, in optics, the inverse of the refractive indices of the different media can be regarded as an analog of the potential barrier. The photons go through the air gap in a way similar to that of the electrons between two conductors.

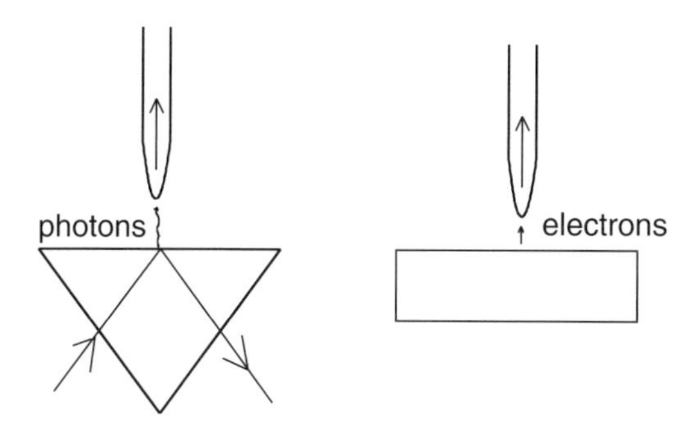

Fig. 9.12. Schematic representation of the analogy between the optical tunneling effect and the electronic tunneling effect

Figure 9.12 illustrates the analogy between the optical tunneling effect and the electronic tunneling effect. In either case, the detected intensity is equal to

$$I = I_0 \exp(-z/2d_\mathrm{p}). \tag{9.1}$$

In the case of the electronic tunneling effect, d_p is given by

$$d_\mathrm{p} = (2m\phi)^{-1/2}/2h, \tag{9.2}$$

where ϕ is the height of the potential barrier between the metallic substrate and the probe.

In the case of the optical tunneling effect, the equation for the penetration depth d_p is

$$d_\mathrm{p} = (\lambda/2\pi)(\sin^2\theta - n^2)^{-1/2}, \tag{9.3}$$

where

$$n = n_2/n_1. \tag{9.4}$$

Here n_1 and n_2 are the refractive indices of air and the prism respectively, λ_1 is the wavelength of light in the first medium and θ is the angle of incidence.

The relation between the electromagnetic field and the photons can therefore be regarded as similar to that of the electrons with the wave function associated with them. However, the equivalence remains approximate and it might even be misleading in certain cases.

As stated in Chap. 1, the field of the evanescent wave generated at total internal reflection can be expressed as a function of the distance from the surface where total internal reflection arises

$$E_{\text{evanescent}} = E(z) = E_{\text{i}} \exp\left(-z/d_{\text{p}}\right), \tag{9.5}$$

where d_p is the penetration depth of the evanescent field. The penetration depth of the evanescent field corresponds to the distance for which the intensity of the evanescent field is equal to the intensity of the field at the interface where total reflection arises, divided by e^2.

The penetration depth of the evanescent field is related to the wavelength: when the value of the wavelength increases, the resolving power of the microscope decreases. Similarly, if the extent of the numerical aperture of the incident beam rises, the value of the penetration depth decreases and the resolution of the microscope is enhanced.

If a medium of index n_3 is brought near the surface illuminated at total internal reflection, then the evanescent wave will be frustrated, provided that the refractive index of this medium is higher than the index of medium 2 and that the distance between this medium and medium 1 is smaller than the penetration depth. Under these conditions, an important part of the light energy will transfer into this medium, as illustrated in Fig. 9.13.

An essential element in the photon tunneling microscope is the Koehler condenser, by means of which the sample can be illuminated at total internal

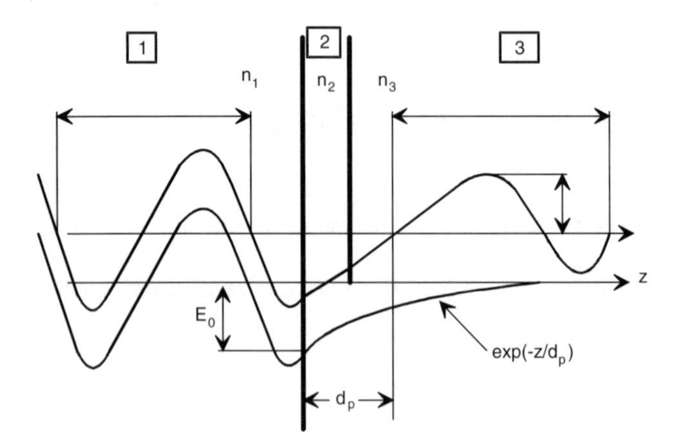

Fig. 9.13. Frustration of an evanescent wave from a third medium

reflection. Unlike dark-field microscopy, where the measurements are performed in transmission, this microscope works in reflection. The microscope employs an oil-immersed objective with a large aperture (N.A. > 1), the oil being deposited upon a transparent plate called a transducer.

By adjusting the focusing, total internal reflection is produced on the lower side of the plate. In the absence of a sample, the light is integrally reflected, as can be seen in Fig. 9.14. The resulting image is therefore uniform. In contrast, if an object is brought into the vicinity of the internal surface, a part of the evanescent wave will be collected by the sample. At distances between the surface of the object and the condenser less than the penetration depth of the evanescent field, the reflectance is weaker. The camera provides an image

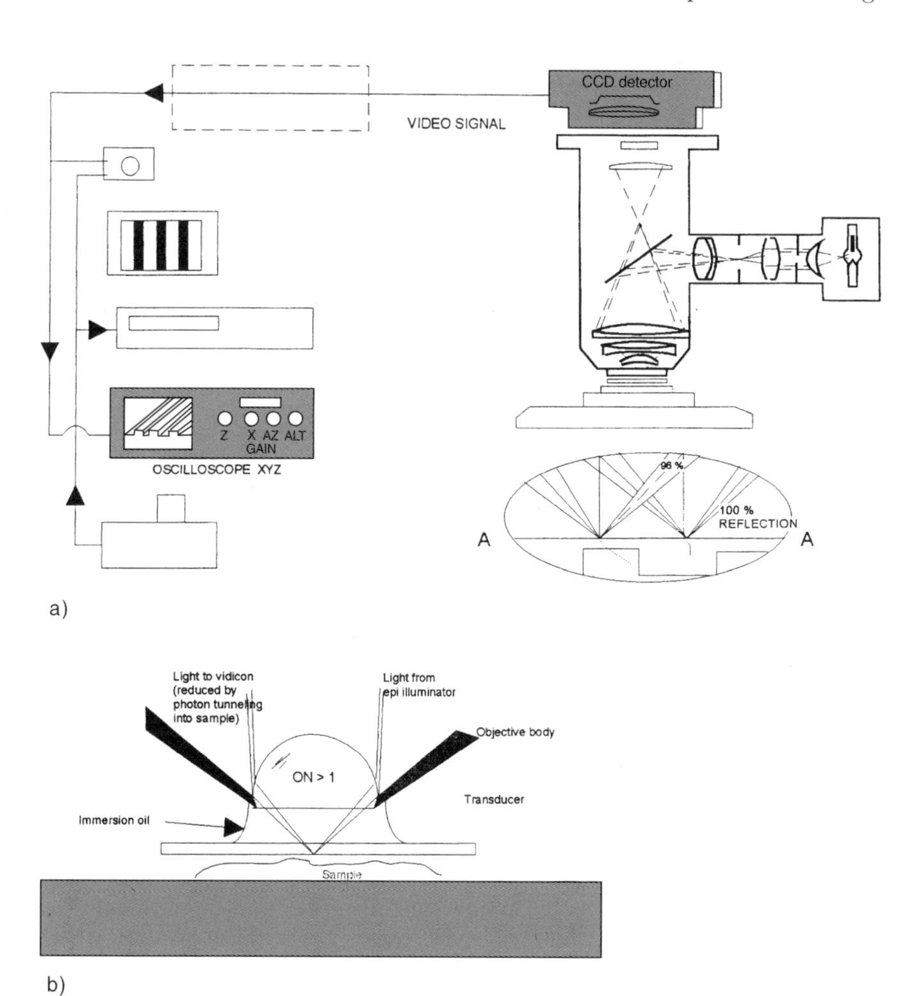

Fig. 9.14. Photon tunneling microscope (**a**) overall arrangement, (**b**) detailed view of the lens, sample, transducer and objective

of these signal variations corresponding to the variations of the height of the object.

The source used for the photon tunneling microscope is a white light source. Enhancement of the resolution of the photon tunneling microscope can be obtained by employing a filter in order to select short wavelengths. The illumination at total internal reflection is accompanied by a Goos–Hänchen shift. However, as the lateral resolution of the microscope is limited by the Rayleigh limit on resolution, this shift is not perceptible.

Examples of images obtained with the photon tunneling microscope are presented in Fig. 9.15 [Guerra 1993b]. The selected samples were homogeneous, so that their shape can be more easily observed.

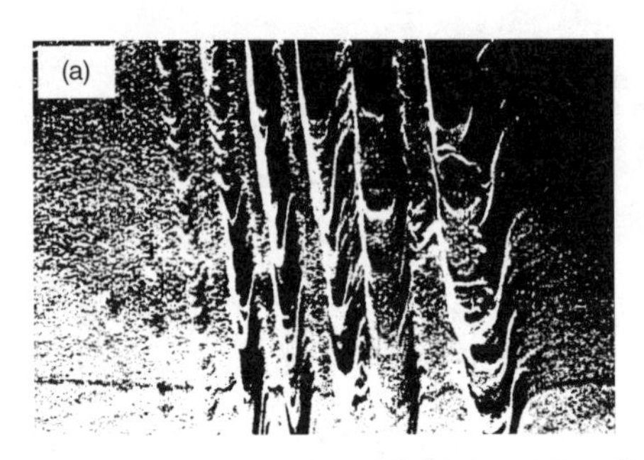

Fig. 9.15. Examples of PTM images (**a**) image of a photoresist layer deposited at the surface of a silica substrate, (**b**) Image of a diamond-turned acrylic [Guerra 1993b, with the permission of the Optical Society of America]

In order to determine the actual geometry of the sample, it is necessary to calibrate the PTM. To this end, a sample whose surface has a calibrated degree of curvature is used for determining the relation between the measured optical density and the width of the corresponding gap. This relation thus defines a calibration function. Evidently, a given such function can be applied only to a single type of material.

If the sample is a dielectric object, the calibration of the microscope can be achieved quite easily. Indeed, in this case, the transmitted intensity is a monotonic function of the distance between the surface and the sample (Fig. 9.16). This is not the case if the sample is a metallic object, for the maximal value of the transmission then corresponds to a distance from the surface of some tens of nanometers. Therefore, the transmitted intensity is not a monotonic function of the distance [Rather 1976]. A possible solution consists in placing between the sample and the condenser, and outside the observation field, a wedge whose thickness corresponds to the position where the transmission reaches its maximal value (Fig. 9.16). The measured distance is therefore unequivocally determined.

The basic features reflect the variations of intensity related to the variations of the height of the sample. In order to obtain a three-dimensional

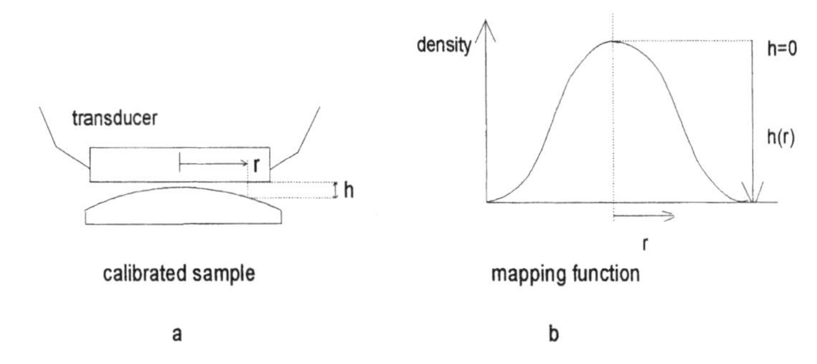

Fig. 9.16. Determination of the calibration function: variation of optical density–variation of height. The curve represented here applies only to dielectric objects

image, the effective variations of height of the object are restored on the basis of the calibration function. As an example, the image of a compact disk obtained with this method is represented in Fig. 9.17 [Guerra 1988].

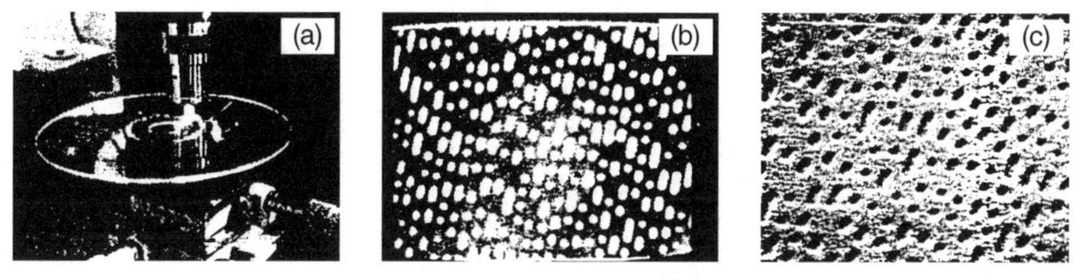

Fig. 9.17. Restoration of the image of a compact disk (**a**) PTM image, (**b**) direct image, (**c**) image of the disk calculated from the mapping function [Guerra 1988]

The lateral resolution of the system can be estimated as equal to 0.5 μm [Guerra 1995]. The value of the resolution is related to the numerical aperture of the condenser (N.A. = 1.25). The vertical resolution, estimated at 1 nm, depends on the sensitivity and on the noise caused by the optical detector.

The set of images shown in Fig. 9.18 represents the evolution of the surface of a polystyrene film heated at 165°C during the condensation phase [Guerra 1993a].

Recently Marchman and Novembre have succeeded in imaging the latent image of a photoresist layer (Fig. 9.19). The formation of the latent image during the sensitization of the photoresist corresponds to the variations of the value of the refractive index during this process. The photon tunneling microscope permits us to observe the transformations that the photoresist undergoes during the formation of the latent lines. This phenomenon is not detectable with an atomic-force microscope [Marchman 1995].

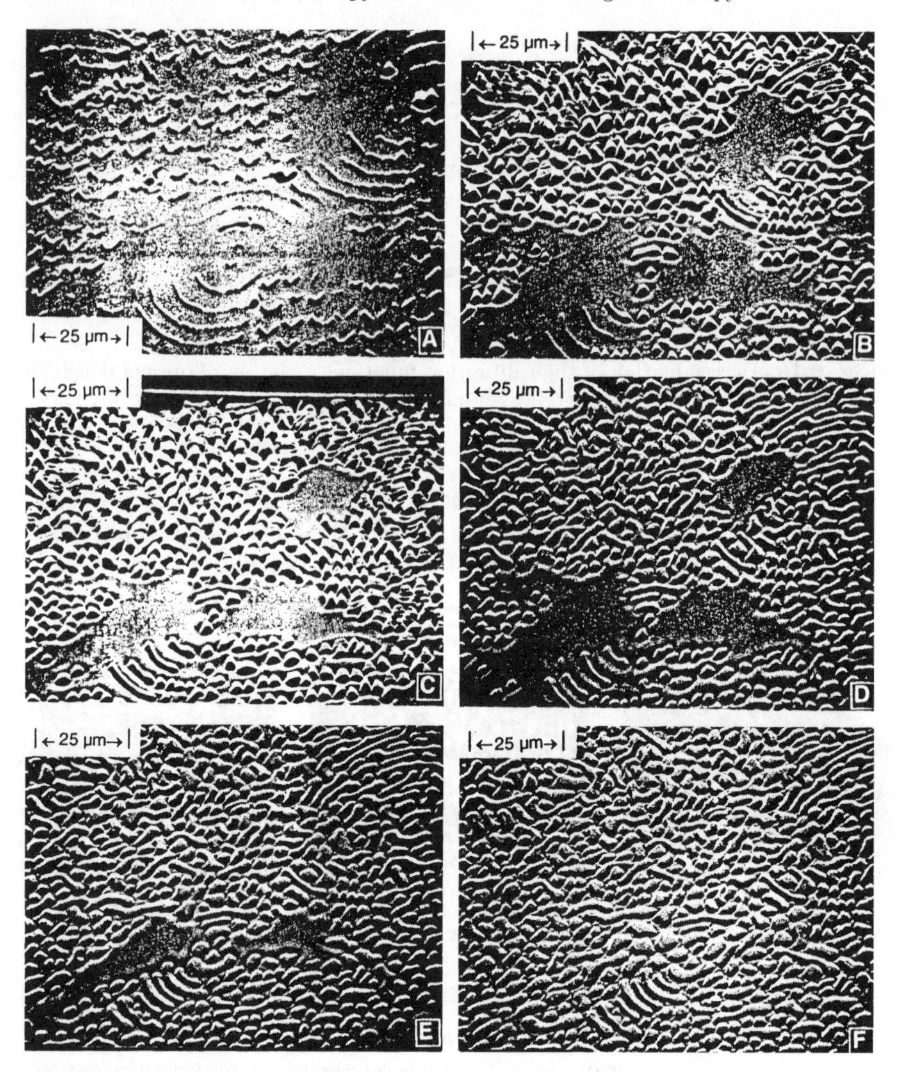

Fig. 9.18. Images of a polystyrene film at different stages of condensation (**a**) 60 s, (**b**) 456 s, (**c**) 540 s, (**d**) 564 s, (**e**) 864 s, (**f**) 1200 s. The difference of height between the most contrasted regions is of the order of 0.2 μm [Guerra 1993a]

9.3 Conclusion

The intent of this chapter was to present two types of microscopes where the properties of the evanescent field are being used: the dark-field microscope and the photon tunneling microscope. A common feature of these microscopes is that, unlike the microscopes which will be presented in the following chapters, their lateral resolution remains within the Rayleigh limit. Nevertheless,

Fig. 9.19. PTM image of the latent features in a photoresist layer [Marchman 1995]

even if these microscopes are based on the use of properties of the evanescent field, the principles on which they rest are markedly different.

The dark-field microscope is based on the diffusion of light from objects with subwavelength extent, and uses the propagative waves generated from the diffraction of the object. As such, this microscope can be employed only for localizing subwavelength samples or details. An extension of dark-field microscopy is the total internal reflection fluorescence microscope. As we have seen, the illumination of a fluorescent material in the conditions of total internal reflection enables a more effective observation of the fluorescent regions in the sample. Similarly, by adjusting the incidence of the source beam, the fluorescence of molecules more or less deeply 'buried' inside the sample can be detected with this microscope.

Historically, the photon tunneling microscope was, after the dark-field microscope, the second type of microscope based on the use of the evanescent properties of the field. Like the dark-field microscope, its lateral resolution is limited by the Rayleigh criterion. The photon tunneling microscope can be used for obtaining three-dimensional images and for gaining information about the height of the sample, with resolution as good as 1 nm. The physical principle which lies behind this microscope is not the diffraction of light from the sample, but the frustration of the evanescent field born in the vicinity of the sample from a medium with high index located near the sample. This requires that the objects to be studied are only slightly rough.

The last decade has seen the development of local-probe microscopies. These microscopies are based on the use of interactions between the surface of a sample and a probe, the latter being assumed to be point-like. Examples of such interactions are:

- the electronic tunneling effect, i.e. the electronic interaction between a probe and a metallic surface,
- the van der Waals effect, i.e. the interaction between the extremity of a flexible cantilever and a surface, and

- the optical tunneling effect, i.e. the photonic interaction between a dielectric probe and a transparent surface.

The microscopies based on interactions of this type are also referred to as proximity microscopies. The remainder of this book will be essentially devoted to near-field optical microscopies.

Conclusion of Part II

As described in the preceding chapters, there exist a large range of devices based on the delocalized use of properties of the evanescent field. The examples that we presented here show that the instruments based on this principle have applications in a number of different fields.

Couplers are devices used for separating or summing different signals. The use of evanescent-field coupling allows us to achieve any coupling or demultiplexing ratio. The characteristics inherent in this type of coupling lead to the possibility of separating the light with respect to certain criteria, for example the polarization, the modal order, or the wavelength. These functions are defined when the coupler is devised, whether in the case of integrated-optical devices or in the case of couplers fabricated from optical fibers.

The propagation in waveguides also depends on properties of the external medium. This dependence can therefore be used for producing sensors designed for the detection of a given parameter. The perturbations, whether they be localized or not, that the variations of a given physical parameter induce on the evanescent field associated with the guided mode(s) of the guiding structure can thus be used for measuring this parameter. Because of the large amplitude of the evanescent field, the level of sensitivity of the sensors based on this principle can be very high.

A prior application of the sensitivity of the evanescent field to the parameters of the medium in which it lies is the technique of internal reflection spectroscopy. Internal reflection spectroscopy systems are based on the fact that the ratio of the reflected light depends on the nature of the medium within which the evanescent field propagates. These spectroscopies permit a spectral analysis of materials deposited in the form of thin layers, and in particular, of materials present in very small quantity.

The evanescent-field manipulation of atoms is an application which was hardly thinkable a few years ago. The results we have described here have shown that the concept of an atom mirror has become a reality and a part of the instrumentation of atom optics. Recent years have seen the very beginnings of the use of hollow dielectric waveguides for the production of atom waveguides.

As the last part of this book is devoted to local-probe near-field optical microscopies, it seemed to us of interest to present first the characteristics of two microscopy systems based on the use of the evanescent field: the dark-field microscope and the photon tunneling microscope, developed a few years before the local-probe microscopy systems. The dark-field microscope can be used for localizing subwavelength structures. The photon tunneling microscope, even if its lateral resolution is limited, presents an axial sensitivity of the order of a few nanometers.

In the next three chapters we shall describe a class of instruments which are presently rapidly expanding and still evolving: scanning near-field optical microscopes.The name 'scanning near-field microscopy' refers generally to local probe or local source microscopes, where the probe or the source is smaller than the wavelength of the light used for illumination. This new generation of microscopes forms the main subject of the third part of this book.

Part III

Localized Interaction with the Evanescent Field

Introduction to Part III

In the last part of this book, we describe microscopes based on the use of the local information contained within the near-field of an object to be observed. By the term 'local' we here understand a domain ranging over an area smaller than the half-wavelength. As described in the first chapters, the near-field contains information both in the form of propagative and of evanescent waves.

It is well known that the resolution power of classical microscopies, which are based on the use of propagative waves, either emitted, reflected or transmitted by the objects to be analyzed, is necessarily half the wavelength of the light used. On the other hand, evanescent waves had not been used in microscopy except in the case of the dark-field microscope developed in 1830 by Nachet. This microscope, however, could be used only for localizing objects with subwavelength extent.

In an article of 1928, E.H. Synge presented what was the principle of a microscope overcoming the Rayleigh limit on resolution. Nevertheless, the first experimental results date only from 1956 and were obtained in the field of microwaves. The development of these techniques for frequencies in the visible spectrum really began during the 1980s.

The fundamental principle of near-field optical microscopy is the use of the interaction between a dielectric or metallic tip and a sample, this interaction arising through the near-fields of the tip and the sample. Several different systems have been developed for generating this interaction. Figure III.1 represents some more or less classical systems that will be detailed at length in the last part of this book.

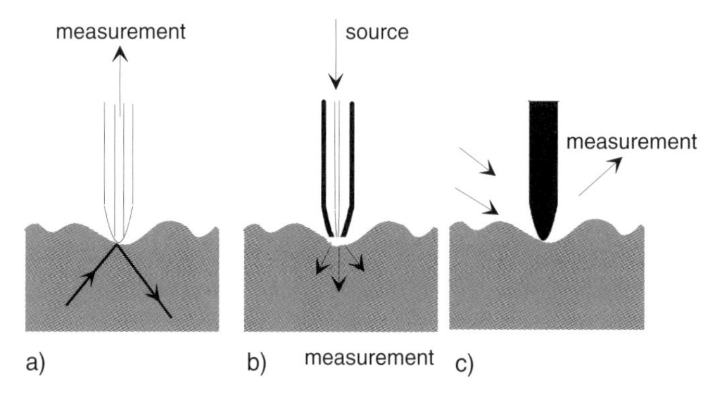

Fig. III.1. Schematic representation of the principles of (**a**) photon scanning tunneling microscopy, (**b**) scanning near-field optical microscopy, (**c**) apertureless microscopy

The different systems of near-field microscopy can be divided into different classes, each being treated in a separate chapter. We first examine the scanning tunneling optical microscopy, then micro-aperture microscopy and finally the apertureless microscopies.

10. Scanning Tunneling Optical Microscopy

We begin the description of local probe microscopies with a description of the scanning tunneling optical microscope. This microscope descends directly from the tunneling optical microscope, which, as we described in Chap. 9, is based on the frustration of light from a glass plate parallel to the surface of the sample. In the scanning tunneling optical microscope, this plate is replaced by a dielectric probe whose end frustrates locally the evanescent field born in the vicinity of the sample. The energy transfer arises essentially at the end of the tip, thereby ensuring the lateral resolution of the system. As the use of local probes in physics is quite recent, we shall treat these elements in detail.

An alternative denomination for the scanning tunneling optical microscope is that of 'photon scanning tunneling microscope' (PSTM), not to be confused with the photon tunneling microscope (PTM) developed by Guerra and described in the preceding chapter.

10.1 Fundamental Principles of the Scanning Tunneling Optical Microscope

Before turning to the analysis of the scanning tunneling optical microscope, the principle of this microscope has to be briefly explained. Figure 10.1 represents schematically the physics underlying this microscope.

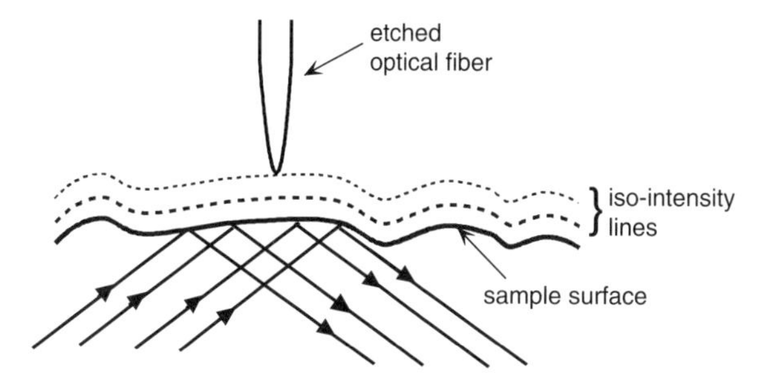

Fig. 10.1. Principle of the scanning tunneling optical microscope

As illustrated in Fig. 10.1, the surface of a sample is illuminated at such an incidence angle that the conditions for total internal reflection are fulfilled. If the surface is perfectly plane, the field is purely evanescent and remains constant within a plane parallel to the scanned surface. If the surface exhibits even a small roughness, or if the sample placed on this surface is small compared with the wavelength, the variations of the electric field correspond approximately to the topography of the scanned surface. Therefore, if a dielectric probe, for example the extremity of a sharpened optical fiber, is moved close to the sample, it will detect a part of the light intensity of the electromagnetic field. Figure 10.2 reproduces the first published decrease curves [Reddick 1989].

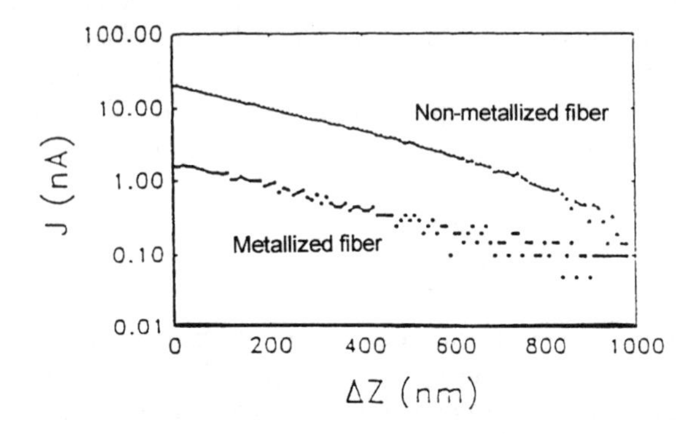

Fig. 10.2. Value of the current detected by the photomultiplier tube as a function of the distance between the fiber and the surface of a thin quartz plate illuminated in total internal reflection under an incidence angle of 45 degrees. The source was a He-Ne laser with power 7 mW [Reddick 1989]

In the presence of a perfectly plane surface, the intensity detected by the probe is a decreasing exponential function of the distance between the fiber and the surface. Hence, as the signal detected by the probe is a monotonic function of the distance, the feedback of the position of the probe can be achieved relative to the surface of the sample. The imaging principle is the same as for the electronic tunneling microscope.

The scanning of the sample can be achieved in two different ways. The two operating modes of the photon scanning tunneling microscope, referred to respectively as constant-height and constant-intensity modes, are illustrated in Fig. 10.3.

In the constant-height mode, the fiber scans the sample along a plane which in general is parallel to the average plane of the sample. The resulting image thus restores the variations of the optical signal as they are detected by the probe during the scanning process.

In the constant intensity mode, the vertical position of the fiber is controlled, while the intensity is kept constant during the scan. The images

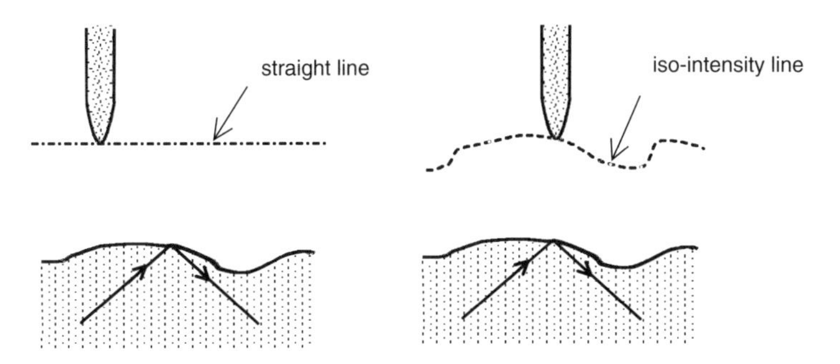

Fig. 10.3. Operating modes of the PSTM (**a**) constant-height mode, (**b**) constant-intensity mode

obtained with this mode reproduce the displacements of the probe normal to the surface of the sample. This mode restores as a first approximation the form of slightly rough samples. In the case of an object with such characteristics that the signal detected by the fiber presents important variations, such as those caused by absorption phenomena, the constant-height mode is more appropriate.

Let us examine more precisely the detection by the probe of the light in the vicinity of the interface.

10.2 Detection of the Near-Field in the Vicinity of a Plane Surface

We have seen in the first chapter how a semiinfinite medium could be used for detecting the evanescent field in the vicinity of a surface illuminated under total internal reflection. The first curves displaying the decrease of the light intensity collected by an optical fiber were presented by Reddick [1989]. These curves are reported here with a semi-logarithmic scale in Fig. 10.2. They indicate that the dependence of the intensity detected by a tip in front of a surface is roughly exponential.

In fact, further measurements have shown that the curve of the intensity is not exactly exponential. Indeed, as can be seen from Fig. 10.4, the curve bends as the distance between the fiber and the surface tends towards zero [Salomon 1991b].

These experimental results have been analyzed on the basis of a three-layer model similar to the one described in Fig. 10.5.

Under these conditions, the curves displaying the energy detected by the third medium representing the fiber were compared to the experimental curves of the energy detected by the optical fiber. The comparison was carried by varying the incidence angle (Fig. 10.6).

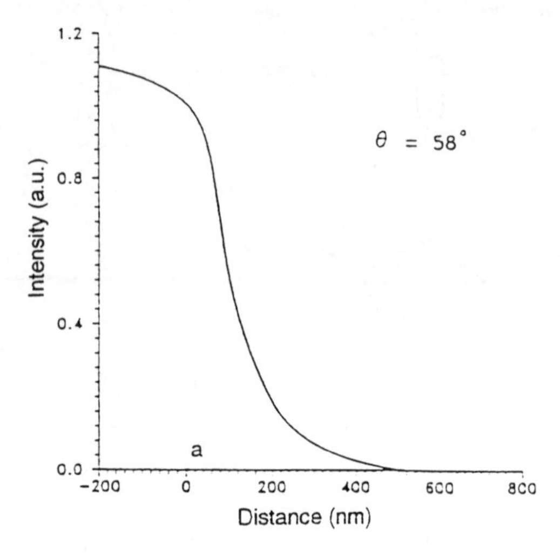

Fig. 10.4. Intensity detected by the optical fiber as a function of the distance between the fiber and the surface [Salomon 1991b]

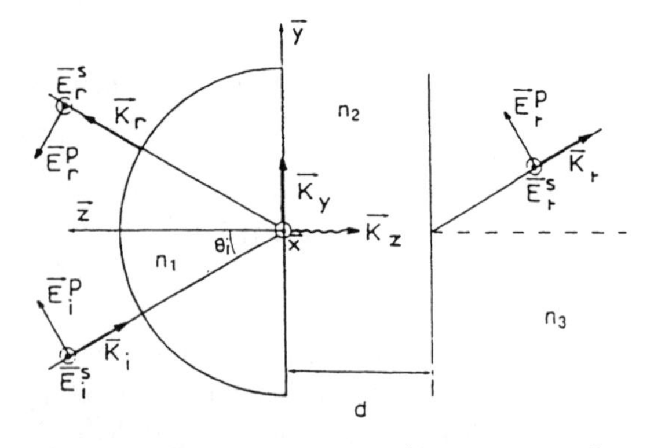

Fig. 10.5. Three-layer system [Salomon 1991a]

Despite the great simplicity of the model where the probe is represented by a semiinfinite plane, this model allows a description of the effect of the distance on the frustration of the evanescent field by the probe, as can be seen in Fig. 10.7.

10.3 Early Results in Scanning Tunneling Microscopy

Several different schemes of the photon scanning tunneling microscope have been designed [Reddick 1989, Courjon 1989]. Figure 10.8 represents one of these arrangements.

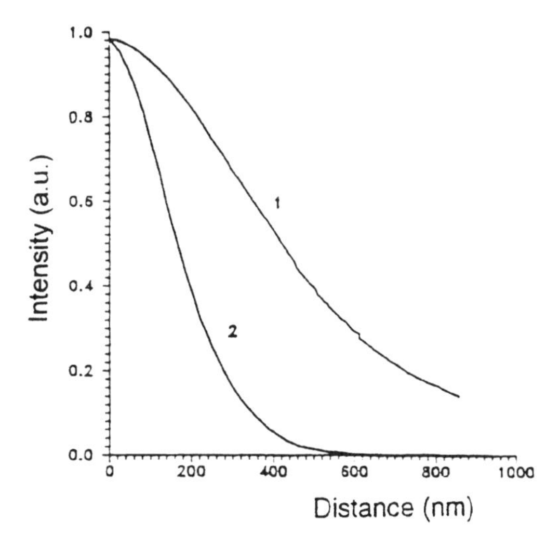

Fig. 10.6. Intensity detected by the probe as a function of the incidence angle of the beam. $\lambda = 1.3$ µm. The value of the refractive index of the fiber and of the prism is equal to 1.447; (**a**) $\theta = 50°$, (**b**) $\theta = 60°$

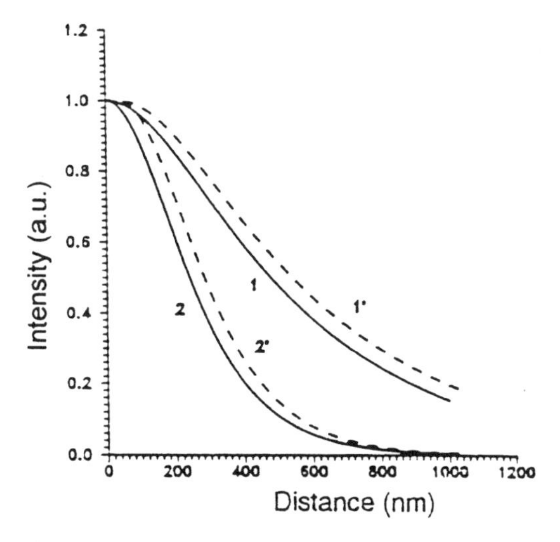

Fig. 10.7. Calculated intensity of the field transmitted in the semiinfinite medium representing the tip. These curves are consistent with experimental measurements (the *curves in dashed lines* correspond to the case of a prism covered with 50 nm of water)

The light source can be either a He-Ne laser with wavelength λ=633 nm or 542 nm [Reddick 1989], a laser diode with wavelength $\lambda = 670$ nm or 1300 nm [Jiang 1991, de Fornel 1992], a white source [Adam 1993, Chabrier1994] or an infrared source [Piednoir 1992, de Fornel 1993, Quartel 1999]. This subject will be discussed at length in the following section.

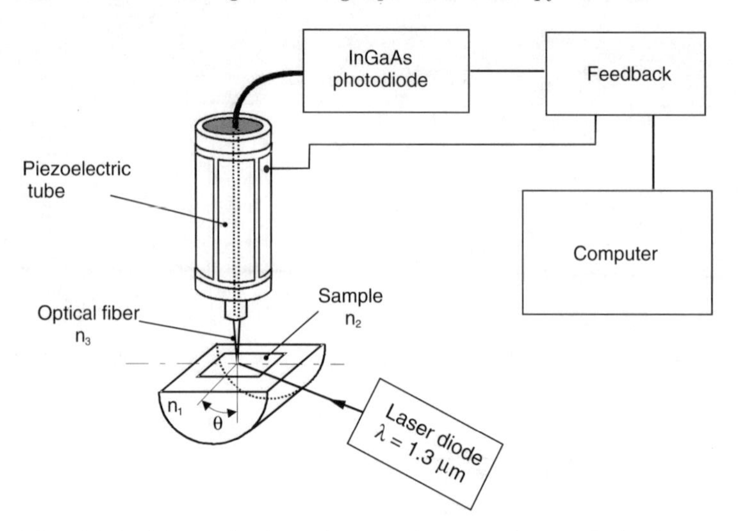

Fig. 10.8. Schematic view of a scanning tunneling optical microscope

The fiber used as probe must be transparent to the excitation wavelength. For applications extending from the visible spectrum to near infrared, silica fibers are well suited [Jiang 1991, Courjon 1989, de Fornel 1989]. For infrared observations, it is necessary to use fibers of a different type, such as for example fluorized or chalcogenure glass fibers [Piednoir 1992].

The fabrication technique of the probe depends evidently on the raw material of the fiber. Examples of techniques used for producing classical silica fibers are chemical etching of the fiber or simply pulling of the fiber with use of a CO_2 laser for fusing the fiber [Garcia–Parajo 1995]. The fabrication of fluorized glass fibers is more complex: indeed, these fibers cannot be chemically etched, and the fusion process is made more difficult by problems related to low-temperature re-crystallization effects which render the tip opaque at the usual wavelengths. Other types of probes have been designed from tips produced for atomic-force microscopes [van Hulst 1993, Fillard 1996]. The latter technique allows a simultaneous utilization of this microscope and the AFM.

Probes fabricated from silica single-mode or multimode fibers with current characteristics present, after having been chemically etched, a radius of curvature of the order of 50 nm. It has been shown that, under certain conditions, the use of highly-doped single-mode fibers permits the production of tips with diameters as small as some tens of nanometers [Pangaribuan 1992, Ohtsu 1998]. Figure 10.9 represents examples of different probes.

The scanning of the probe in front of the sample is generally controlled from a piezo-electric tube, similar to those a scanning tunneling microscope is equipped with. The scanned surfaces, which of course are larger than in the

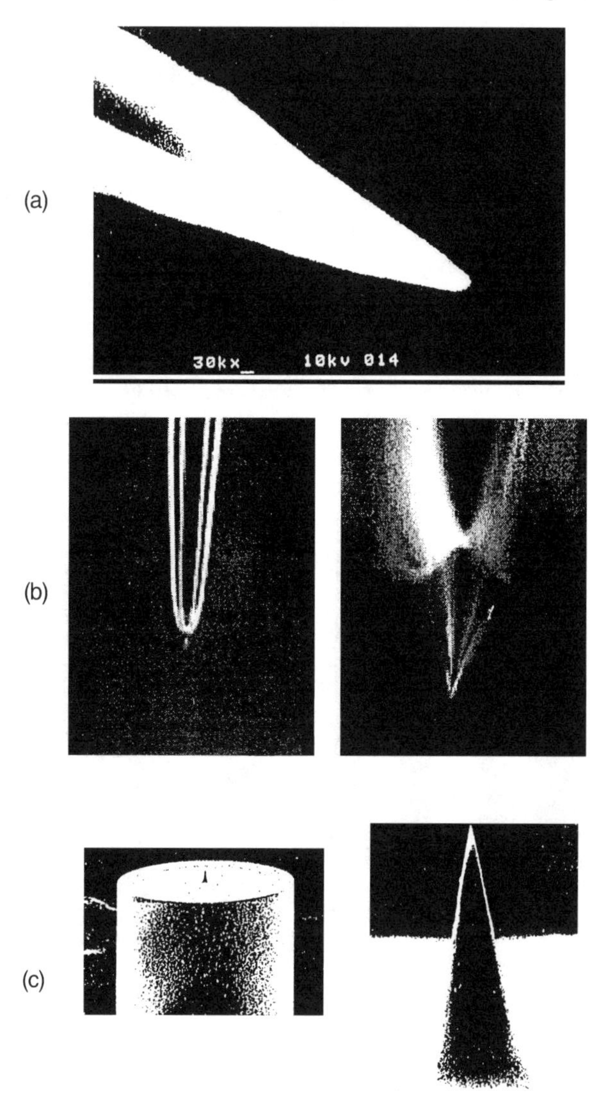

Fig. 10.9. Micrographs of probes (**a**) 50 nm diameter tip [Salomon 1991b], (**b**) 10 nm diameter tip [Pangaribuan 1992], (**c**) pulled tip [Garcia–Parajo 1994]

case of a STM, range between a few square microns and some hundred square microns. The problems that one encounters when using piezo-electric tubes, namely hysteresis, non-linearity or drifts, have been extensively described in the literature devoted to scanning tunneling microscopy and atomic force microscopy. In the analysis of the images obtained with any near-field microscopy system, the possibility that some of these phenomena had arisen must always be borne in mind.

Let us now return to the description of the photon scanning tunneling microscope. The detector used can be either a photodiode or a photomultiplier. The light signal is transformed into an electric current recorded by the computer while the probe scans the surface of the sample along a plane parallel to the average plane of the surface. This scanning mode is referred to as constant-height mode. The other type of scanning is performed by controlling the vertical position of the probe so that the intensity detected during the scanning remains constant. The recorded signal corresponds to the variations of the altitude of the probe relative to its position. This mode is referred to as constant-intensity mode.

Among the earliest images obtained in 1989, the images due to Reddick and to Courjon can be mentioned here [Reddick 1989, Courjon 1989]. Figure 10.10 presents the images of a low-amplitude grating (40 nm) with a large period (1.3 μm).

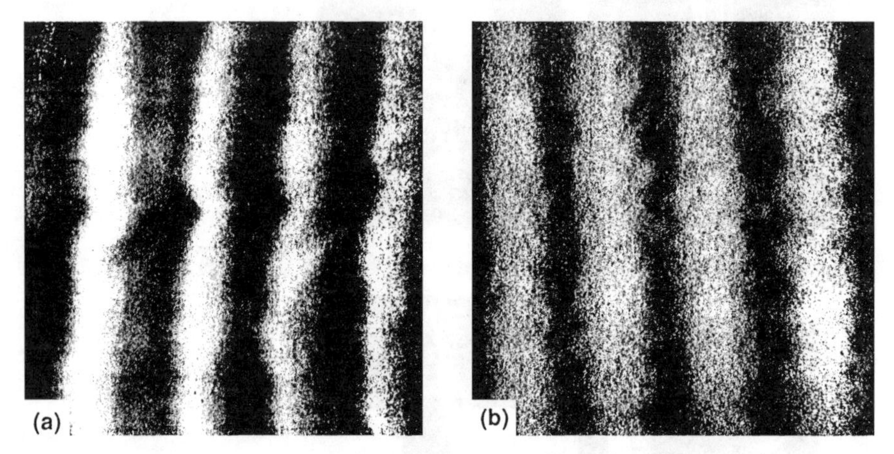

Fig. 10.10. Images of a grating (**a**) STM image, (**b**) PSTM image [Reddick 1989]

Figure 10.11 reproduces the image of a step on a mica sample obtained by Courjon. The step is schematized against the image.

At first sight the correspondence between the near-field image and the effective surface of the sample may appear to be obvious. However, further researches have demonstrated that the image obtained does not actually represent the surface of the sample, but rather the iso-intensity lines, i.e. the lines along which the light intensity detected by the probe is constant [de Fornel 1992, van Hulst 1992a].

In reality, the PSTM images of a sample can just be said to represent the diffraction of the incident electromagnetic field from the surface and the perturbation induced by the probe. The effects of the respective illumination systems have been investigated by different authors. Researches carried out

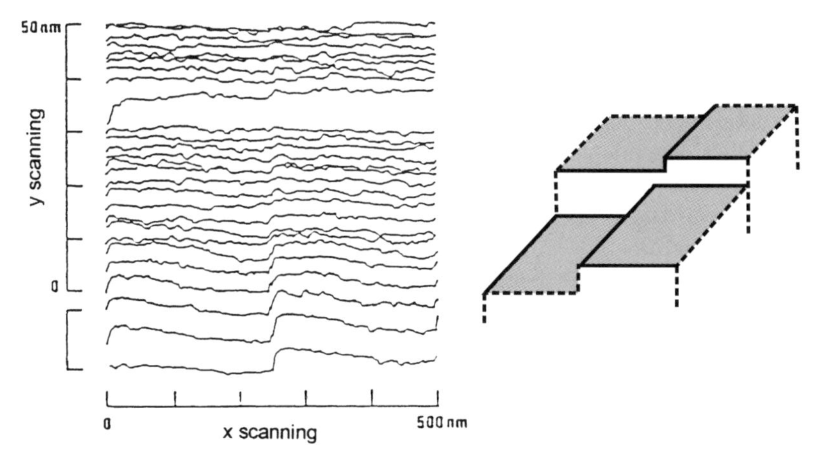

Fig. 10.11. Image of a mica step [Courjon1989]

by van Hulst, Courjon and Ferrell have shown that the distance between the probe and the sample and the polarization of light plays an essential part [van Hulst 1992b, Ferrell 1991]. As soon as the probe is at a distance from the sample of the order of $\lambda/2$, there appear huge corrugation phenomena, which reflect the phase summation of the different propagative signals diffused by the different parts of the sample. If the fiber is extremely close to the sample, the part of the evanescent field frustrated by the fiber dominates over all other parts.

The effect of the shape of the probe on the images has not been entirely analyzed. Using traditional probes, details of the order of 20 nm have been imaged for wavelengths in the visible spectrum (400 nm, 800 nm) [Courjon 1991, Salomon 1991a]. Infrared measurements (λ = 7 μm), have shown a resolution beyond the Rayleigh limit [Piednoir 1993].

Nevertheless, the resolution of such a system remains a controversial subject. Indeed, the information provided by a PSTM image is related not only to the sample itself, but also to the coupling of the probe and the sample. Only if the sample is almost even does the PSTM image correspond to the effective topography of the sample. If this is not the case, a more accurate analysis of the image is required.

10.4 Near-Field Study of Homogeneous Samples

As mentioned earlier, the images obtained with a photon scanning tunneling microscope depend on several different parameters. The respective effects of these parameters will be examined in the following subsections. In classical optics, the resolution limit of a given system depends on the value of the wavelength of the light used for illuminating the sample. The resolution limit can therefore be determined in each particular case.

In optical near-field microscopy, the resolution limit has still not been exactly determined. This comes in particular from the fact that the formation of images depends not only on the characteristics of the incident waves (wavelength, polarization, coherence, incidence angle) but also on the interactions between the probe and the near-field of the sample. The effect of some of these parameters on the formation of PSTM images will thereafter be addressed in detail. We begin with the description of the effects of the polarization and of the orientation of the source.

10.4.1 Effects of the Polarization and Orientation of the Source

Let us examine the case of a sample with a fairly simple geometry, that is a single step on a surface (Fig. 10.12).

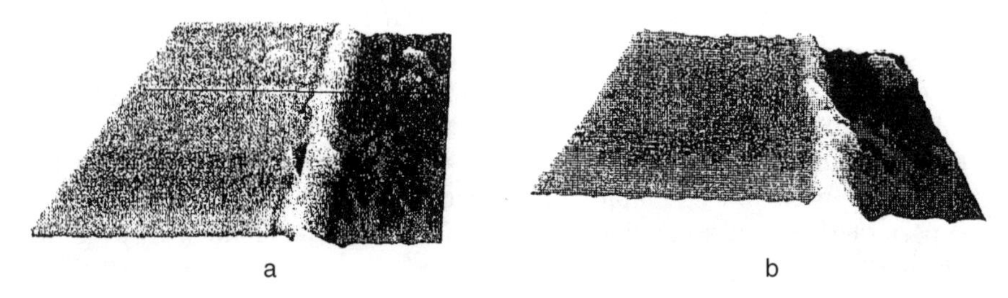

Fig. 10.12. Images of a geometrical step (**a**) in p polarization, (**b**) in s polarization

The results presented in Fig. 10.12 are to be compared with the related theoretical calculations. As an example, a numerical simulation carried out by Santenac *et al.* for the same step is represented in Fig. 10.13. These results were obtained with two different methods, the finite element method and perturbation theory. In the case of this example, the values obtained with the two methods turn out to be identical.

Even if these calculations remain approximate, the effect of polarization is clearly visible in this simulation. The model used here does not take into account the presence of the probe. This simplification is justified in the present case in so far as the probe is located at a distance of several tens of nanometers from the surface and hence has a weakly perturbing effect. The effect of polarization on the images of a grating was observed and confirmed by van Hulst [1992a].

The features of the images also depend on the orientation of the sample relative to the incident wave.

It can be seen from the images reproduced in Fig. 10.14 that the geometry of an object appears in a form which directly depends on the direction

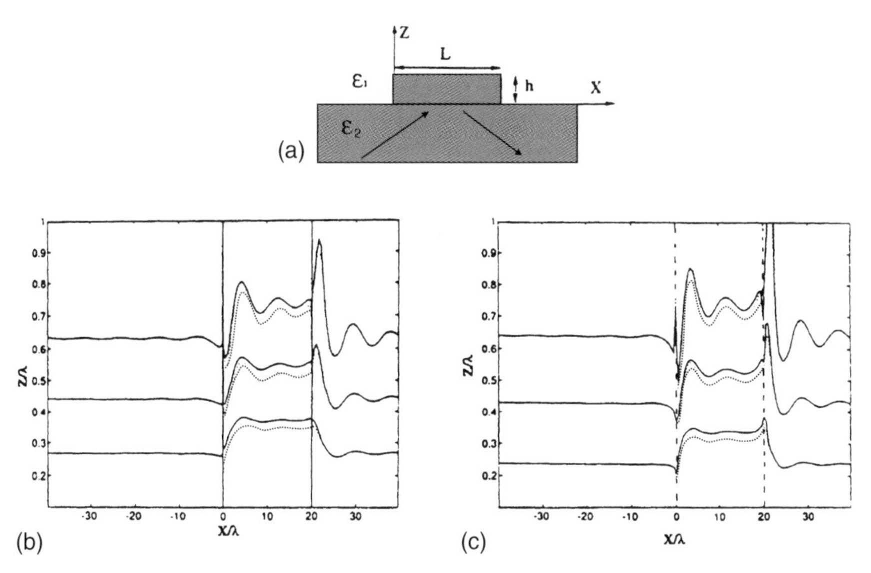

Fig. 10.13. (**a**) Geometry of the sample, $h = 50$ nm, $L = 20\lambda$, (**b**) iso-intensity lines in p polarization, (**c**) iso-intensity lines in s polarization [Santenac 1995]

of illumination. Here again these results are consistent with the numerical simulations reported in Fig. 10.13. This example is rather ideal, and a real sample may simultaneously present the different configurations described above. The shape that a more complex structure assumes on the PSTM image can therefore be very different from the real geometry of the sample [Salomon 1991a, Meixner 1994, Santenac 1995].

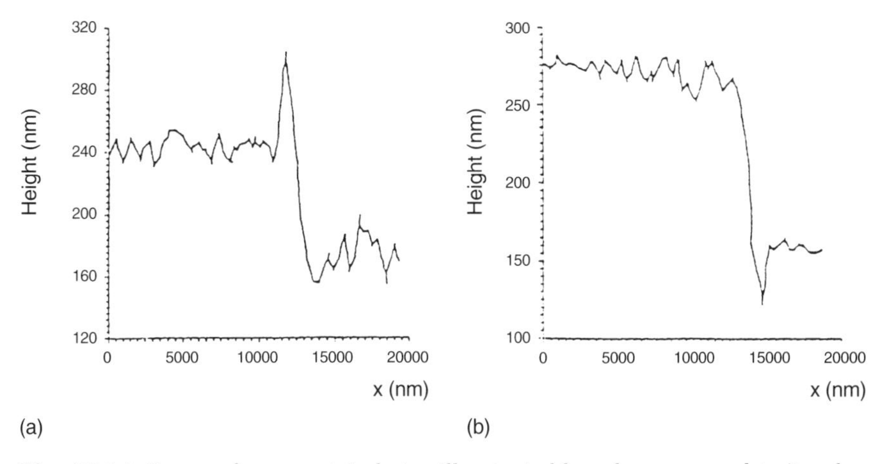

Fig. 10.14. Image of a geometrical step illuminated by a beam normal to its edge (**a**) the source is on the left of the step, (**b**) the source is on the right of the step

10.4.2 Effect of the Distance Between the Probe and the Surface

The effect of the distance on the formation of the images will not be described here in detail. A sample observed with the photon scanning tunneling microscope can be assumed to consist of several multiple radiative secondary sources. If the probe is near the surface, only the light emitted from the surface will be detected. In contrast, if the probe is far from the surface, the light detected is the sum of the light re-emitted by each secondary source. If the source is sufficiently coherent, significant variations of the signal can be observed. These variations result from the amplitude summation of the light emitted by the different parts of the surface (Fig. 10.15) [Nevière 1992].

The theoretical results presented in Fig. 10.15 are consistent with measurements obtained by different authors [Ferrell 1991, van Hulst 1992b]. Figure 10.16 displays the effect of distance on the experimental PSTM images.

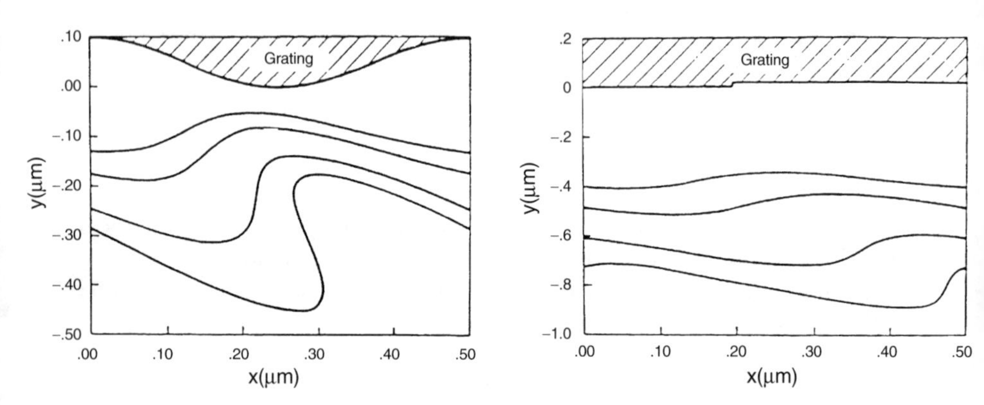

Fig. 10.15. Numerical simulation of the effect of the distance on the form of the iso-intensity lines of the field. $\theta = 53°$, $n = 1.58$, $\lambda = 633$ nm

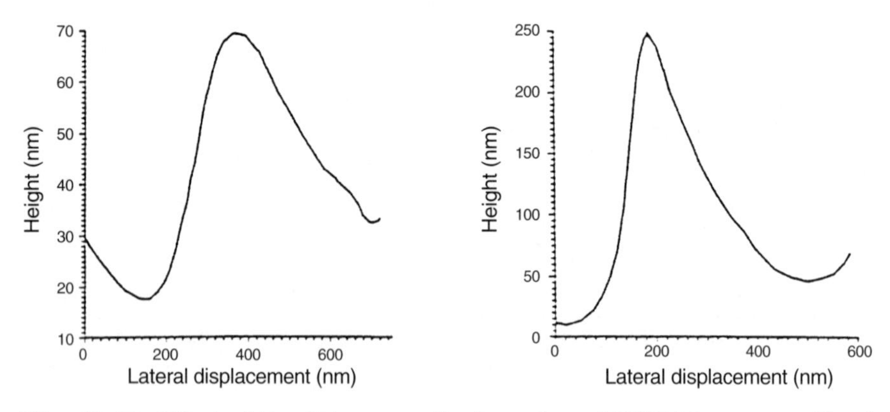

Fig. 10.16. Effect of the distance on the formation of PSTM images: profile of the image of a quartz pad corresponding to the numerical simulation shown in Fig. 10.15, $d = z_0$ and $d = z_0 + 500$ nm

If the probe is located at a distance smaller than half the wavelength from the surface, the image restores the structure of the near-field of the sample.

If the distance between the probe and the surface is larger than this value, the probe remains beyond the near-field of the sample. The image provides the diffraction pattern of the field scattered by the sample, and the information thus recorded does not depend on the local structure of the sample placed in front of the probe. In order to retain the local information related to the sample, the probe has to be located at a very small distance from the sample.

10.4.3 Effect of the Coherence of the Source

The images of a surface with a random roughness exhibit a granular structure [de Fornel 1994]. This type of structure, when observed in far-field, is referred as speckle. The same term will be used hereafter for describing similar near-field structures [Maréchal 1960, Goodman 1972].

If the light used for the illumination of the sample is coherent, the diffraction pattern of each particular detail of the object presents a specific structure. As an example, we shall examine here the case of simple spherical particles deposited upon a plane surface. The image represented in Fig. 10.17 displays the diffraction patterns of latex spheres illuminated with a coherent source [Ferrell 1991, van Hulst 1992b]. If the particles are observed under incoherent illumination, the fringes disappear and the effective geometry of the sample is more effectively restored. Similar results can be obtained with a dielectric step (Fig. 10.18).

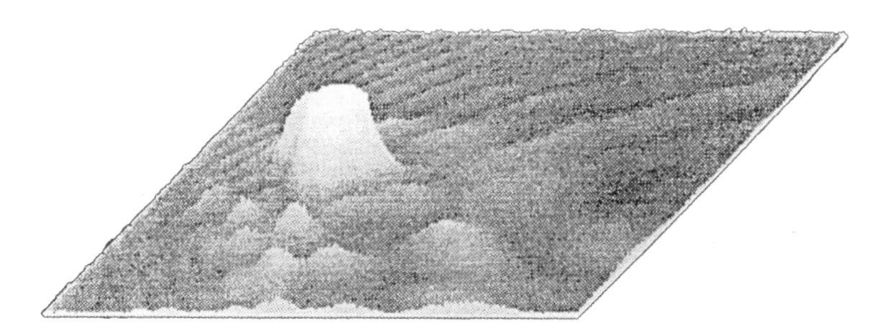

Fig. 10.17. Image of latex spheres with diameter 91 nm under coherent illumination. Scanning domain: 8 μm × 8 μm [van Hulst 1992b]

These results clearly present us with the problem of defining the effect of the coherence of the source in near-field microscopy, and, in the particular case of photon scanning tunneling microscopy, of evaluating this effect for structures separated by very small distances.

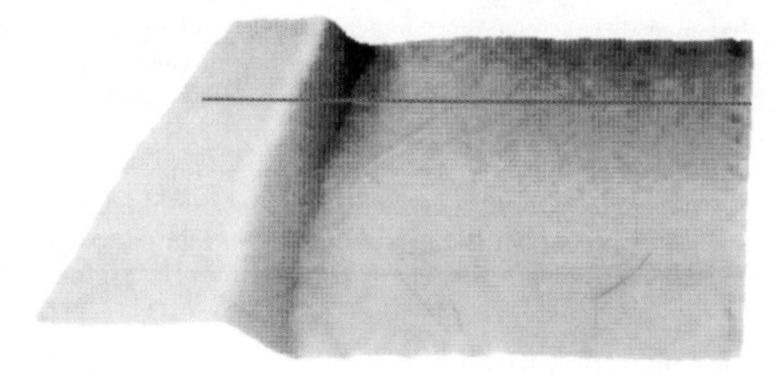

Fig. 10.18. Image of a dielectric step under incoherent illumination [de Fornel 1994]

A theoretical study of this problem was carried out by Greffet *et al.* The effect of the degree of coherence on the formation of PSTM images is apparent in the numerical simulation presented in Fig. 10.19.

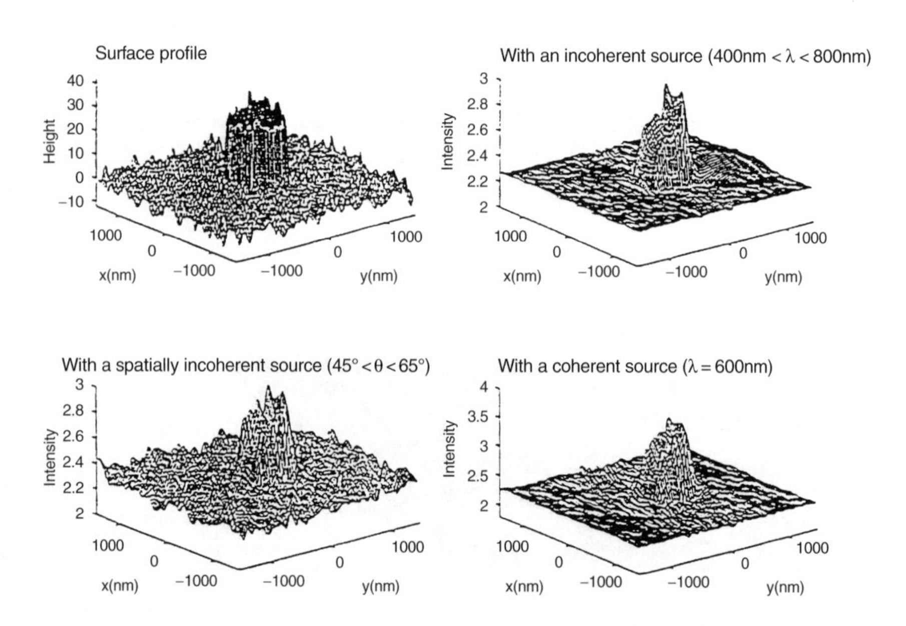

Fig. 10.19. Numerical simulation showing the effect of the degree of coherence and the appearance of speckle-type structures in near-field [Greffet 1995]

Further calculations indicate that if the sample were illuminated from a source with large numerical aperture and low degree of coherence, the resulting images would come very close to the effective shape of the sample

[Greffet 1995, Greffet 1997]. These calculations do not take into account the specific features of the detection of the near-field by the probe. Indeed, depending on the physical parameters of the probe, the detection of the field in the vicinity of the sample is not performed under the same conditions. The response to the degree of coherence depends also on the physical characteristics of the probe [Salomon 1997].

10.4.4 Effect of the Wavelength

To this day, the effect of the incident wavelength on the formation of the images has not been the subject of any exhaustive analysis. In particular, the effect of the wavelength of the incident light on the resolution has not been investigated.

However, the perturbations of the measurements induced by the mechanical or electronic noise have been related to the value of the wavelength used. As demonstrated by Adam, the penetration depth of the evanescent field extends as the value of the incident wavelength rises. Accordingly, the higher the wavelength, the smaller the value of the gradient of the signal [Adam 1993]. A consequence is the relative increase of the mechanical or electronic noise compared with the signal detected by the probe. Therefore, under the same experimental conditions, the images obtained with high wavelengths are more noisy than the images obtained at shorter wavelengths.

10.4.5 Effect of the Probe

Numerical simulations performed by Carminati suggest the possibility of associating a transfer function with the photon scanning tunneling microscope [Carminati 1995]. The experimental system is illustrated in Fig. 10.20, and the transfer function of the probe is defined in Fig. 10.21.

The effect of the probe can be analyzed with a different procedure, by first obtaining an AFM image from a given region of a calibrated sample, while recording the PSTM image. The transfer function of the probe is then determined from the image obtained with the photon scannning tunneling microscope. The PSTM images of a different region are then calculated on the basis of the transfer function thus determined and of the AFM image of this region. The resulting photon scanning tunneling microscope images are in agreement with the numerical simulations [Weeber 1995].

The influence of the probe in the detection of light can also be determined from the dependence of the intensity detected by the probe on the distance, for different types of fibers. The intensity detected by single-mode and multimode fibers is represented in Fig. 10.22 as a function of the distance for three angles of incidence.

From observation of these curves, it is apparent that the detection of the field in the vicinity of the sample depends strongly on the type of fiber

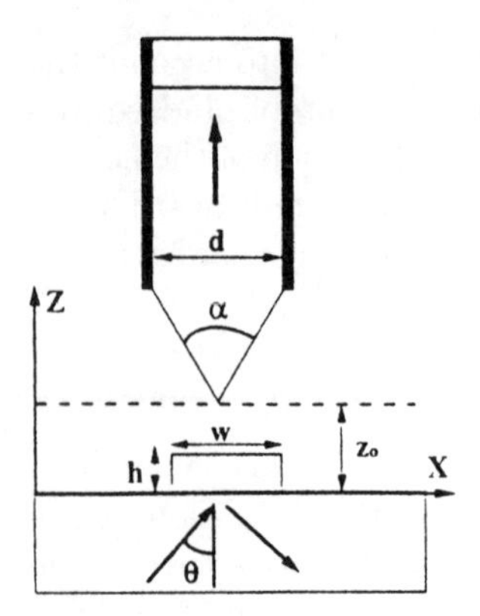

Fig. 10.20. System of the probe and the sample [Carminati 1995]

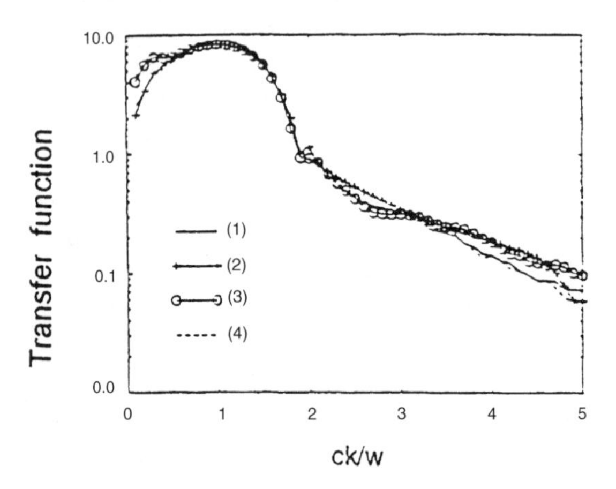

Fig. 10.21. Spectrum of the transfer function of the PSTM [Carminati 1995]

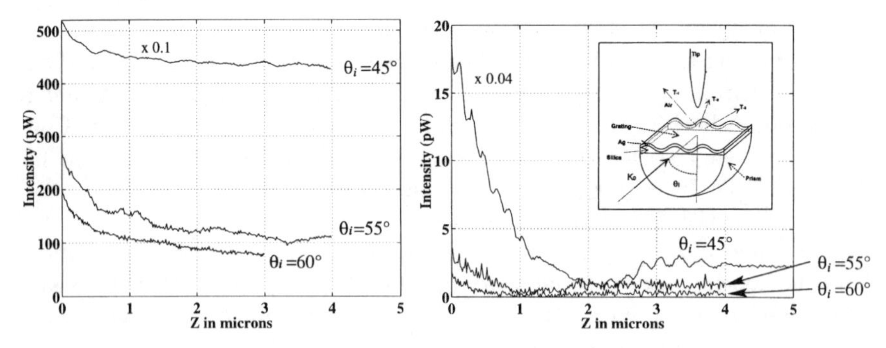

Fig. 10.22. Intensity detected by two fibers, for three different values of the incidence angle (the sample is represented in the insert). The probe is (**a**) multimode with a core of diameter 50 μm, (**b**) single-mode with a core of diameter 3 μm

probe. With a multimode fiber, both the evanescent and propagative waves are detected. This explains the high level of the intensity detected in far-field. With a single-mode fiber, the intensity detected in far-field is very low in comparison with the intensity detected at very small distances from the surface. The difference in the detection of the near-field appears if one compares the two images represented in Fig. 10.23. The first was obtained with a probe fabricated from a multimode fiber, while the other was obtained with a single-mode fiber. It may be mentioned that the PSTM images are in general obtained with single-mode fibers.

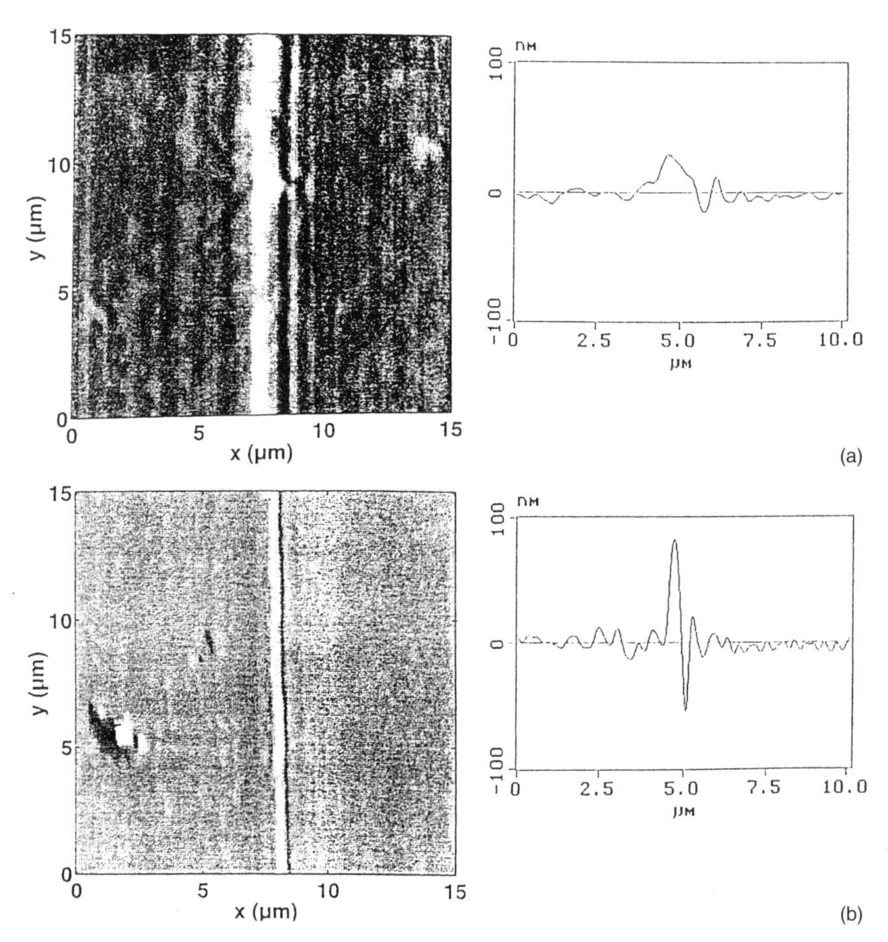

Fig. 10.23. Image of a 50 nm wide and 15 nm high silica bar (**a**) the probe is fabricated from a multimode fiber, (**b**) the probe is fabricated from a single-mode fiber

10.5 Near-Field Study of Non-Homogeneous Samples

The images that the photon scanning tunneling microscope provides are related to the electromagnetic field present near the sample. Therefore it is not surprising that an important application of this microscope is the study of non-homogeneous samples. The PSTM image of a non-homogeneous sample is determined by the product of the index variations (Δn) and of the topographical variations (Δe). The latter can be viewed as the difference between the refractive index of the sample and the refractive index of air. In general, a sample does not present only index variations, and the topography of the sample must be first determined. To this effect, instruments like atomic-force microscopes or scanning tunneling microscopes can be used as complements to the photon scanning tunneling microscope.

Figure 10.24 represents the PSTM image of a lithium niobate plate where protons were locally implanted through the grid of a microscope. Whereas the implanted regions (20 μm × 20 μm) appear in the AFM image with a 5 nm height, the PSTM image displays 30 nm hollows instead. This essentially reflects the difference of the dielectric constant in the implanted regions and in the rest of the sample [Dazzi 1994].

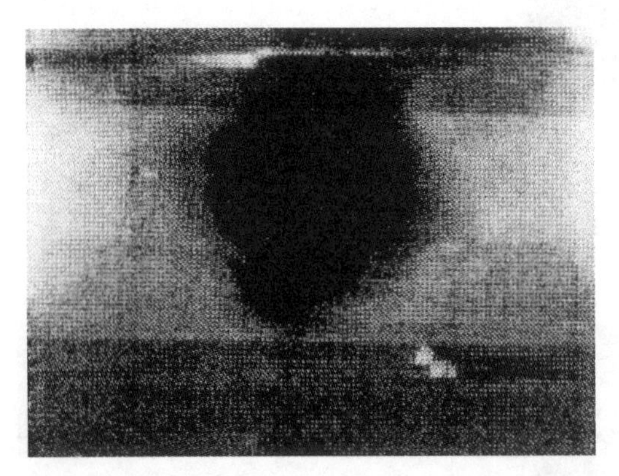

Fig. 10.24. PSTM image of a lithium niobate plate locally implanted with protons [Dazzi 1994]

These images demonstrate that the photon scanning tunneling microscope, far from being only a microscope, can be used as an optical analyzer for detecting localized index variations. This type of measurement is well adapted to the study of integrated-optical devices, as will be described in the following.

10.6 Near-Field Study of Optical Waveguides

An area where the photon scanning tunneling microscope has become a very useful instrument is the study of optical waveguides. Three different types of measurements can be performed using this microscope.

- The observation of index variations.
- The detection of the evanescent field of a guided mode.
- The observation of the structure of guided modes.

These subjects will be successively examined in the following subsections.

10.6.1 Observation of the Index Variations of a Waveguide

Planar optical waveguides can be produced either by ion implantation or by the thermal diffusion of impurities. The PSTM image presented in Fig. 10.25 corresponds to a strip-loaded waveguide fabricated by the local diffusion of silver salts in glass. The profile of the observed waveguide has a Gaussian shape, while the index difference between the core and the cladding is equal to 0.01. The width of the waveguide is of the order of 20 μm, this being consistent with the value determined from the PSTM images. It can be noted that the PSTM images reflect the index variations of the waveguide in all directions, both laterally and transversely [Bourillot 1992, Bourillot 1995].

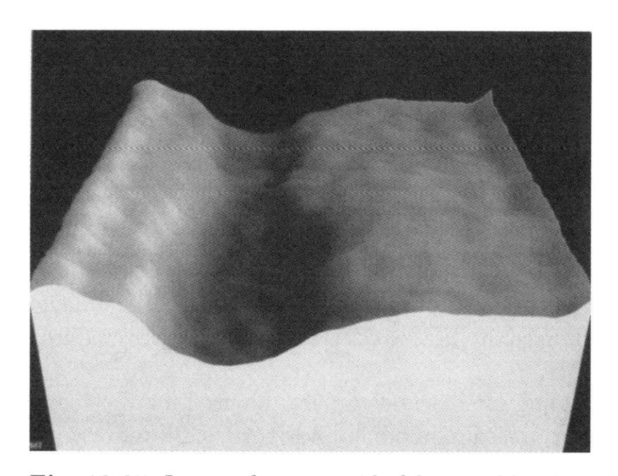

Fig. 10.25. Image of a waveguide fabricated by the diffusion of silver salts in glass [Bourillot 1992]

Any waveguide, whether it be fabricated from dielectric devices or from semiconductors and whatever its dimensions, can be analyzed using a photon scanning tunneling microscope.

10.6.2 Detection of the Evanescent Field of Guided Modes

The modes sustained by a strip-loaded waveguide typically present an evanescent part extending in air. As shown in Fig. 10.26, a part of the evanescent field of a mode can be detected by bringing a tapered fiber into the near-field of the waveguide.

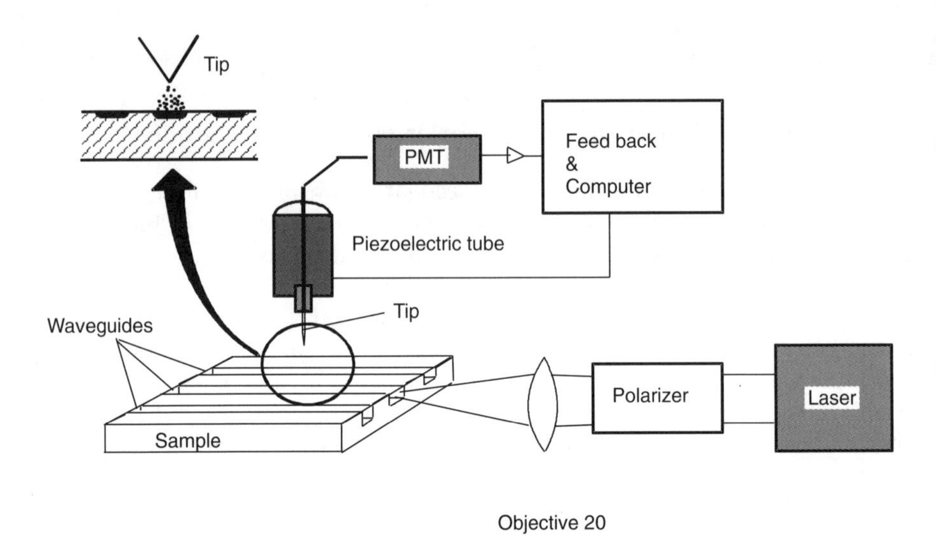

Fig. 10.26. Arrangement used for detecting the evanescent field of the waveguide [Tsai 1990]

The fiber brought into the near-field of the waveguide will detect the evanescent field of the guided mode(s). The decrease curve reported by Tsai of the intensity detected by the probe indicates that the length of decrease associated with the guided modes is equal to 26 nm. This value is identical within about 3 nm with the theoretical predicted value. The PSTM image of the guided mode is reproduced in Fig. 10.27 [Tsai 1990, Jackson 1991, Wang 1992].

The photon scanning tunneling microscope can also be used for the characterization of other integrated-optical components. Figure 10.28 represents a coupler fabricated on a silica substrate [Choo 1994], while Fig. 10.29 displays the variation of intensity detected by the probe along the coupler.

This example shows that the photon scanning tunneling microscope can be used for characterizing integrated-optical devices. Further, the analysis of these components can be carried out without having to subject them to metal coating. As a non-destructive technique, it can therefore be used for testing integrated-optical devices under their actual operating conditions.

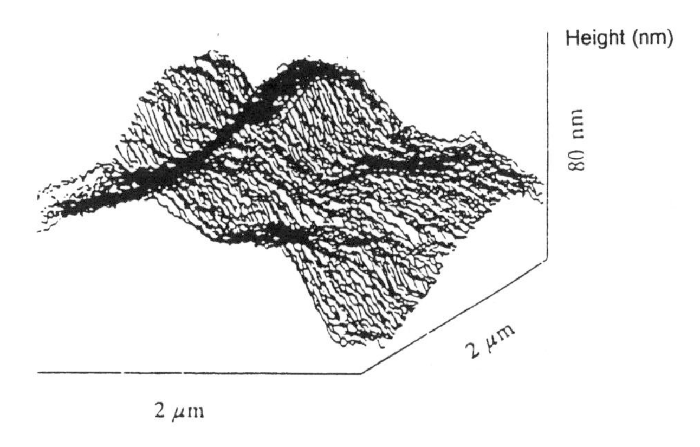

Fig. 10.27. Image of the guided mode [Tsai 1990]

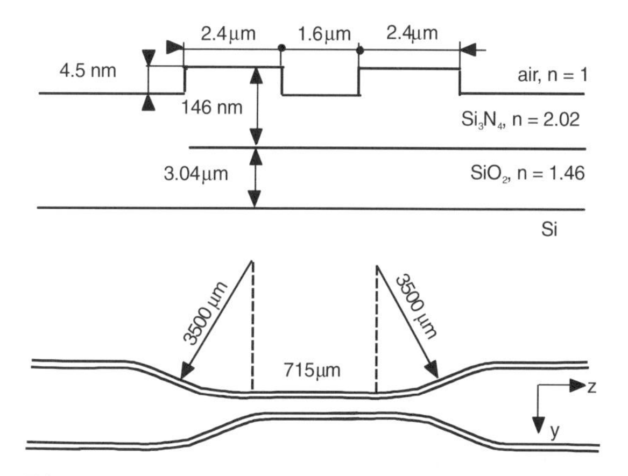

Fig. 10.28. Illustration of a coupler

10.6.3 Near-Field Analysis of the Structure of Guided Modes

The results presented hereafter, if not obtained directly with the photon scanning tunneling microscope, were based on measurements performed in the near-field of the waveguide. Therefore, it seemed of interest to present them in this chapter.

To begin with, we shall resume researches described by Cella [1995]. The waveguide used in these experiments, whose geometry is represented in Fig. 10.30, was made of different layers deposited by MOCVD upon an InP substrate. The 1 μm thick InP epitaxial layer was locally chemically etched using a mask fabricated by photolithography. The width L_{mesa} ranges from 2 mm to 17 μm.

The measurements were carried out using a tapered fiber probe fabricated by fusion and pulling with a CO_2 laser. The sharpened extremity of

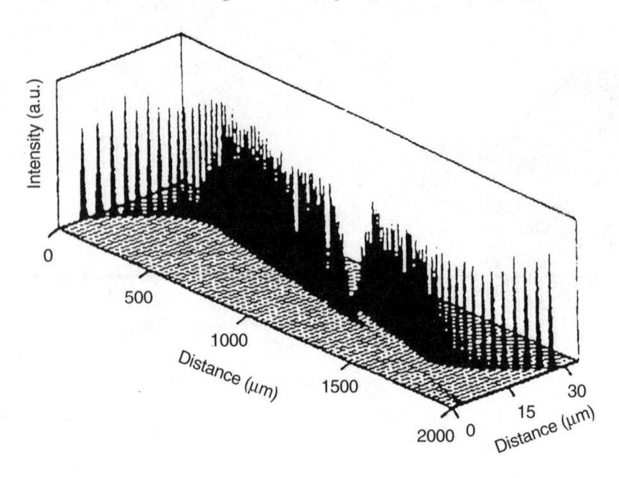

Fig. 10.29. Representation of the intensity detected by the probe along the coupler [Choo 1994]

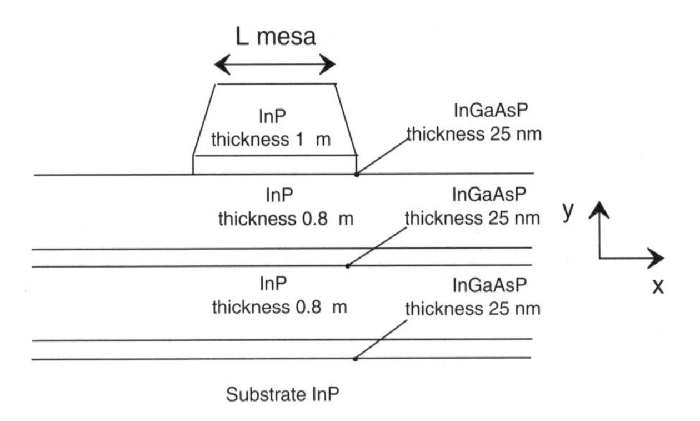

Fig. 10.30. Geometry of the waveguide

the fiber presents a radius of curvature of the order of 50 nm. The fiber is brought to a distance from the output side of the fiber far smaller than the wavelength. The fiber, whose displacements are being piezo-electrically controlled, scans a domain parallel to the output side. Figure 10.31 shows the modal distribution with respect to the ribbon width.

In Fig. 10.31, the variations of the detected intensity were compared with the theoretical determination of the field of the modes obtained from the beam propagation method (BPM) [Feit 1978]. The waveguide used here was single-mode for ribbon widths under 10 μm and 12 μm for TE and TM polarizations respectively. The detection of the field of the modes in near-field is quite consistent for the transverse distribution, but the vertical confinement of the field is weaker than the predicted value calculated from the characteristics of the waveguide.

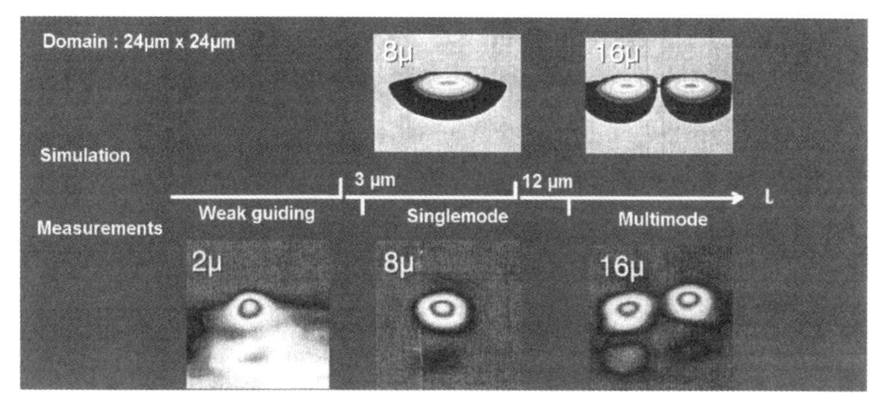

Fig. 10.31. Modal distribution with respect to the ribbon width in p polarization [Cella 1995]

This example indicates that a near-field measurement of the modal distribution of a waveguide allows a control of the modes of the waveguide.

10.7 Local Near-Field Spectroscopies

An advantage of near-field optical microscopy lies in the possibility of carrying out local spectroscopic analyses. The basic principle of these analyses is fairly simple: the end of the fiber, that is, of the local probe, is placed within the near-field of the sample. The optical signal detected by the probe depends on the light emitted or absorbed by the sample. The signal is then transmitted along the fiber onto a detector where its spectrum is analyzed. Until the present, relatively few experiments of this type have been carried out [Moyer 1990, Sharp 1993, Piednoir 1993].

The earliest results on near-field spectroscopy were presented by Moyer in 1990. The excitation laser was a He-Cd laser emitting at wavelength 441.6 nm and the prism was a chromium Cr^{3+} doped ruby crystal Al_2O_3. A part of the signal collected by the fiber was used for the feedback of the PSTM, while the remaining part of the signal was analyzed with a spectrometer. The doublet detected at $\lambda = 692.4$ nm and $\lambda = 694.2$ nm is due to the Cr^{3+} luminescence. Later, results on the local luminescence of an organic film deposited at the surface of a prism were presented by Piednoir (Fig. 10.32).

The film has a very limited thickness, far below the excitation wavelength, and so the signal analyzed corresponds only to the thin organic layer. The different spectra represented in Fig. 10.32 are designed for a comparison between spectra obtained under near-field conditions and spectra obtained in far-field conditions.

The utilization of the PSTM for spectroscopic studies is limited by the necessity of depositing the materials to be studied in the form of thin layers.

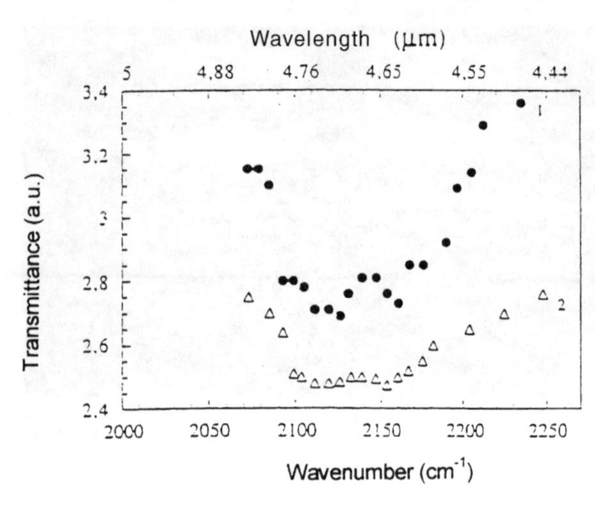

Fig. 10.32. Near-field and far-field luminescence spectra of a thin organic film [Piednoir 1993]

Indeed, if the samples are in a different form, the spectra detected provide the response of the ensemble sample. In this case, it might then be difficult to differentiate the near-field and far-field responses. If a material cannot be prepared in this way, it is necessary either to locally illuminate the sample or to modulate the distance between the tip and the surface, in order to separate the contribution of the evanescent field from the field generated by the whole sample.

10.8 Photon Scanning Tunneling Microscopy and Fluorescence

The detection of the fluorescence of materials has important applications in physics as well as in chemistry and in biology. An example is the localization of fluorescent markers released in a biological material in order to study molecules or chemical reactions.

The fluorescence of small spheres containing fluorescent markers has been observed with the PSTM. As an example, Saiki was able to localize from the detected fluorescence signal fluorescent particles which had been deposited at the surface of a material (Fig. 10.33) [Saiki 1995].

For imaging fluorescent particles, the use of a filter is necessary. As an example, Fig. 10.34 represents an experimental arrangement used for the observation of such particles.

Images of fluorescent polystyrene spheres, reproduced in Fig. 10.35, have been obtained at different wavelengths [Rahmani 1996]. The feedback of the PSTM is performed on the total near-field signal detected by the fiber. The

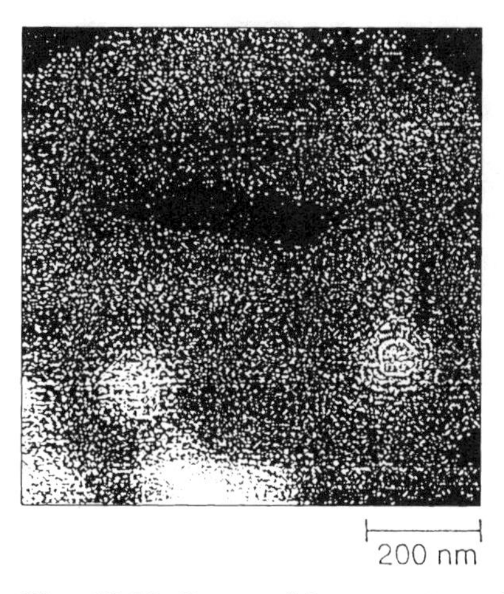

Fig. 10.33. Image of fluorescent particles [Saiki 1995]

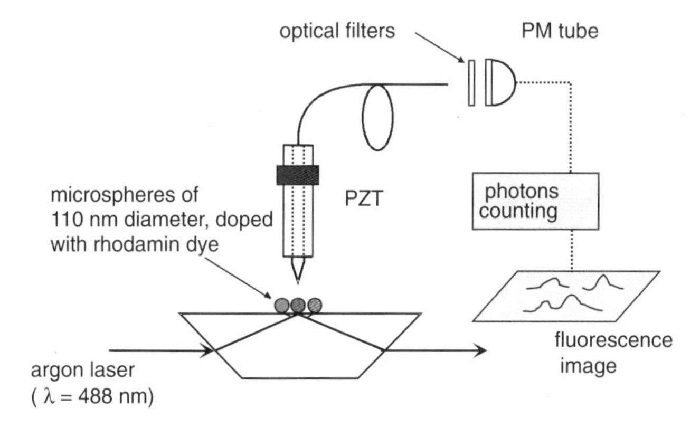

Fig. 10.34. Experimental arrangement used for the observation of fluorescent particles [Saiki 1995]

near-field signal is composed of the pump signal and of the fluorescence signal if the fluorescence is excited [Rahmani 1996]. The resulting images show that the spheres are localized more effectively if the excitation of the fluorescence is maximal. Similar results have been obtained using a scanning near-field optical microscope. Vaez-Iravani has observed the same type of localization phenomena in the case of a porous silicon sample [Rogers 1995].

These researches on optical near-field fluorescence are necessary for extending photon scanning tunneling microscopy, for example for applications of this microscope to biology. The analysis of near-field fluorescence phenomena should be greatly enhanced by employing for the measurements a shear-force feedback system which allows us to place the fiber at a distance of a few

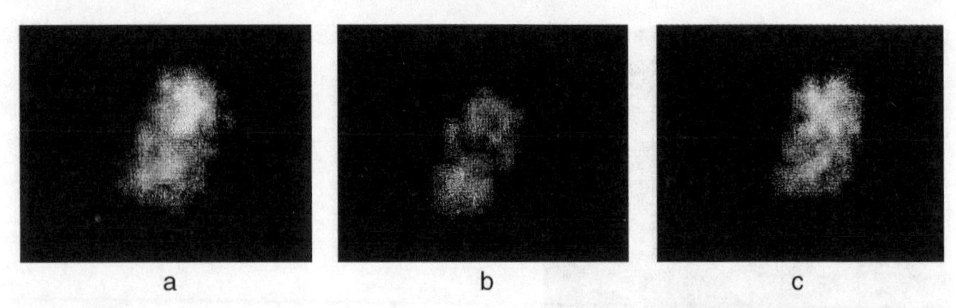

Fig. 10.35. PSTM images of fluorescent particles at different wavelengths (**a**) $\lambda = 612$ nm, (**b**) $\lambda = 535$ nm, (**c**) $\lambda = 532$ nm [Rahmani 1996]

nanometers from the surface. In these conditions, it will be necessary to take into account the interaction between the fiber and the sample.

10.9 Near-Field Study of Surface Plasmons

The present section will be devoted to the near-field observation of surface plasmons. These plasmons had previously been observed only in far-field [Rather 1976]. The physics of surface plasmons cannot be described here in detail. Surface plasmons observed with the photon scanning tunneling microscope are excited in the Kretschmann configuration [Kretschmann 1968]. A thin silver film is deposited at the surface of a prism illuminated in the conditions of total internal reflection, as illustrated in Fig. 10.36.

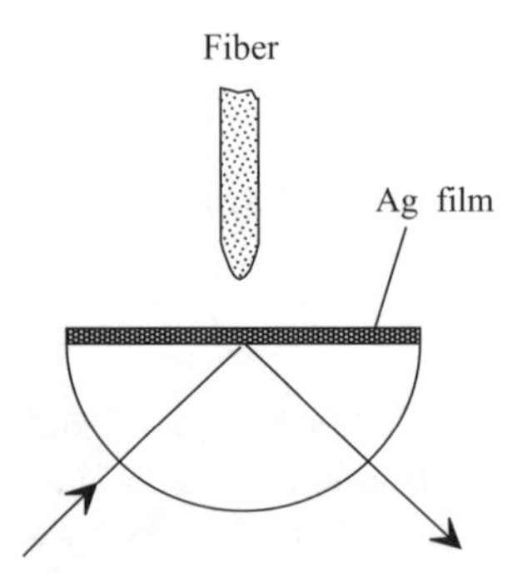

Fig. 10.36. Kretschmann configuration used for the excitation of surface plasmons and detection by the PSTM [Kretschmann 1968]

The dispersion relation of the surface plasmons excited at the air–metal interface is

$$k_y = \sqrt{\varepsilon_1}\frac{\omega}{c}\sin\theta_{\mathrm{p}}, \qquad (10.1)$$

where θ_p is the excitation angle of the plasmon, ε_1 the dielectric constant of the prism, and k_y the component along the y axis of the wave vector associated with the plasmon. The excitation of the plasmon brings the intensity reflected by the prism to a minimal value. The reflectance of a thin silver layer deposited on a prism reaches a minimum when the plasmon is excited at the air–metal interface.

Marti was able to measure the intensity of the field in the vicinity of a metallic film deposited at the surface of a prism [Marti 1993]. The plasmon resonance can be recognized in the decrease curves reproduced in Fig. 10.37. Unfortunately, as the apex of the tip used for the experiment had a radius of curvature of the order of one micron, the far-field signal and the near-field signal were mixed in the response of the system.

In order to correlate the near-field and far-field measurements, the reflectance of the film was determined for different values of the incidence angle. For each value of the incidence angle, the decrease curve of the signal detected by the optical probe was recorded. The probe used in this experiment had a radius of curvature of the order of 50 nm.

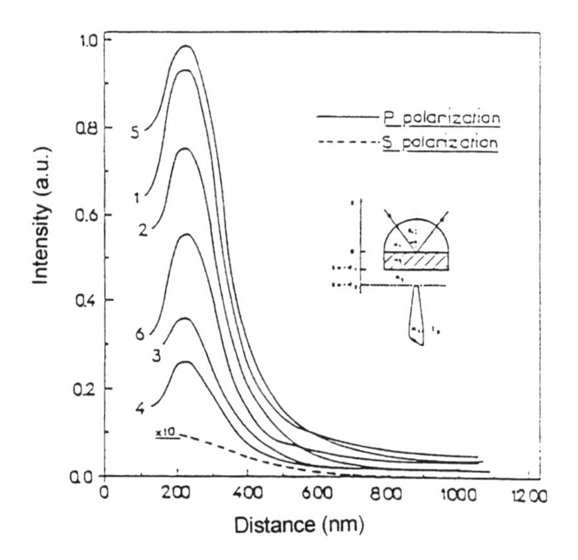

Fig. 10.37. Intensity detected by the probe as a function of the distance between the silver film and the probe

It can be seen in this figure that the maximal value of the intensity collected by the probe is at the θ_p plasmon resonance for fiber–surface distances larger than several hundred nanometers. In contrast, when the probe gets closer to the sample, the angle corresponding to the maximal value is shifted a fraction of degree. This shift indicates that the probe becomes active at small distances. The presence of the probe locally modifies the conditions of the resonance of the plasmon [Adam 1993]. The effect of the probe can be understood using the multilayer model for describing the system.

Other experiments have been conducted on surface plasmons. An example is the experiment described by Dawson, where a thin film is illuminated with a light spot with diameter 15 μm and with the beam emitted by a laser with diameter 2 mm. The latter configuration, described in Fig. 10.38, can be compared with the illumination of the film by a plane wave. The two sources have different wavelengths, so that the feedback is carried out over the intensity detected by the tip in the evanescent field associated with the plane wave. The variations of the signal detected at other wavelengths are simultaneously recorded. This type of measurement is referred to under the name of pilot-wavelength mode.

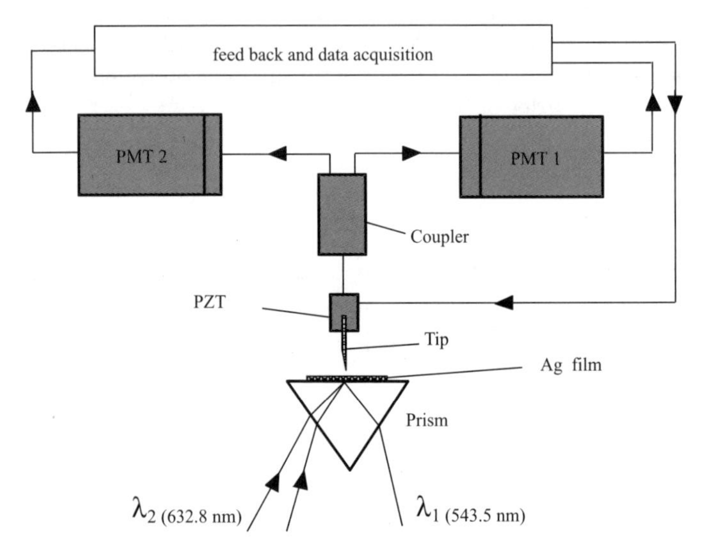

Fig. 10.38. Experimental arrangement used for the near-field measurement of the lateral de-excitation of the surface plasmon

If the value of the incidence angle is different from the angle of the plasmon resonance, the image corresponding to the Gaussian spot reveals a symmetrical structure. In contrast, if the incidence angle is within the range of the plasmon resonance, an asymmetry in the form of the spot can be observed, reflecting the lateral propagation of the plasmon [Dawson 1994, Dawson 1999]. The difference between the images of the spot obtained at the resonance and beyond the resonance can be seen in the images represented in Fig. 10.39.

Thus far, few researches have investigated the physics of surface plasmons [Bielefeldt 1993, Tsai 1994, Bolzhevolny 1995a, Bolzhevolny 1995b], [Krenn 1996, Salomon 1999]. Let us mention here results described by Krenn. Using an aluminum-metallized probe, Krenn was able to adjust the distance

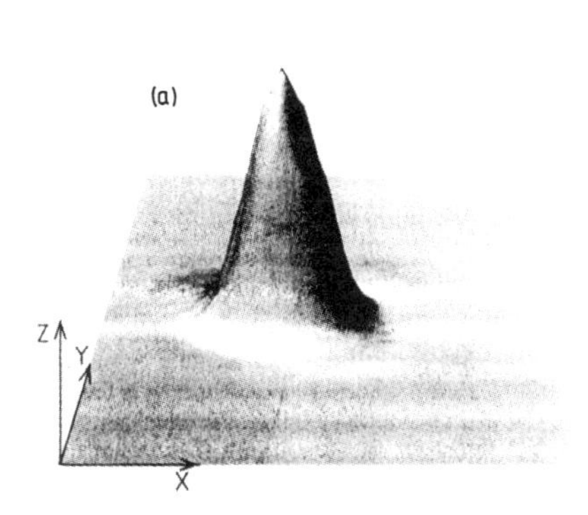

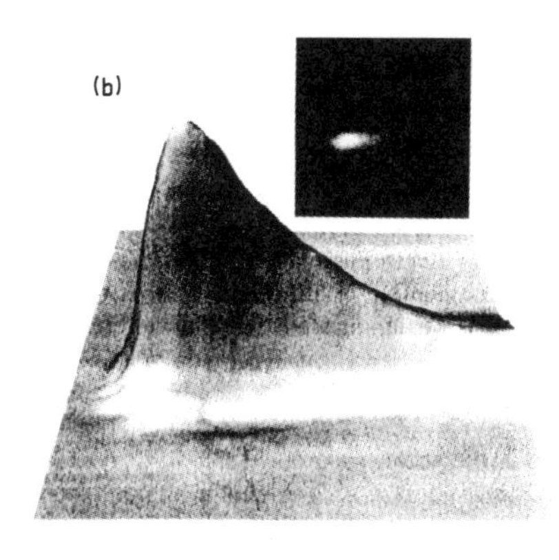

Fig. 10.39. Images of the spot at the resonance and beyond the resonance. The scanning range is 40 μm × 40 μm [Dawson 1994]

between the probe and the surface from the electric intensity of the tunneling current detected by the probe [Krenn 1996]. He thus demonstrated that the response of a sample consisting of gold dots with different sizes and shapes deposited upon a quartz plate depends on the characteristics of the dots.

10.10 Conclusion

The photon scanning tunneling microscope has shown the capability of resolving objects with a subwavelength extent. The analysis of the images obtained with this microscope is quite complex, for the presence of the field diffracted by the sample and the interaction between the probe and the field have both to be taken into account.

The photon scanning tunneling microscope has turned out to be a complement to the microscopy systems used for obtaining topographical images of samples. Indeed, the photon scanning tunneling microscope can be used for obtaining maps of the index variations of the objects under consideration. Further, the photon scanning tunneling microscope can be used for studying the modes of optical waveguides. An example of application of this microscope in this field is the observation of the coupling between two integrated waveguides. For these reasons, the photon scanning tunneling microscope has a large range of applications related to the characterization of integrated-optical devices.

The use of near-field detection enables one to obtain localized spectroscopic information. Among the topics related to local near-field spectroscopy, the researches presently carried on near-field fluorescence are directly connected with the possibility of extending photon scanning tunneling microscopy, for example to biology. Enhancement of these analyses could be achieved by coupling the near-field measurement with a shear-force feedback.

The study of the formation of the optical near-field and of its coupling with a dielectric probe is an emerging field of research. Further researches are presently in process, addressing the theoretical analysis of these phenomena as well as the collection of experimental data.

11. Micro-Aperture Microscopy

The principle underlying the scanning near-field optical microscope (SNOM) is the use of an aperture with subwavelength extent [1928, O'Keefe 1956]. The idea of a microscope based on such a principle was first advanced by E.H. Synge. He suggested as early as 1928 in a letter to Einstein, and then in an article, to illuminate a sample with an aperture perforated in an opaque screen [1928]. He believed that with an aperture with subwavelength extent, the resolution limit that results from the diffraction of the field could be broken.

These ideas were vindicated more than ten years later with the analysis developed by Bethe and Bouwkamp of the field present in the vicinity of an aperture with a limited extent [1944, Bouwkamp 1950a, Bouwkamp 1950b]. A similar analysis was presented for microwave frequencies by Toraldo di Francia [1942a, 1942b].

The *Gedankenexperiments* which had been described by Synge were effectively performed only during the 1970s, first in the field of centimetric waves, and then for wavelengths of the visible spectrum.

It may be pointed out that the theoretical analysis of this problem was addressed before experimental results had been produced. The analysis of the diffraction of light from an aperture with subwavelength extent has been presented in Chap. 2.

11.1 Fundamental Principles of the Scanning Near-Field Optical Microscope

The scanning near-field optical microscope employs an aperture smaller than the wavelength used for the illumination of the system. Depending on the arrangement used for the scanning near-field optical microscope, this aperture acts either as a localized source or as a local detector [Pohl 1991]. This microscope can work either in transmission or in reflection. These two modes are represented in Figs. 11.1 and 11.2 respectively.

Let us first examine the case of a transmission SNOM. The probe serves either for the illumination of the sample or for the detection of the optical signal. In the first system, an incident wave strikes the aperture. Since

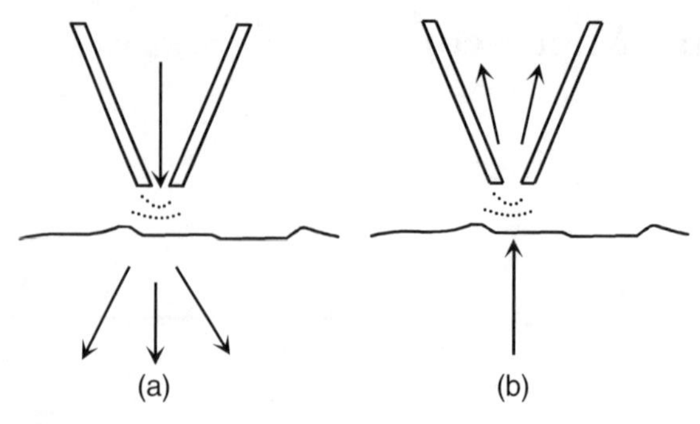

Fig. 11.1. Transmission scanning near-field optical microscope (**a**) illumination from the probe, (**b**) detection by the probe

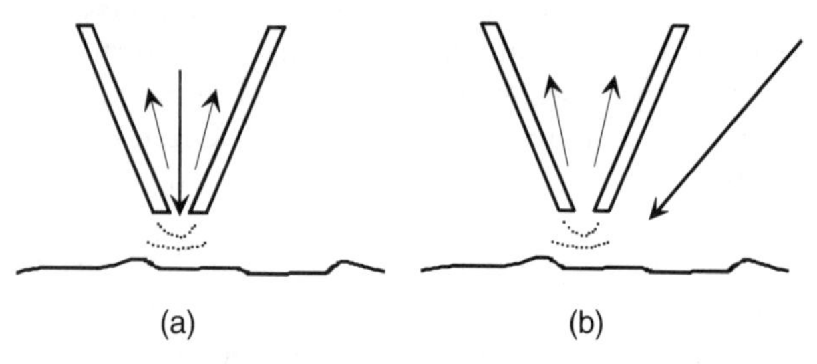

Fig. 11.2. Reflection scanning near-field optical microscope (**a**) illumination and detection by the probe, (**b**) external illumination and detection by the probe

the aperture is smaller than the wavelength, it generates an evanescent field highly confined in the vicinity of the aperture. The aperture is then brought near the sample, thereby perturbing the evanescent waves generated by the aperture. The variations of the light intensity of the propagative waves can thus be detected in far-field. These variations express the near-field coupling between the sample and the aperture [Massey 1983].

The roughness or inhomogeneities of the surface of the sample can be regarded as generating an evanescent field. Therefore, with a scanning near-field optical microscope where the aperture is used as a probe, a part of the evanescent field is transformed into a propagative field which then propagates into the detector.

Different arrangements of reflection scanning near-field optical microscopes have been developed. As an example, the aperture can be used simultaneously as a source and as a detector. The sample can also be illuminated from an external source while the aperture detects the light reflected by the object. Here again, the interaction between the object and the sample is detectable in far-field.

The different systems of scanning near-field optical microscopy, and the different types of probes used in these arrangements, will be described in more detail in the following sections.

11.2 The Breaking of the Rayleigh Limit on Resolution for Microwave and Optical Frequencies

As stated earlier, the very principle of local probe optical near-field microscopy dates back to Synge [1928]. The first experimental evidence for a resolution going beyond the Rayleigh limit is much more recent and came from experiments where centimetric waves ($\lambda = 3$ cm) were used [Ash 1972]. The scheme of the first such experiment, where a $\lambda/60$ resolution was achieved, is presented in Fig. 11.3.

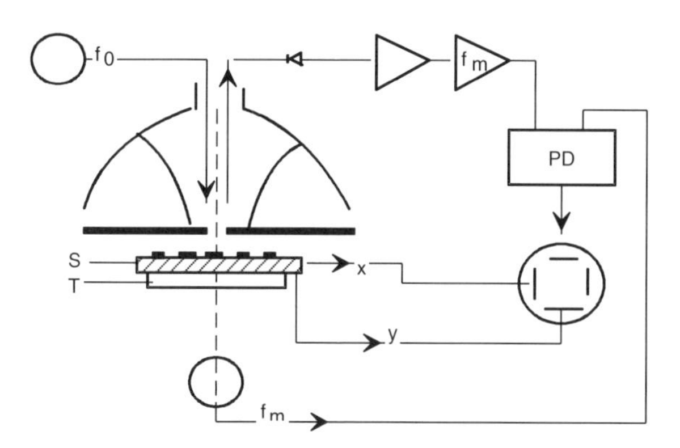

Fig. 11.3. Scheme of Ash's experiment. The source is an open resonator with frequency equal to 10 GHz ($\lambda_0 = 3$ cm) and radius of curvature 10 cm, located at 9.3 cm from the aperture. The sample vibrates at the frequency f_{m}. Only the signal reflected by the sample is modulated at the same frequency f_{m} and restored by an oscilloscope [Ash 1972]

Figure 11.4 is designed for a comparison between the image obtained with the system illustrated in Fig. 11.3 and the sample. The 2 μm high ($\lambda/15$) letters were formed by an aluminum deposit on glass. The possibility of reaching a resolution of the order of $\lambda/60$ has been suggested from other experiments [Ash 1972].

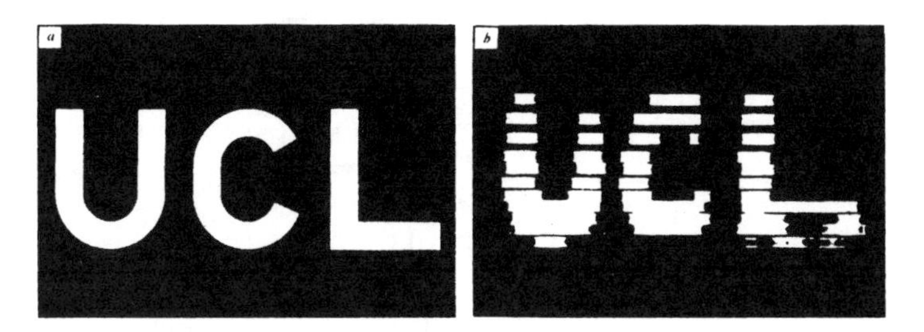

Fig. 11.4. Comparison between the sample and its near-field image [Ash 1972]

The proposal of a source or of a detector with subwavelength extent has been made at several times. In optics, the problem of the fabrication of such elements began to be effectively addressed during the 1980s with the researches carried at IBM by Pohl, and independently at Cornell University by Harootunian, Lewis and later Betzig [Lewis 1984]. It might be remembered here that the diffraction of light from a small aperture had been theoretically analyzed a few decades earlier by Bethe at Cornell University.

The theory of the diffraction of light from subwavelength apertures has been developed in the first part of this book, and therefore we shall not return to the theoretical problems underlying this question. Instead, the present chapter will be concerned essentially with the experimental results obtained in this field and the related applications.

Whereas at centimetric frequencies, the production of apertures with extent far below the wavelength is not prevented by any significant difficulties, if the wavelength of the light used for illuminating the sample belongs to the visible spectrum, the size of a subwavelength aperture must be a hundred times smaller than for microwave frequencies. For this reason, different stages of development were necessary for the perfecting of apertures smaller than the wavelengths of the visible spectrum.

Several techniques have been proposed by Pohl and Fischer. The earliest one consisted in coating a crystal with metal, and then in bringing the crystal in contact with a plane surface. The metal layer was locally pierced during the contact, and thus an aperture with very small extent was created.

In his 1982 patent, Pohl explains that the key element in the microscope is the aperture, whose size he estimates to be of the order of 20 nm [Pohl 1982].

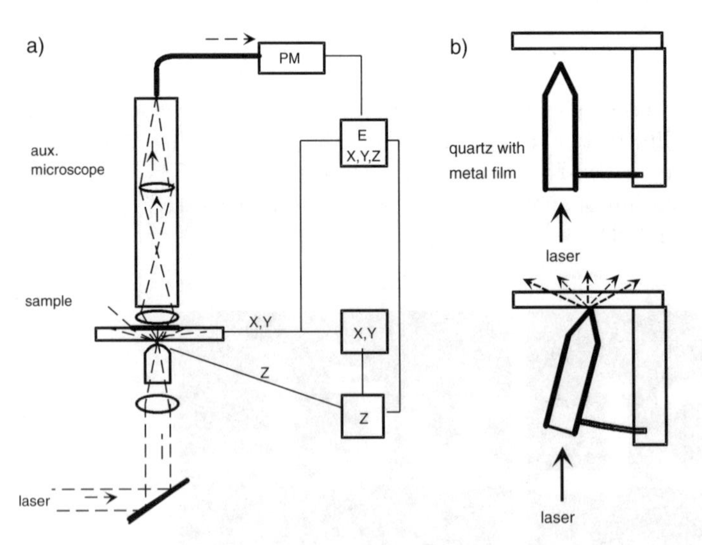

Fig. 11.5. Microscope developed at IBM Zurich [Pohl 1982] (**a**) overall arrangement, (**b**) details of the method used for the fabrication of the aperture

The main disadvantage of this technique is the lack of stability of the geometry of the aperture thus created. Indeed, the surface tension exerted by the metallic film tends to widen the size of the aperture. Figure 11.5 illustrates the principle of this microscope.

The first results obtained already indicated a resolution of the order of 10 nm, as can be observed in Fig. 11.6 [Pohl 1984, Pohl 1985].

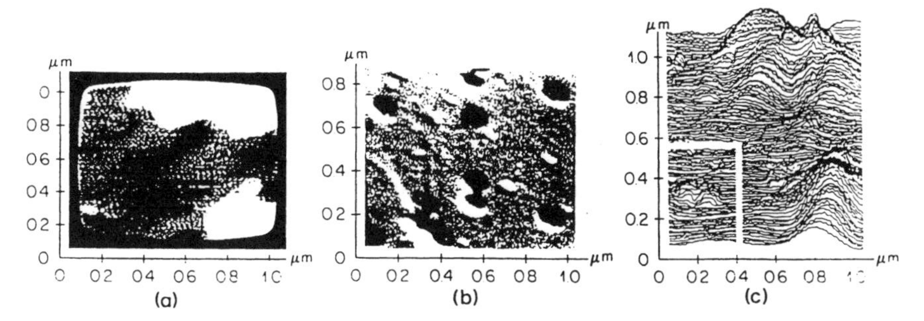

Fig. 11.6. Images of a tantalum film containing holes with diameter 100 nm (**a**) and (**b**) images obtained in optical near-field in transmission, (**c**) image obtained with a scanning electron microscope [Pohl 1985]

At the same time, similar experiments were performed at Cornell University, with the difference that the source apertures used there were formed from micropipettes. An example of an aperture of this kind is represented in Fig. 11.7.

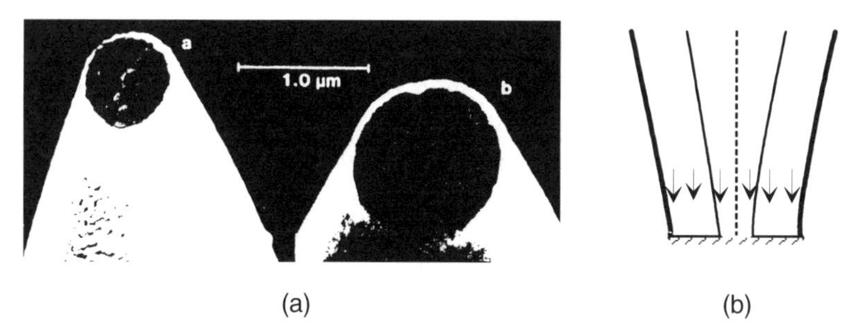

Fig. 11.7. Apertures created from micropipettes (**a**) SEM image, (**b**) schematic view of the tip [Harootunian 1986]

The resolution of the images obtained with such apertures was already very high, as can be seen in Fig. 11.8 [Harootunian 1986].

The technique of scanning near-field optical microscopy has rapidly advanced in the following years. Present-day systems will be described in the next sections.

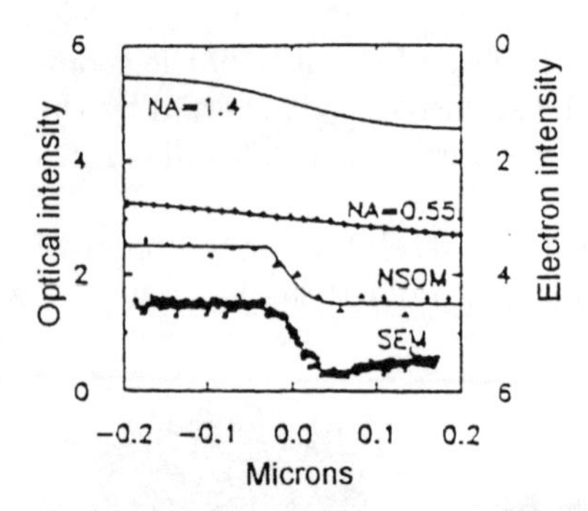

Fig. 11.8. Comparison of image profiles of the edge of a 50 nm thick chromium deposit [Harootunian 1986] (**a**) measurement with a densitometer of a profile of a SEM image, (**b**) variation of the optical signal during the displacement of the source ($\lambda = 570$ nm), (**c**) and (**d**) responses of a traditional microscope with a 0.55 and 1.4 numerical aperture. The curves are normalized in order that the variation of the signal for each curve is unitary

11.3 Description of the Scanning Near-Field Optical Microscope

Several different schemes of scanning near-field optical microscopes have been demonstrated in recent years. The microscopes presently used differ from the early systems in two major respects: the aperture and the scanning process. We shall describe here a typical, modern scanning near-field optical microscope.

As was already advocated by Pohl, the probes are fabricated from silica single-mode optical fibers [Pohl 1982]. For scanning near-field infrared optical microscopy the probes are fabricated from single-mode chalcogenide fibers [Schaafsma 1999]. The fiber is strongly pulled in order to obtain a diameter of the end of the order of a few tens of nanometers. As represented in Fig. 11.9, the fusion and pulling process is generally performed in the beam of a CO_2 laser. The fiber is then metallized using a masking technique in order not to cover its extremity. The aperture of the probe is formed by the non-metallized region. Under these conditions, we obtain apertures whose diameter can be of the order of a few tens of nanometers (Fig. 11.9).

The fiber probe must then be brought inside the near-field of a sample at a distance of a few nanometers from the latter. In order to prevent damage to the sample during the scan, the vertical position needs to be controlled. The first solution to this technical problem consisted in metallizing the sample. The feedback of the position of the probe was then achieved by keeping constant the tunneling current between the probe and the sample during the scan [Fisher 1988]. A few years later, systems for the feedback of the probe based on the use of the contact forces between the probe and the surface of the sample were at the same time developed by Betzig and Vaez-

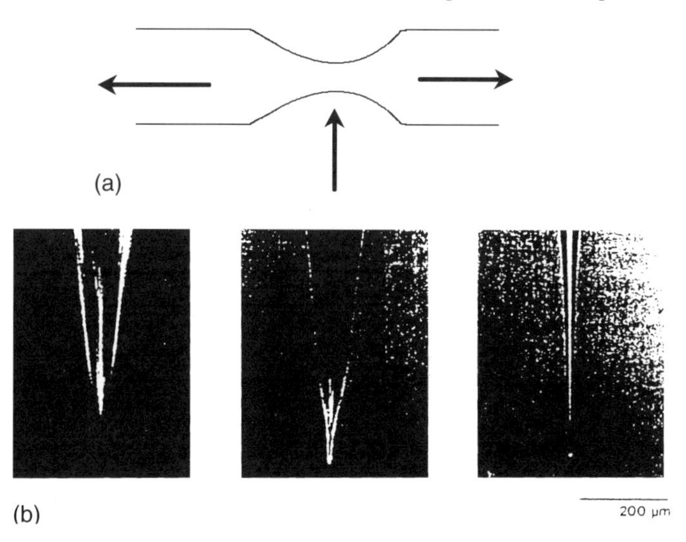

Fig. 11.9. (**a**) Schematic illustration of the fabrication technique of probes from metallized fused-tapered fibers, (**b**) and (**c**) images of two probes fabricated from pipettes, (**c**) image of a probe fabricated from a single-mode fiber. The difference in the guiding parameters between these probes can be noted

Iravani [Vaez-Iravani 1992, Betzig 1991]. Figure 11.10 represents the system developed by the latter at Rochester University.

The probe is mounted upon a piezo-electric element, here a parallelepipedal bimorph fabricated from a piezo-electric material. A piezo-electric tube vibrating at a frequency ω_s can also be used in different experiments. The light emitted by a laser travels through a Wollaston prism, is reflected from the fiber, returns onto the separating slide and is finally detected. By using synchronous detection, only the modulated signal will be detected. The frequency ω_s corresponds to the resonance frequency of the fiber or of the pipette fixed on the piezo-electric element.

When the fiber is brought near the surface of the sample, an effect of the interaction which arises between the probe and the surface is that the amplitude of the oscillation of the probe at the point of contact with the surface becomes almost zero. During the scan, the distance between the probe and the surface is controlled so that the amplitude of the vibrations remains constant (Fig. 11.11). This feedback system is referred to as shear-force feedback. Under these conditions, the images corresponding to variations of the distance at each point of the scanning domain reproduce roughly the topography of the sample. The image simultaneously recorded which restores the variations of the detected light intensity provides the optical response of the sample [Betzig 1993b].

In the descriptions of their experiments, most of the authors employing shear-force regulation agree on the fact that the probe is moved laterally

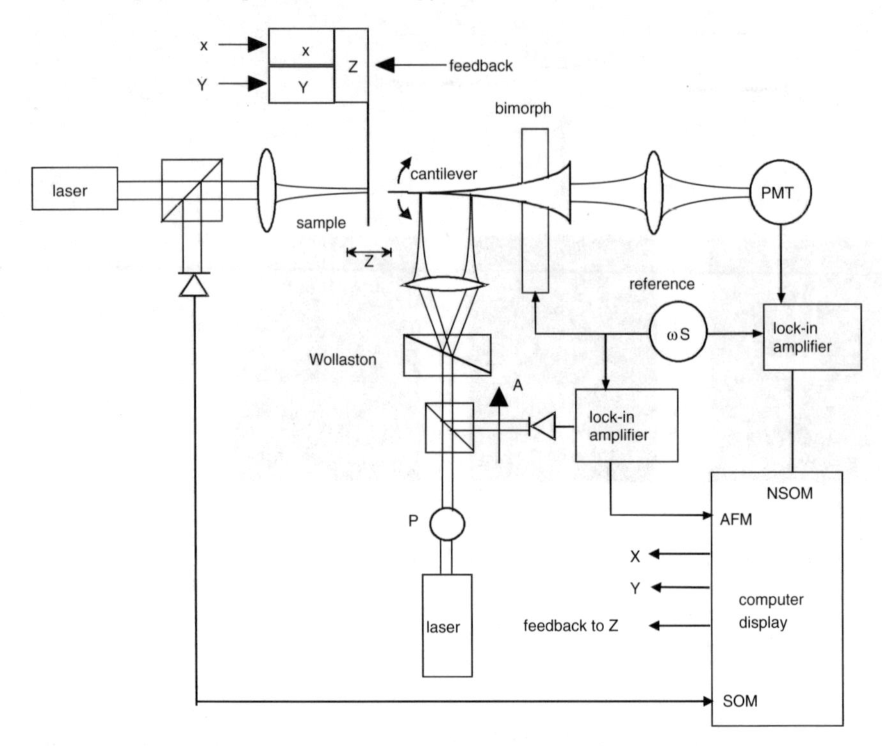

Fig. 11.10. Principle of the feedback system developed at Rochester University

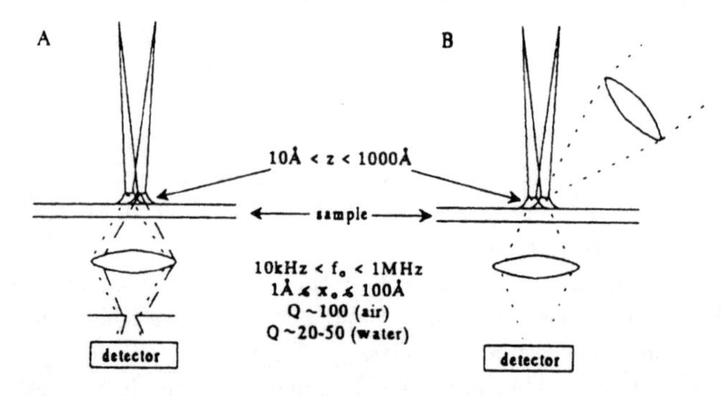

Fig. 11.11. Schematized illustration of the detection of the displacements of the probe. Two examples (**a**) by the detection of the image through an aperture, (**b**) by the measurement of diffusion [Betzig 1993b]

with an amplitude of a few nanometers. The curves presented in Fig. 11.12 illustrate the effect of the vibration amplitude on the determination of the distance between the probe and the surface [Moers 1995].

The samples which are typically observed with the SNOM can be grouped into two classes: metallic or absorbing patterns and fluorescent samples. The

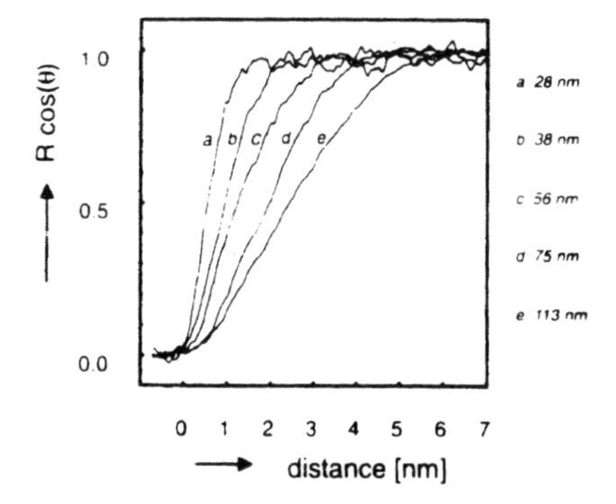

Fig. 11.12. Effect of the modulation amplitude of the probe on the response measured with synchronous detection for different values of the distance between the probe and the surface [Moers 1995]

latter will be treated in more detail near the end of the chapter. As far as we know, for samples presenting only index variations, the 'optical' images obtained with this microscope have never been analyzed in regard to the effective variations of the optical index of the sample.

11.4 Effects of the Physical Parameters on the Formation of the Images

As in scanning tunneling optical microscopy, all the physical parameters play a part in the formation of the images. Therefore, the images obtained with a scanning near-field optical microscope depend in particular on the coherence of the light used for illuminating the sample, on the polarization, and clearly on the wavelength. Since the orientation of the incident wave does not have any effect on the images, the number of variable parameters of the system will be more limited than for photon scanning tunneling microscopy.

In the present section, we shall examine separately the different parameters involved in the formation of the images by considering results presented by various authors. A difference between the photon scanning tunneling microscope and the scanning near-field optical microscope is that in the latter case the respective effects of these parameters have not been theoretically investigated to the same degree.

11.4.1 Effect of the Polarization

The effect of the polarization of the source has been extensively analyzed with the experimental results presented by Betzig [1992]. The arrangement used in these experiments was similar to ordinary scanning near-field optical

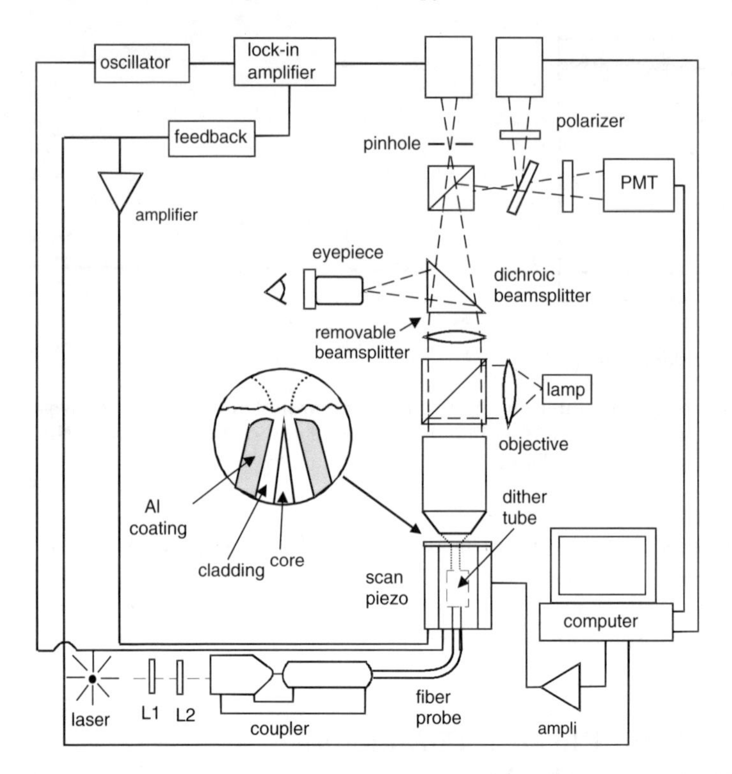

Fig. 11.13. Experimental arrangement used for determining the effect of polarization [Betzig 1992]

microscopes. The scheme of the experimental setup used here is represented in Fig. 11.13.

The polarization of the light depends on the optical fiber placed between the source and the sample and is selected from the orientation of a $\lambda/2$ slide and of a $\lambda/4$ slide placed at the input of the fiber. Even if the rotation of polarization is not the same for all fibers, the variations of the polarization can be effectively determined in each particular case.

The sample that the authors used in these studies was a set of four aluminum rings placed upon a glass plate. The aluminum layer had a thickness of 20 nm for a diameter of 1 μm. The following images (Fig. 11.14) illustrate some of the cases that might be encountered.

Fewer studies are presently available for the reflection scanning near-field optical microscope. Figure 11.15 is designed for a comparison between the shear-force image and the SNOM image of a sample formed by an aluminum deposit on glass [Cline 1995].

The absence of a systematic theoretical analysis of the effect of the polarization on the images results in a difficulty of discussing the results presented

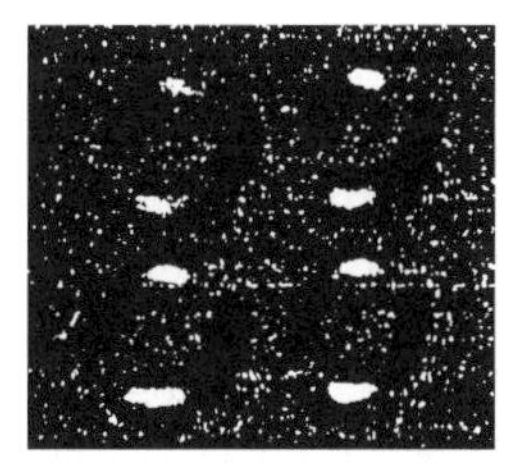

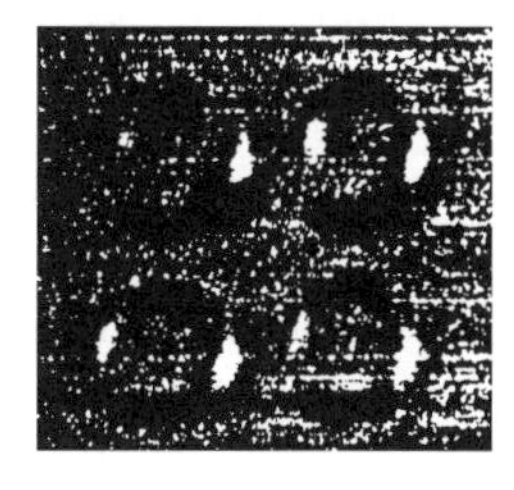

Fig. 11.14. Images of the rings obtained for different polarizations in transmission mode: the polarizations of the source and of the detector were (**a**) vertical, (**b**) horizontal [Betzig 1992]

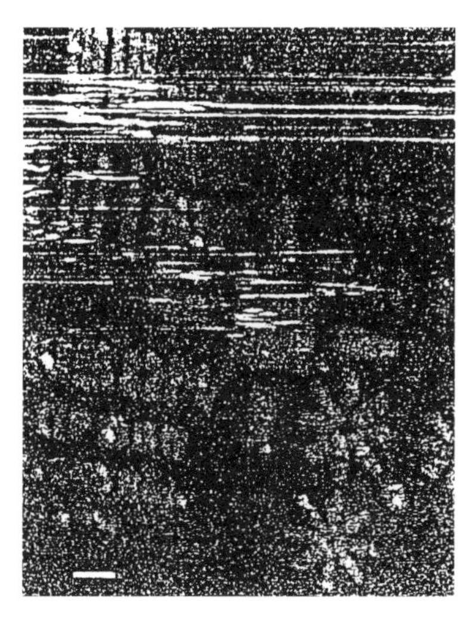

Fig. 11.15. Images obtained in reflection mode and in shear-force mode respectively [Cline 1995]

herein. In fact, the effects of certain parameters, such as for example the effects induced by the polarization at the very end of the probe, are still not entirely determined.The studies about effect of the polarization are still in progress [Higgins 1996].

11.4.2 Effect of the Wavelength

The images obtained with scanning near-field optical microscopes apparently depend also on the wavelength of the light used for illuminating the sample, but the exact relation between this parameter and the images remains to be determined. As far as we know, the effect of the wavelength of the source on the formation of the images has not been the subject of any exhaustive analysis.

11.4.3 Effect of the Coherence of the Source

In his PhD thesis, E. Betzig presented images that demonstrate the influence of the degree of coherence of the source on the formation of the images [Betzig 1988]. The weaker the coherence of the source, the higher the resolution of the images (Fig. 11.16). In transmission, the effect of the coherence of the source is relatively limited. To a large extent, the weak effect of the coherence results from the fact that the illumination has a localized character. Hence the light is diffracted only from the elements contained within the spot, each of these elements acting as a secondary source.

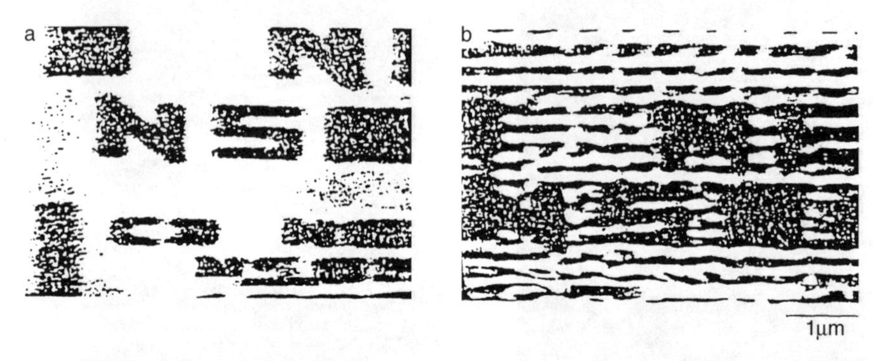

Fig. 11.16. Effect of the coherence on the formation of images (**a**) image obtained with an incoherent source, (**b**) image obtained with a laser [Betzig 1988]

11.4.4 Effect of the Distance Between the Probe and the Surface

The effect of the distance on the formation of images has also be investigated by Betzig [1992]. The images represented in Fig. 11.17 demonstrate that the resolving power of the microscope decreases slowly when the probe is more than some tens of nanometers from the surface of the sample.

The use of shear-force regulation allows us to bring the probe to a few nanometers from the surface under investigation. The results obtained with this technique can therefore be reproduced, this being a prerequisite for the development of any commercial instrument. Up to the present, the resolution limit of the scanning near-field optical microscope has not been precisely determined. The resulting optical images must be analyzed cautiously since they contain not only optical data, but also purely topographical data which are dependent on the regulation mode, and which therefore are called artefacts by some authors [Hecht 1997].

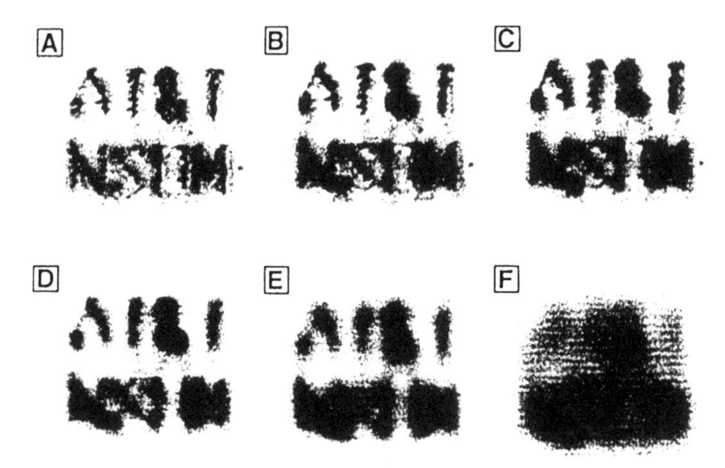

Fig. 11.17. Images obtained for different values of the distance between the fiber and the surface of the sample (**a**) the two are in contact, (**b**) 5 nm, (**c**) 10 nm, (**d**) 25 nm, (**e**) 100 nm, (**f**) 400 nm [Betzig 1992]

11.5 Local Fluorescence Detection

By attaching a shear-force type regulation for restoring the topography of the sample, complementary optical information can be drawn from the optical signal. An example is the detection of the fluorescence of the sample, which has great importance in a number of extensions of scanning near-field optical microscopy.

In classical microscopy, the measurement of the fluorescence had always been a complementary instrument of analysis of the observations. The phenomena of near-field fluorescence had been the subject of researches carried out by Drexhage [1968, 1970 and 1974]. At the time when the first researches on optical near-field were carried out, images of fluorescent samples had already been obtained, both by the Cornell group and by the IBM Zurich group (Fig. 11.18) [Harootunian 1986].

The early scanning near-field optical microscopes lacked an effective feedback system, and so the position of the probe relative to the surface of the sample could not be controlled. The results obtained with these microscopes were therefore not reproducible. From around 1992, the observation of the sample could be carried out while simultaneously detecting the local variations of the fluorescence signal. As an example, we may mention here recent researches carried out at Twente University in the Netherlands [Jalocha 1995].

The possibility of accurately localizing fluorescent elements is of great interest not only for biology, but also for physics, for example in the study of optical fibers doped with rare earth elements. Betzig has been able to image the field emitted by a single molecule [Betzig 1993a]. Figure 11.20a reproduces the image of carbocyanine molecules diluted in a resist of PMMA

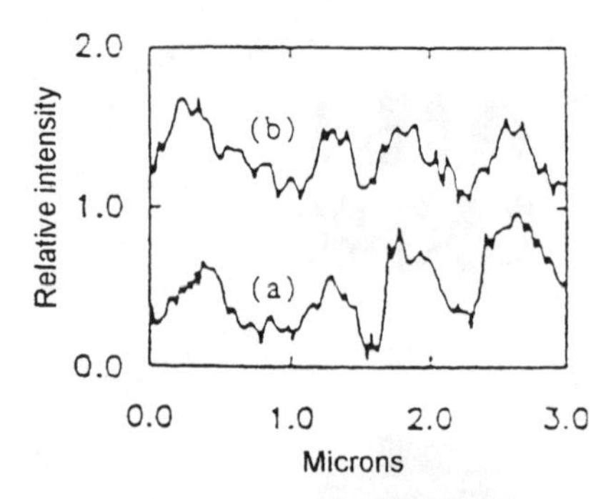

Fig. 11.18. Profiles of different images of a fluorescent sample (**a**) image obtained from the fluorescence signal, (**b**) SEM image [Harootunian 1986]

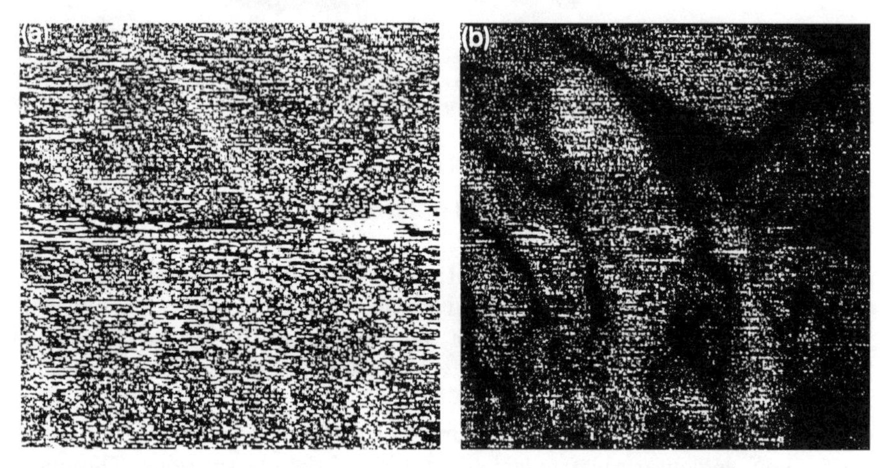

Fig. 11.19. Comparison between (**a**) the shear-force type image and (**b**) the fluorescence image of a DPDA film. The scanning range is equal to $12 \times 12\ \mu m^2$. The domains are clearly visible in the fluorescence response [Jalocha 1995]

(polymethylmethacrylate) type. The apparent shape of the molecules is not the same throughout the entire sample. Indeed, the shape of the image of each molecule depends on the orientation of its active dipole. Figure 11.20b represents a numerical simulation where the possible orientations of the dipoles coherent with the image reproduced in Fig. 11.20a are taken into account.

The use of spectroscopy associated with near-field optical measurements enables a thorough study of fluorescent molecules. Following these researches, local spectroscopic measurements have thus been obtained [Trautman 1994, Jahncke 1995, Moerner 1999].

As an example, we hereafter examine works by Trautman *et al.* which were carried out at AT&T Bell [Trautman 1994]. 1,1'-dioctadecyl-3,3,3',3'-tetramethylindocarbocyanine (Dil) molecules are released in a thin film of polymethylmethacrylate (PMMA). Single molecules are first localized from

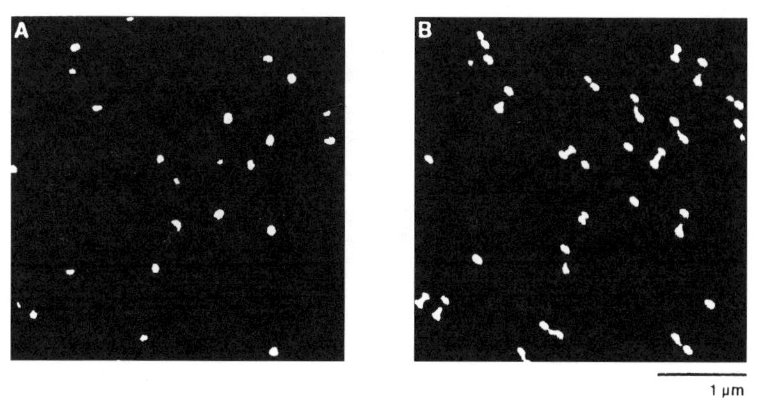

Fig. 11.20. Molecules released in PMMA (**a**) SNOM image of the molecules, (**b**) numerical simulation corresponding to the previous image

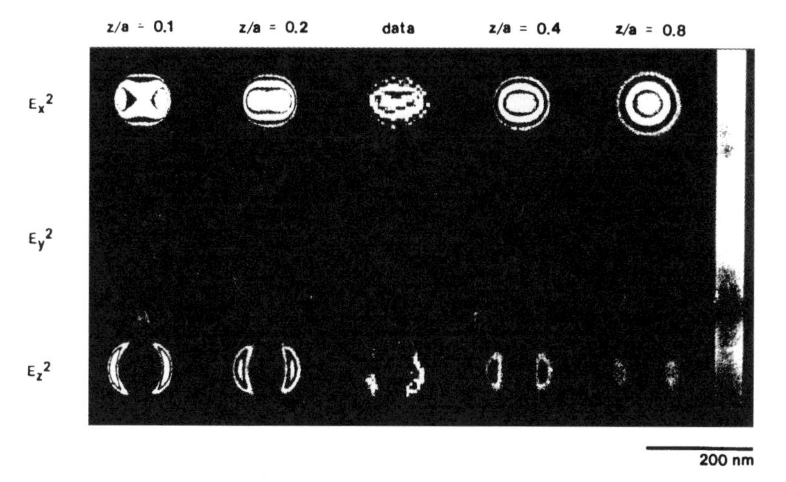

Fig. 11.21. Model of the electromagnetic field in the vicinity of molecules with distinct orientations

the measurement of the fluorescence of these molecules, as represented in Fig. 11.22. The fluorescence signal is then analyzed by a spectroscope. The spectra obtained in the vicinity of two molecules are reproduced in Fig. 11.23.

The photobleaching of single molecules can also be observed using near-field scanning optical microscopy (Fig. 11.24).

A large number of studies in optical near-field microscopy have been concerned with observations of the fluorescence of single molecules [Trautman 1994, Ambrose 1994]. Fluorescent molecules are used as markers, for example for the monitoring of chemical reactions. Even if these researches extend largely beyond the scope of this book, the potential applications of near-field optical fluorescence detection have such an importance that it was necessary to briefly mention this subject.

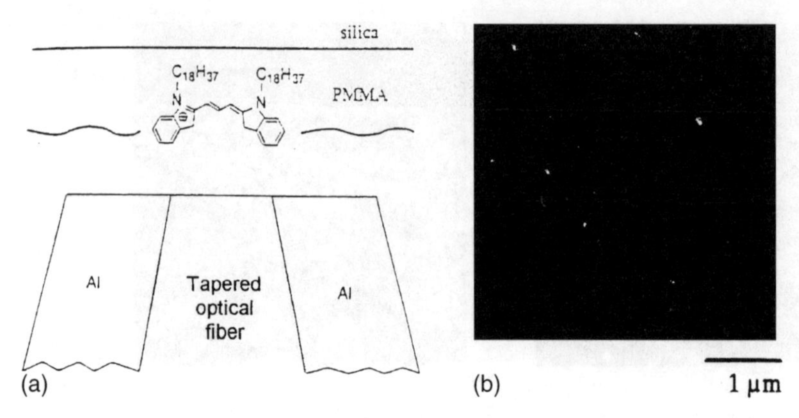

Fig. 11.22. Schematic view of the probe in front of a fluorescent molecule. Image of the fluorescence signal displaying a few isolated molecules

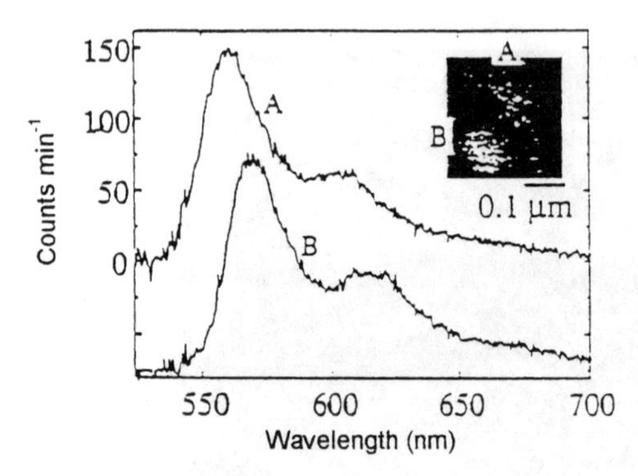

Fig. 11.23. Spectra and images of two molecules [Trautman 1994]

11.6 Near-Field Optics and Photolithography

From the very beginnings of researches on near-field microscopy, the possibility of using the confinement of light by probes for carrying out photolithographies had been apparent to all. An arrangement used for this purpose is represented in Fig. 11.25. As can be seen in this figure, it resembles the arrangement of the scanning near-field optical microscope.

The authors have compared the photolithographies obtained with a non-metallized fiber and with an aluminum-coated fiber. Here again, the difference between the two is apparent from observation of the resulting images. Figure 11.26 represents the profile of the AFM image of lines produced in a photoresist using chemically etched fibers of each type [Rugar 1990, Krausch 1995].

As can be seen from these profiles, the modulation induced in the photoresist is close to a Gaussian curve. The half-width of the modulation is about 254 nm for the uncovered fiber and 82 nm for the metallized fiber. In the first

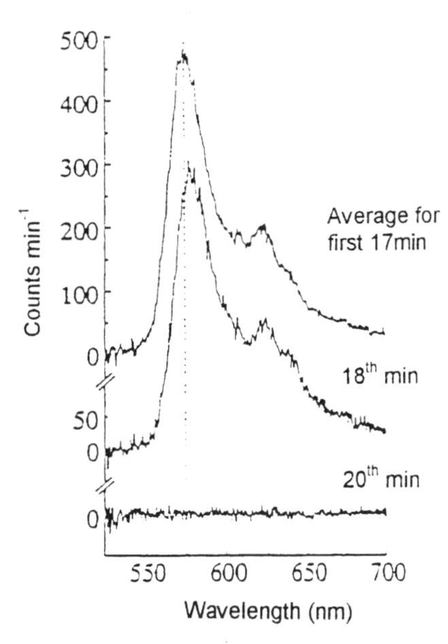

Fig. 11.24. Evolution of the fluorescence spectrum of a single molecule. The upper spectrum corresponds to the average of seventeen spectra recorded during the seventeen first minutes. The second spectrum was obtained one minute later and exhibits a 5 nm shift. At the nineteenth minute, the molecule was photobleached. The authors have determined that during these measurements around 2×10^8 photons have been emitted by the molecule [Trautman 1994]

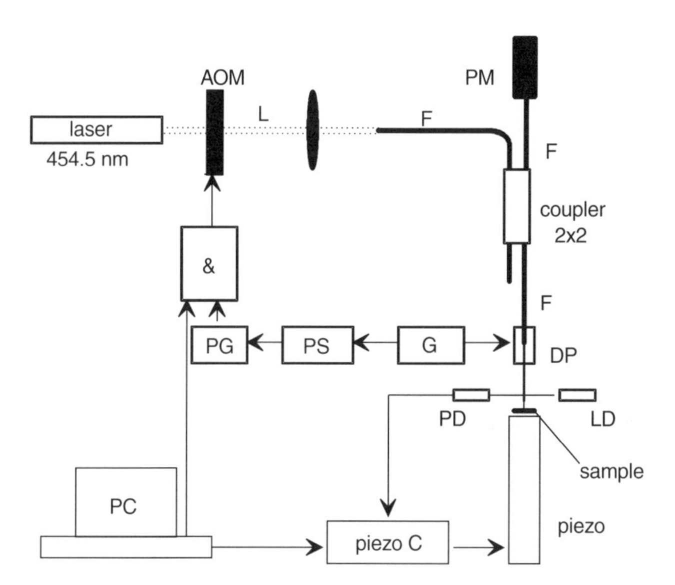

Fig. 11.25. Experimental arrangement used for carrying out photolithographies [Krausch 1995]

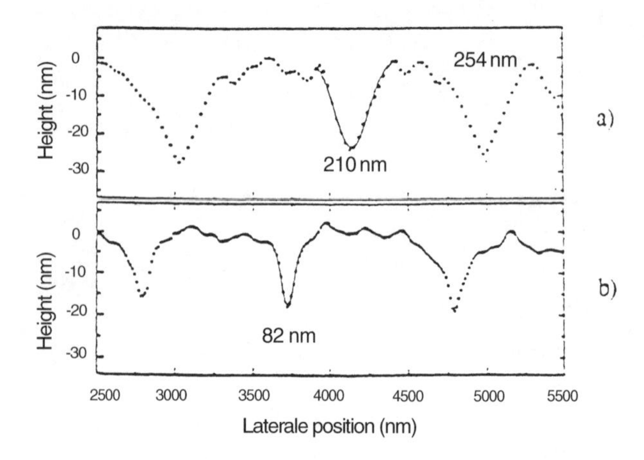

Fig. 11.26. Profiles of AFM images of lines produced in a photoresist. The lines were engraved with (**a**) a non-metallized fiber, (**b**) a fiber coated with 65 nm aluminum. The *solid curve* corresponds to an adjustment with a Gaussian curve [Krausch 1995]

case, the Gaussian shape is consistent with the intensity distribution of the beam emitted from the fiber. In contrast, for the metallized fiber, one would have expected a more complicated shape, because of the distribution of the p-polarized field at the very end of the probe [Novotny 1994]. The fiber used there was not polarization preserving, and hence it is difficult to determine exactly the state of polarization near the aperture.

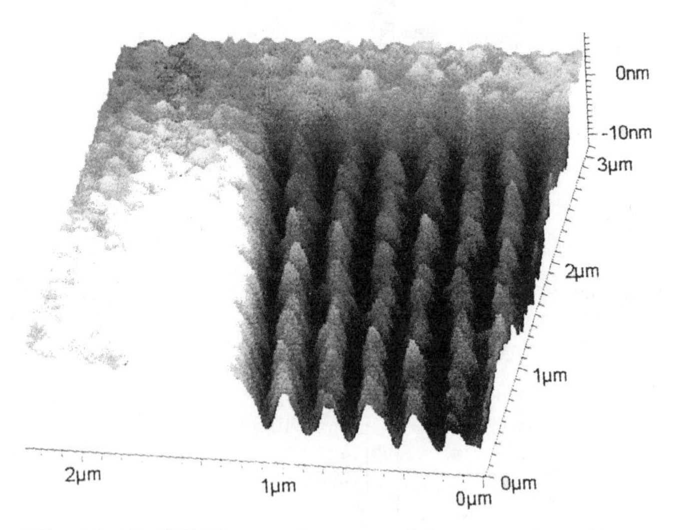

Fig. 11.27. AFM image of an area of the resist photolithographied with a metallic tip [Krausch 1995]

11.7 Conclusion

A resolution in near-field microscopy equal to $\lambda/60$ was attained as early as 1972. These early results were obtained in the field of microwave frequencies. From 1982, the interest raised by photon scanning tunneling microscopy has led to the development of local-probe microscopes where a resolution of the same order as the value claimed by Ash in 1972 was reached. In the initial scanning near-field optical microscopes, however, the probe had to be brought to within a few nanometers of the surface, and this often resulted in mechanical deterioration of the probe. Therefore, this restricted the imaging capabilities of this microscope.

These defects were settled when an accurate feedback was associated with the optical measurements. The feedback of the microscope, based on the use of the interactions of the probe and the surface, permits us to control precisely the distance between the probe and the surface of the sample. The topography of the sample can therefore be imaged with a very high degree of resolution, while the optical data are simultaneously detected.

Another extension of the scanning near-field optical microscope is in the field of fluorescence studies. Using this microscope, very effective fluorescence measurements were performed. An example concerns the detection of single molecules. Further, the orientation of a molecule can be determined from the form of the intensity detected by the probe. The spectrum of a single molecule has also become measurable with this microscope.

There still remain a few unresolved technical problems. Among these are the control of the size of the aperture at the end of the probe and the reduction of the overheating of the extremity of the probe caused by the high concentration of light.

Finally, certain researches indicate the possibility of using the tunneling optical effect for carrying out photolithography. The arrangement used in this case would then be quite similar to the scanning near-field optical microscope. Resolutions of engraved lines of the order of 80 nm have been obtained.

12. Apertureless Microscopies

There exist several different techniques for detecting the information contained within the evanescent field of an object. In the two previous chapters we have described the techniques referred to as photon scanning tunneling microscopy and scanning near-field optical microscopy. We shall end this survey of local probe microscopies with the presentation of some other systems of local probe microscopy. All the systems that will be described here are presently in an earlier state than photon scanning tunneling microscopy or scanning near-field microscopy, but they nevertheless can be regarded as promising.

The common feature of the systems that we are to examine in this chapter is that they all are apertureless microscopes. This means that the essential parameter to be considered in them is the fact that the probe consists of an opaque element used for inducing a perturbation on the evanescent field present in the vicinity of the object.

12.1 Near-Field Optical Microscope Based on the Local Perturbation of a Diffraction Spot

As we have seen in the preceding chapter, a subwavelength aperture can be used as a local probe. This is actually the principle of the scanning near-field optical microscope. The scheme of the system developed by Boccara and Bachelot can be regarded as symmetrical to the system of a subwavelength aperture suggested by Synge [1928], provided however that the opaque probe used in it also has a subwavelength extent. This symmetry can be found in the other microscopies described in this chapter.

The principle of this microscope can be briefly stated as follows [Gleyzes 1995]: a thin metallic probe is brought within the near-field of a locally illuminated sample placed at the focus of an objective. The diffraction spot has an extent of the order of one micrometer. The presence of the probe inside the near-field of the sample induces a perturbation of the field in the vicinity of the sample. Besides, the probe has a periodic motion of low amplitude, i.e. of the order of 100 nm and vibrates normally to the surface of the sample.

During the oscillation of the probe, and as long as it remains located at a great distance from the surface, the perturbation induced on the ob-

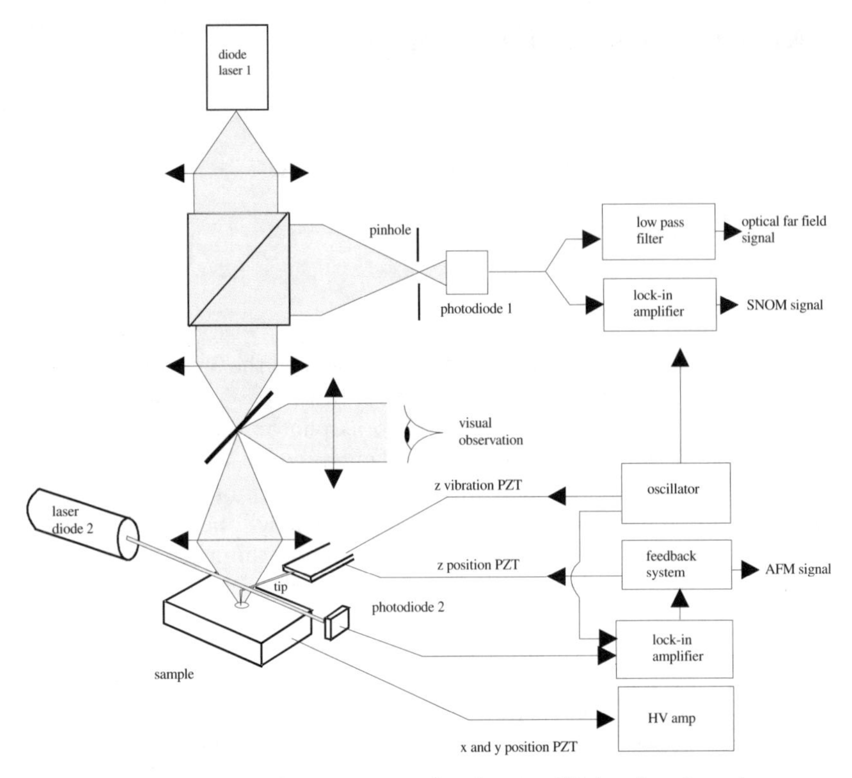

Fig. 12.1. Experimental arrangement for the metallic local-probe microscope

ject under study is very weak and is therefore not detectable in far-field. In contrast, as the probe moves closer to the sample, the perturbation induced on the electromagnetic field in the vicinity the sample becomes important enough to be measured in far-field.

As demonstrated by the authors, the signal reflected and detected in far-field has two components: the first term is related to the dimension of the diffraction spot, while the second term depends on the intrinsic characteristics of the sample placed near the probe [Bachelot 1995].

The system uses the objective of a microscope with a magnification factor of 100 and a numerical aperture equal to 0.85, while the source is a laser diode emitting at wavelength 0.670 μm. The objective is used to focus light onto the sample and to collect the reflected light, which is then detected by the photodiode D1. The probe is a tungsten cone with an extremity of diameter 0.1 μm, bent and attached to a piezoelectric ceramic tube exciting it at resonance frequency (5 kHz). The amplitude of the oscillation is measured by a second photodiode D2.

As can be seen in the images represented in Fig. 12.2, the resolution of this microscope extends beyond the Rayleigh limit. The sample examined here was a gold layer in which grooves had been cut and coated internally with silicon. The topographical information of the sample is obtained using atomic force microscopy detection. The optical parameters are simultaneously measured with the optical detection system. The images obtained with this system present quite a high degree of agreement with the geometrical characteristics of the sample.

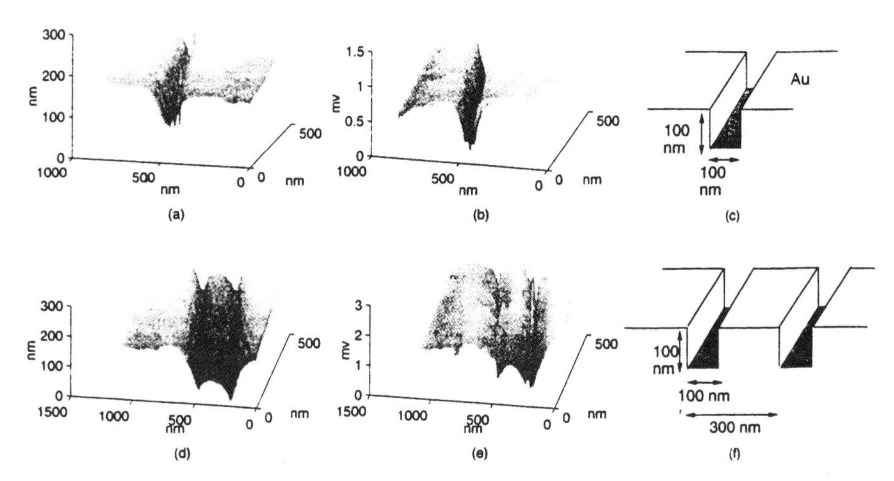

Fig. 12.2. (**a**), (**d**) AFM and (**b**), (**e**) optical images of different areas (**c**), (**f**) of the sample

The comparison between the profiles, represented in Fig. 12.3, of images of a non-homogeneous plane sample obtained with different types of microscopes is more instructive. As can be seen in this figure, while the profile of the image obtained with an atomic-force microscope is planar, the profile of the image obtained with the near-field optical microscope exhibits a step which reveals a change in the optical characteristics of the sample. The enhancement of the resolution is clearly visible when this image is compared with measurements performed with a traditional optical microscope.

The lateral resolution of this microscope is apparently of the order of about 15 nm for wavelengths in the visible spectrum. Similar observations carried out in the infrared ($\lambda = 10.6$ μm) have shown a resolution of the order of 20 nm [Lahrech 1996]. Therefore, the resolution obtained for infrared observations turns out to be of the same order as for observations in the visible spectrum. This suggests that the resolution is directly related to the size of the tip. The respective effects of the size of the spot and of polarization on resolution have been reported by Adam [1998a, 1998b]. The direct effect of wavelength on resolution still needs to be further inquired into.

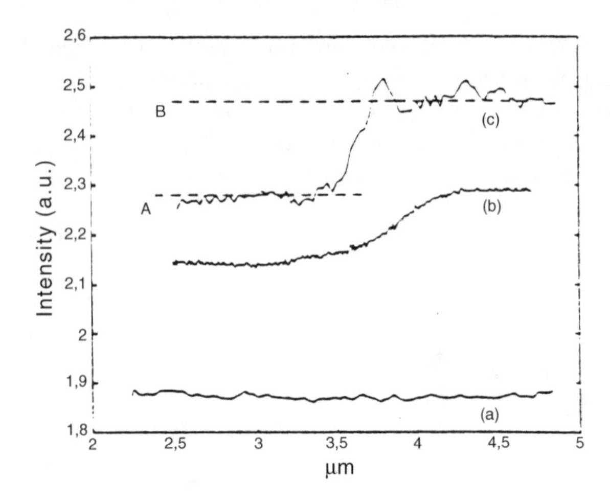

Fig. 12.3. Image profiles of a non-homogeneous planar sample (the left and right sides are respectively formed of GaAs and $Ga_xAl_{1-x}As$) obtained with three different microscopy systems (**a**) atomic-force microscope, (**b**) classical optical microscope, (**c**) near-field microscope

12.2 Scanning Interferometric Apertureless Microscope

Recently a different type of apertureless microscope has been devised by Wickramasinghe *et al.* The principle of this microscope rests on the interaction between the sample and a metallic tip and on the perturbation of the field present in the vicinity of the tip and of the sample [Wickramasinghe 1987, Zenhausern 1994, Wickramasinghe 1995]. The scheme of this microscope, referred to under the name of scanning interferometric apertureless microscope (SIAM) is represented in Fig. 12.4.

A laser beam focused onto the surface of the sample travels through the sample. A probe, whose position is controlled from an atomic-force microscope, is then brought near the sample. The probe oscillates vertically with a frequency of the order of 250 kHz and an amplitude comprised between 6 and 10 nm, while the spring constant is equal to 20 nN/m. Finally, the extent of the extremity of the tip can be estimated as 5 nm. As the feedback of the probe is achieved using an AFM in attractive mode, the position of the probe with respect to the surface can be controlled within about one or two nanometers.

The question of the physical significance of the optical signal detected by the system may be raised. The reflected signal is the sum of $E_{r'}$ and E_s, which respectively denote the signal reflected by the sample surface and the signal corresponding to the perturbation induced by the vibrating tip on the evanescent waves born in the vicinity of the sample. The reflected signal is then added to a reference signal E_r. Using such an interferometric arrangement, the phase and amplitude of the reflected signal can be determined.

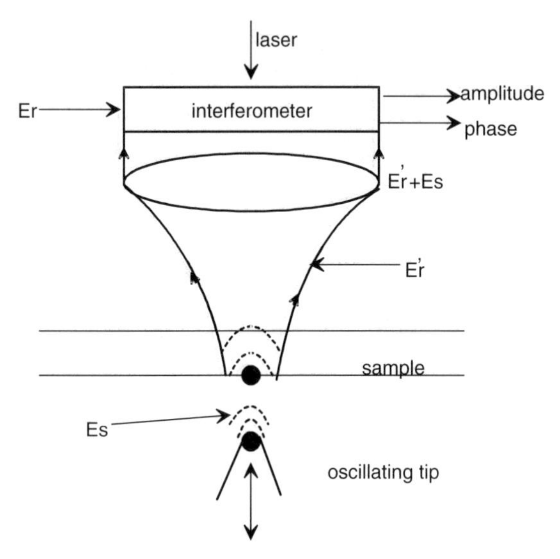

Fig. 12.4. Principle of the scanning interferometric apertureless microscope [Wickramasinghe 1987]

Figure 12.5 displays a comparison between the atomic-force microscope image and the image corresponding to the variations of the light intensity diffracted by the probe. Comparing the two images clearly shows the high sensitivity of the image obtained with the scanning interferometric apertureless microscope. We use here the term of sensitivity, rather than that of resolution, which supposes the existence of a reference sample with a well-defined topography. Even if a relative correlation between the AFM images and the optical images can be observed, the question of the physical origin of this correlation is presently left unanswered.

The question of the origin of the resolution in the scanning interferometric apertureless microscope is a controversial one. The variations of the optical signal have sometimes been analyzed as an effect of perturbations, for example of interferences, affecting the diffraction spot.

We here review the theoretical analysis developed by Wickramasinghe in the article that he published in *Science* in 1995 [Zenhausern 1995]. The fact that the images obtained with this microscope reflect a local perturbation is explained by Wickramasinghe in the following way: according to him, the interaction between the tip and the sample can be described in terms of an interaction between the dipole of the tip and the dipole of the surface of the sample placed in front of the tip [Zenhausern 1995].

The end of the tip and the section of the sample in front of the tip can be schematically represented by two spheres with radius a and polarizabilities α_{t} and α_{f} respectively, contained within the electric field E_{i}.

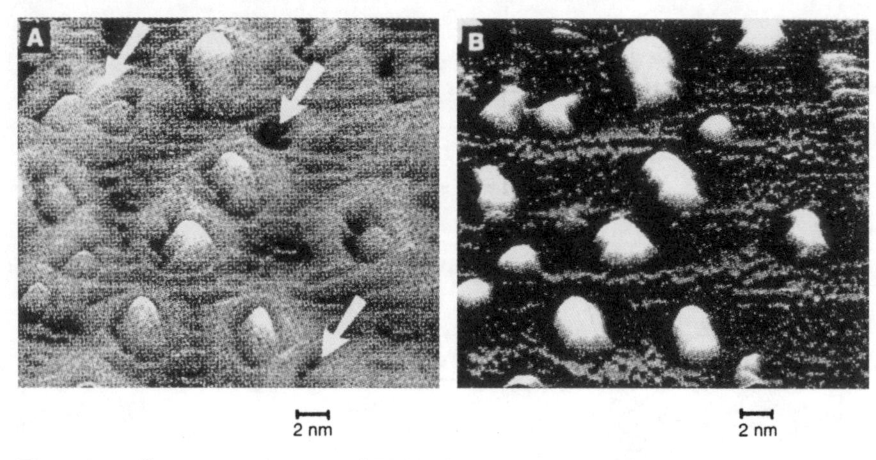

Fig. 12.5. Comparison between AFM and optical images of oil drops spread out on a mica plate (**a**) AFM image: the drops appear in general in the form of bumps, but, as an effect of electrostatic forces, some of them are imaged in inverted contrast, (**b**) simultaneous images of the variations of the optical signal: some details (indicated with an arrow) are reversed in comparison with the AFM image. The smaller details are of the order of a few nanometers

The coupling between the two spheres induces a variation of their polarizabilities. This is expressed by the equation

$$\Delta\alpha = \frac{2\alpha_{\mathrm{t}}\alpha_{\mathrm{f}}}{(r^2 + a^2)^{3/2}}. \tag{12.1}$$

In far-field, the amplitude of the wave scattered by a particle of extent smaller than 50 nm is given by the equations

$$E_{\mathrm{s}} = \frac{E_{\mathrm{i}}}{d} k^2 \alpha \tag{12.2}$$

and

$$\Delta E_{\mathrm{s}} = \frac{E_{\mathrm{i}}}{d} k^2 \Delta\alpha. \tag{12.3}$$

Equation (12.3) expresses the modulation of the field induced by the coupling of the two dipoles. The modulation of the field is represented in Fig. 12.6. From observation of these curves, it can be seen that to a large extent the modulation depends on the distance between the tip and the surface. A very high degree of agreement can be observed between the theoretical model and the experimental values which in this case were obtained with a probe placed within the near-field of a planar chromium surface.

This microscope is still in an early experimental state, and up to the present the limits of its performances have not been determined. Nevertheless, the authors believe that an atomic resolution can be reached with this system. If this were the case, the problem of the physical meaning of the optical signal

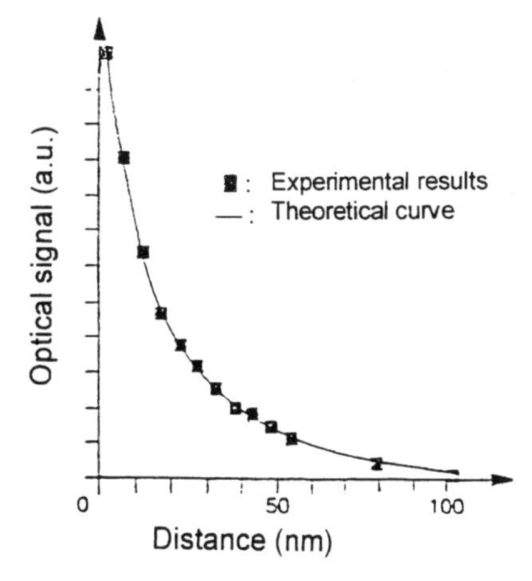

Fig. 12.6. The detected optical signal as a function of the distance between the probe and the surface of a chromium film deposited on a glass plate. The theoretical curve is in *solid lines* and the experimental values are represented by *squares* [Wickramasinghe 1987]

would have to be addressed. This in fact implies that further researches are to be conducted on this subject.

12.3 Tetrahedral Probe Microscope

A microscope based on a different principle has been developed by U. Fisher at the University of Münster [Koglin 1996, Ferber 1999]. The feedback of the position of the probe is achieved by using the electric tunneling effect. A tetrahedron is partially coated with metal, as illustrated in Fig. 12.7. The metal used is gold, and is deposited by evaporation in the form of a 50 nm layer.

The light emitted by a laser is focused onto the non-metallized side of the tetrahedron. Then, after it has propagated inside the layer, the light is emitted by the vertex at A and is detected by the photodiode, as represented in Fig. 12.8. The feedback of the system can be achieved by using the end of this probe in the same way as in a scanning tunneling microscope arrangement.

From profiles effected through different parts of the scanning near-field optical microscope image, the resolving power of this microscope has been estimated to be of the order of a few nanometers. Images of samples obtained with this technique have shown a resolution of the same magnitude (Fig. 12.9).

The origin of the resolving power of this microscope is still not entirely understood. It has been suggested that a surface plasmon is excited on the

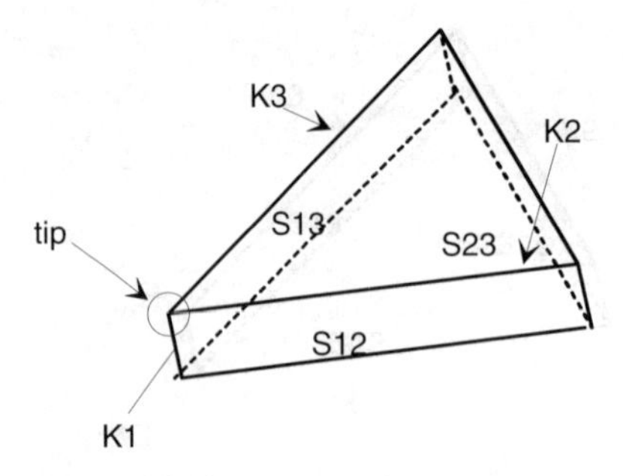

Fig. 12.7. Scheme of the tetrahedron used as a probe. The sides S_{12}, S_{13} and S_{23}, edges K_2 and K_3 and the vertex at A are covered with gold, while the edge K_1 remains free

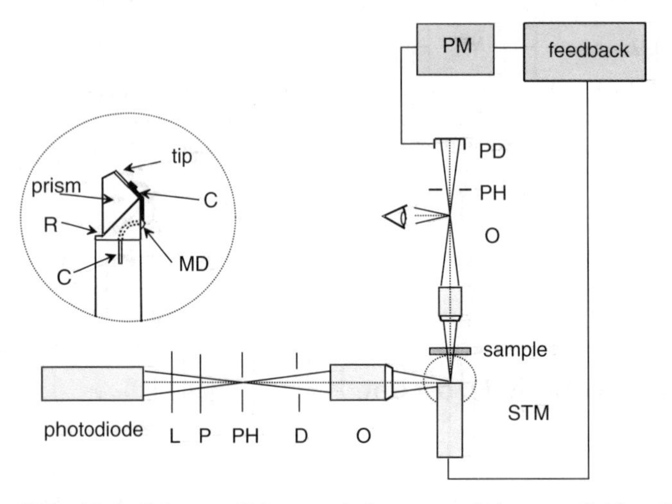

Fig. 12.8. Scheme of the coupled system of the near-field optical microscope and the scanning tunneling electron microscope; C: contact, R: mirror, D: diaphragm, PH: hole, CE: control electronics, PD: photodiode [Koglin 1996]

metallic side of the tip and perturbed by the sample. Up to now this hypothesis has not been completely confirmed.

As earlier, the question arises whether the optical images are related to the interaction between the tip and the surface or to a variation of the signal due to the spot as a whole. To answer this question, further researches are necessary.

12.4 Local Probe Microscope Derived from the PSTM

In this section, we examine another system of optical near-field microscopy. Here a thin metallic tip is used for perturbing the evanescent field generated in the vicinity of the structure to be studied, and for partially transforming

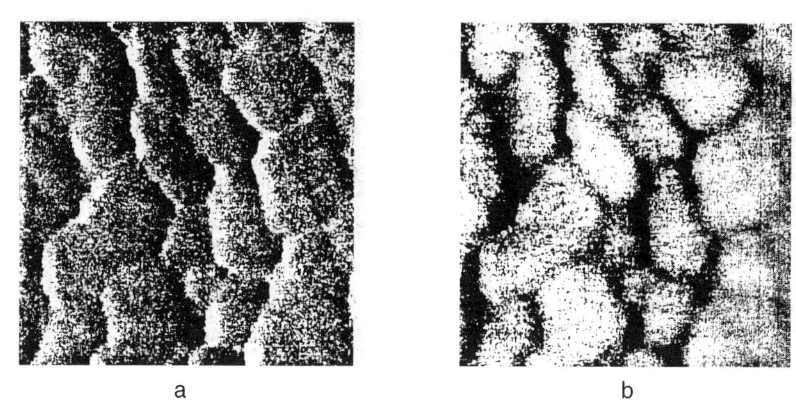

Fig. 12.9. Image of a thin silver film (thickness 0.5 nm), evaporated upon an ITO substrate. (**a**) Image of the variations of the optical intensity, (**b**) STM images obtained simultaneously [Koglin 1996]

it into a propagative field. Only a few studies on this subject have been published so far.

Various configurations can be used for generating an evanescent field near the sample. If the object is transparent, it is possible to use an arrangement derived from the PSTM. In this case, the sample is illuminated at total internal reflection. Figure 12.10 represents the microscope developed by Inouye [1994]. The metallic probe is coated with a thin aluminum oxide (Al_2O_3) layer. The source used here is a 10 mW laser diode emitting at wavelength 670 nm.

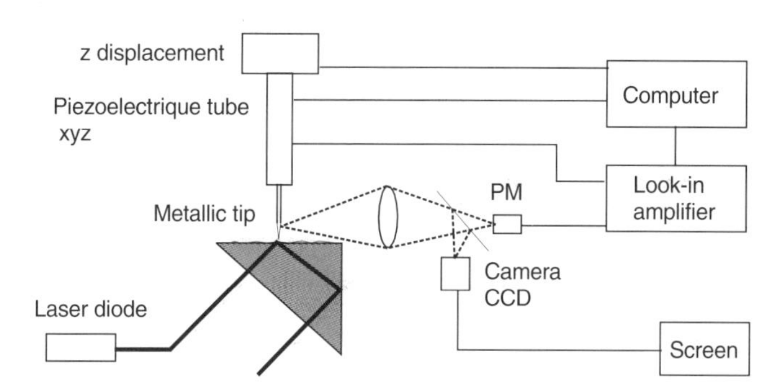

Fig. 12.10. Scheme of the arrangement used

The photograph in Fig. 12.11 shows the extremity of the tip used for scattering the evanescent field. The distance between the tip and the surface is within about ten nanometers.

The measurements of the dependence of the signal on the distance between the probe and the surface have revealed a decrease length of the order of

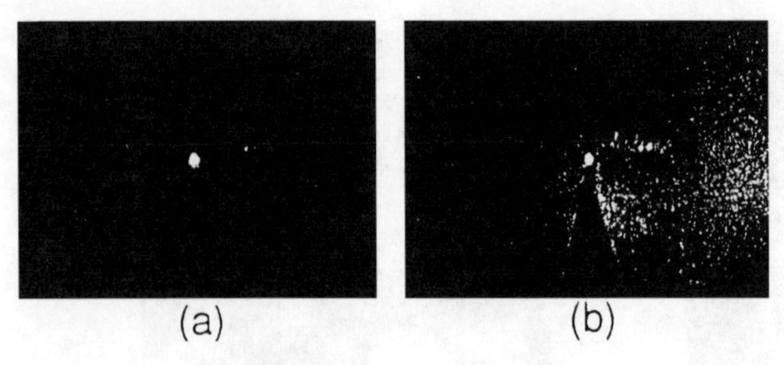

Fig. 12.11. Photograph of the probe in front of the sample (**a**) the probe is illuminated while the evanescent field is present, (**b**) the probe is illuminated from an external source [Inouye 1994, with permission of O.S.A.]

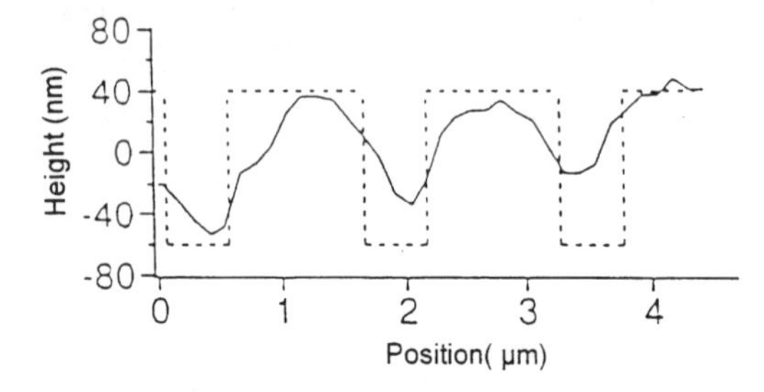

Fig. 12.12. Image profile of a compact disk. The curve in dashed lines corresponds to the theoretical geometry of the sample [Inouye 1994, with permission of O.S.A.]

200 nm. These results are consistent with the value of the penetration depth of the evanescent field for an angle of incidence equal to 47°. Figure 12.12 represents the profile of the image of a compact disk.

The microscopes described thus far in this chapter were all based on the idea of exploiting the far-field effects of a perturbation induced by a thin metallic tip on the near-field of the sample.

The microscope that we shall describe hereafter does not belong to this class of microscopes, and relies on an entirely different principle. Its distinctive feature is the use of radiation forces for displacing a probe inside a sample.

12.5 Radiation Pressure Scanning Microscope

The last type of microscope that will be described is based on the use of the radiation pressure exerted by light beams for displacing a particle, for example a microsphere [Kawata 1994]. The sphere acts either as a source or as a scattering center. The displacements of the sphere are achieved with a laser beam focused by the objective of a microscope with a 0.85 numerical

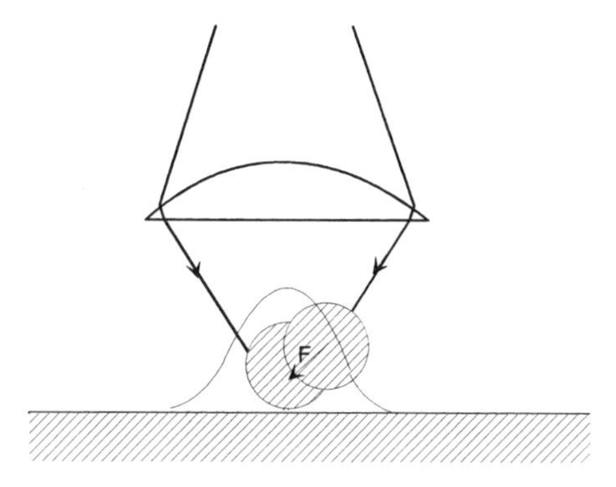

Fig. 12.13. Radiation pressure of a Gaussian light beam focused onto a particle [Kawata 1994, with permission of J.A.P.]

aperture and a magnification factor 40. The power of the Nd:YAG laser is equal to 150 mW (Fig. 12.13).

The interest raised by this microscope is linked with the fact that the particle can be displaced along three directions. In order to scatter the evanescent field generated in the vicinity of the surface of the sample illuminated at total internal reflection, a particle with a small extent was used (Fig. 12.14). The sample was illuminated with an argon laser at an angle greater than the critical angle. In order not to perturb the conditions of the trapping, the power of the illumination laser was 10 000 times weaker than that of the trapping laser.

The following images were obtained using a polystyrene probe with diameter 1 μm. The probe sphere was lying in a film of water, while the sample

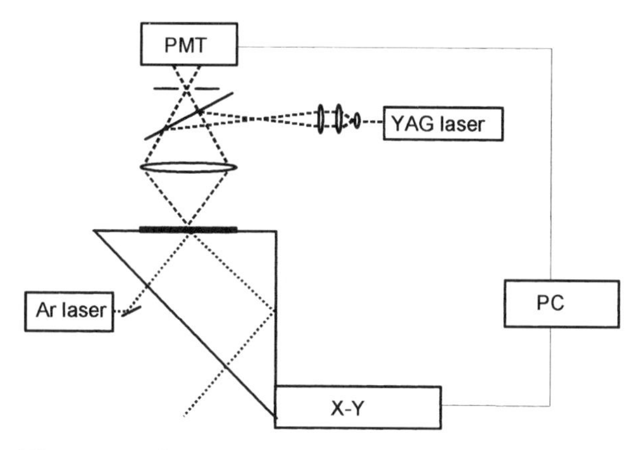

Fig. 12.14. System where the radiation pressure of a focused photon beam is used for displacing a particle [Kawata 1994, with permission of J.A.P.]

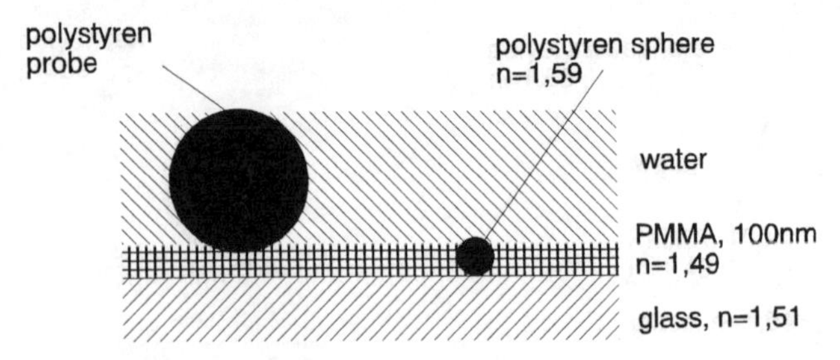

Fig. 12.15. Schematic representation of a polystyrene sphere in PMMA resist

consisted of polystyrene spheres with diameter 100 nm released inside a layer of PMMA resist (Fig. 12.15).

Images of fluorescent spheres with a resolution of the order of a few hundred nanometers have also been obtained (Fig. 12.16).

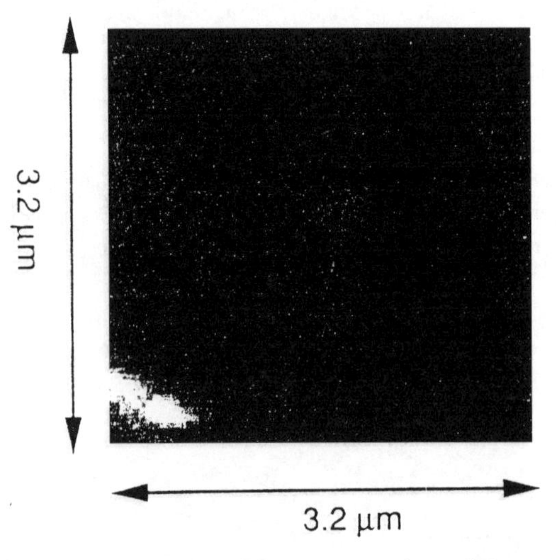

Fig. 12.16. Image of fluorescent spheres [Kawata 1994, with permission of J.A.P.]

This microscope presently has a low resolution. The displacement of the sphere is carried out at the surface of the sample. The authors expect that in the next phase of the development of this technique, the sphere will be displaced inside the sample itself. This would open perspectives of application of this microscope to the observation of biological phenomena.

12.6 Conclusion

This chapter has been devoted to different apertureless microscopy systems where a non-transparent tip is used for perturbing locally the near-field of the sample. Evidence of the possibility of reaching a very high resolution with microscopes using a metallic probe tip has been experimentally obtained recently.

The purely optical character of the information detected by these microscopes is still disputed by some authors. According to them, the images obtained with this microscope would be due at least partially to a correlation with the feedback system (STM or AFM). The results presented by Wickramasinghe, which have been compared with numerical results, demonstrate that the detected information is effectively related to the optical near-field of the sample. On the other hand, the question whether an atomic resolution can be reached with such a system remains presently unanswered.

The experiments on the possibility of using microspheres as scattering centers are still in their beginnings. These researches are directed at the observation of biological elements using the displacements of a probe inside the sample to be studied. Such a system, where the probe is displaced due to the radiation pressure exerted by the evanescent field, would have several applications in biology.

Conclusion of Part III

The presence within evanescent waves of information related to details smaller than half the wavelength was demonstrated at the beginning of the twentieth century with the researches carried out in ultramicroscopy. This information, which is not contained within propagative waves, could be obtained only through a perturbation of the evanescent waves. In dark-field microscopy, both the illumination of the sample and the detection had a delocalized character.

Local-probe microscopies have introduced a localized element, which can be either the source, as in the scanning near-field optical microscope, or the probe, as in the photon scanning tunneling microscope. This localized character has given these microscopies a lateral resolution which goes beyond the Rayleigh limit on resolution.

The development of local-probe microscopies has benefitted from advances in different fields. In particular, the progress in the technologies of piezoelectric microdisplacements, of the fabrication of microstructures and of data processing were essential in the development of these microscopies.

Local-probe microscopies are based on the exploitation of the information contained within the near-field of the sample. This has led to the possibility of observing objects smaller than $\lambda/60$. The two most representative local-probe microscopes are the photon scanning tunneling microscope (PSTM) and the scanning near-field optical microscope (SNOM), where a subwavelength source is used. More recently, the possibility of achieving a high resolution with apertureless microscopy systems has been demonstrated.

Further, the combined utilization of several microscopy systems (STM-PSTM, AFM-SNOM, etc.) allows the simultaneous detection of different types of information (topography of the sample, local absorption, etc.). This clearly is one of the advantages of these new microscopies. The use of near-field measurements also allows the characterization of integrated-optical devices, for example waveguides, from the detection of the evanescent part of the guided mode(s).

Local-probe microscopes will certainly evolve in the next few years, as regards both the construction of the probes and the analysis of the coupling between the sample and the probe.

The physics of the near-field is still a recent field of research, and notwithstanding dark-field microscopy and internal reflection spectroscopy, has been investigated only during the last ten years. The concept of near-field will certainly become more and more employed. The trend towards the miniaturization of integrated-optical components and the researches on microcavities are but two examples of the extension of the use of the concept of near-field.

References

Abramowitz H., Stegun I. (1964): "Handbook of Mathematical Functions", Nat. Bur. Stand. U.S., GPO, Washington
Adam P.M., Salomon L., de Fornel F., Goudonnet J.P. (1993): "Determination of the spatial extension of the surface-plasmon evanescent field of a silver film with a photon scanning tunneling microscope", Physical Review B, 48, 4, 2680
Adam P.M., Royer P., Laddada R., Bijeon J.L. (1998a): "Polarization contrast with an apertureless near-field optical microscope", Ultramicroscopy, 71, 1-4, 327
Adam P.M., Royer P., Laddada R., Bijeon J.L. (1998b): "Apertureless near-field optical microscopy: influence of the illumination conditions on the image contrast", Applied Optics, 37, 1814
Adams M.J. (1981): "An Introduction to Optical Waveguides", Wiley, New-York
Albiol F., Navas S., Andreas M. (1993): "Microwave experiments on electromagnetic waves and tunneling effect", Am. J. Phys., 61, 2,165
Ambrose W.P., Goodwin P.M., Martin J.C., Keller R.A. (1994): "Alteration of single molecule fluorescence lifetimes in near-field optical microscopy", Science, 265, 364
Aminoff C.G., Steane A.M., Bouyer P., Desbiolles P., Dalibard J., Cohen-Tannoudji C. (1993): "Cesium atoms bouncing in a stable gravitational cavity ", Phys. Rev. Lett., 71, 3083
Arnaud J. A. (1976): "Beam and Fiber Optics", Academic Press, New York
Arnaud J. A. (1977): "Applications des techniques hamiltoniennes aux fibres multimodes ", Ann. Télécom., 32, 135
Ash E. A., Nicholls G. (1972): "Super-resolution aperture scanning microscope", Nature, 237, 510
Aspect A., Dalibart J. (1994): "Le refroidissement des atomes par laser", La Recherche, 261, 25, 30
Aspect A., Kaiser R., Vansteenkiste N., Vignolo P., Westbrook C.I. (1995): "Nondestructive detection of atoms bouncing on an evanescent wave", Phys. Rev. A, 52, 6, 4704
Axelrod D., Hellen E.H, Fullbright R.M. (1992): "Total internal reflection fluorescence", in Topics in Fluorescence Spectroscopy, J.Lakowicz, Plenum Press, New york, 3, 289
Bachelot R., Gleyzes P., Boccara A.C. (1995): "Near-field optical microscope based on local perturbation of a diffraction spot", Optics Lett., 20, 18, 1924
Balykin V.I., Letokhov V.S., Ovchininnikov Yu. B., Sidorov A.I. (1988): "Quantum-state-selective mirror reflection of atoms by laser light", Phys. Rev. Lett., 60, 21, 2137
Bennamane B. (1985):"Couplage entre deux fibres optiques multimodes", DEA Report, Limoges, France
Berger V. (1996): "Cristaux photoniques", La Recherche, 290, 74
Bergh R. A., Koitler G., Shaw H.J. (1980): "Single-mode fibre optic directional coupler", Electronics Lett., 16, 7, 260

Bethe H.A. (1944): "Theory of diffraction by small holes", Physical Rev., 66, 7 and 8, 163

Betzig E., Lewis A., Harootunian A., Isaacson M, Kratschmer E. (1986): "Near-field scanning optical microscopy (NSOM), development and biophysical applications", Biophys. J., 49, pp.269

Betzig E., Harootunian A., Lewis A., Isaacson M. (1986): "Near-field diffraction by a slit: implications for superresolution microscopy", Appl. Optics, 25, 12

Betzig E. (1988) PhD, New York

Betzig E., Trautman J.K., Harris T.D., Wetner J.S., Kostelak R.L. (1991):"Breaking the diffraction barrier: optical microscopy on a nanometric scale", Science, 251, 1468

Betzig E., Trautman J.K., Weiner J.S., Harris T.D., Wolfe R. (1992): "Polarization contrast in near-field scanning optical microscopy", Applied Optics, 31, 22, 4563

Betzig E., Chichester R.J. (1993): "Single molecules observed by near-field scanning microscopy", Science, 262, 1422

Betzig E. (1993): "Principles and applications of near-field scanning optical microscopy (SNOM)" in Near Field Optics, D. Pohl and D. Courjon, NATO ASI Series, 7, Kluwer, Dordrecht

Bielefeldt H., Hecht B., Herminghaus S., Mlynek J., Marti O. (1993): "Direct measurement of the field enhancement caused by surface plasmons with scanning tunneling optical microscope", in Near Field Optics, D. Pohl and D. Courjon, NATO ASI Series, 242, Kluwer, Dordrecht

Binning G., Rohrer H. (1982):"Scanning tunneling microscope", Helv. Phys. Acta, 55, 726

Bluestein B.I., Walczak I.M., Chen S.-Y. (1990): "Fiber optic evanescent wave immunosensors for medical diagnostics", Tibbtech, 8, 161

Bolzhevolny S.I., Vohnsen B., SmolyaninovI.I., Zayats A. (1995) "Direct observation of surface polariton localization caused by surface roughness", Opt. Comm., 117, 417

Bolzhevolny S.I., SmolyaninovI.I., Zayats A. (1995) "Near-field microscopy of surface plasmon polaritons: localization and internal interface imaging", Phys. Rev. B, 51, 17916

Born M., Wolf E. (1959): "Principles of Optics", Pergamon Press, London

Bose J.C. (1897): "On the influence of the thickness of air-space on total reflection of electric radiation", Proc. R. Soc. London, 62, 300

Bourillot E., de Fornel F., Salomon L., Adam P., Goudonnet J.P. (1992): "Observation de structures guidantes en microscopie à effet tunnel optique", J. Optics, (Paris), 23, 57

Bourillot E., Hosain S.I., Goudonnet J.P., Voirin G., Kotrotsios G. (1995): "Determination of mode cutoff wavelengths and refractive index profile of planar optical waveguides with a photon scanning tunneling microscope", Phys. Rev. B, 51, 16, 11 225

Boutry G.A. (1946): "Optique instrumentale", Masson, Paris

Bouwkamp C.J. (1950): "On Bethe's theory of diffraction by small holes", Philips Recs. Dep., 5, 321

Bouwkamp C.J. (1950): "On the diffraction of electromagnetic waves by small circular disks and holes", Philips Recs. Dep., 5, 401

Bretenaker F., Le Floch A., Dutriaux L. (1992): "Direct measurement of the Goos–Hänchen effect in lasers", Phys. Rev. Lett., 68, 7, 931

Broeng J., Mogilevstev, Barkou S.E., Bjarklev A. (1999): "Photonic crystal fibers: a new class of optical waveguides", Optical Fiber Technology, 5, 3, 305

Brunner R., Doupovec J., Suchý F., Berta M. (1995): "Evanescent wave penetration depth in capillary optical fibres: challenges for liquids sensing", Acta Physica Slovaca, 45, 4, 491
Buckle P.E., Davies R.J., Kinning T., Yeung D., Edwards P.R., Pollard-Knight D., Lowe C.R. (1993) :"The resonnant mirror: A novel optical biosensor for direct sensing of biomolecular interactions. Part II: Applications", Biosens. Bioelectronics, 8, 355
Bures J., Lacroix S., Lapierre J. (1983):"Analyse d'un coupleur bidirectionnel à fibres optiques monomodes fusionnées", Appl. Optics, 22, 12, 1918-1922
Carenco A. (1983): "Le coupleur directif optique: un composant universel", L'Echo des Recherches, 112, 15
Carminati R., Greffet J.J. (1995): "Two dimensional numerical simulation of the photon scanning tunneling microscope. Concept of transfer function", Optics Comm., 116, 316
Carnal O., Mlynek J. (1992): "L'Optique Atomique", La Recherche, 247, 23, 1134
Carniglia C.K., Mandel L., Drexhage K.H. (1972): "Absorption and emission of evanescent photons", J.O.S.A., 62, 4, 479
Cella R., Mersali B., Barthe F., Licoppe C. (1995): "Etude modale de guides à puits quantiques dilués par la microscopie en champ proche", XV èmes J.N.O.G., Palaiseau
Centeno E., Guizal B., Felbacq D. (1999): "Multiplexing and demultiplexing with photonic crystal", J. Opt. A: Pure Appl. Opt., 1, L10
Chabrier G., de Fornel F., Bourillot E., Salomon L., Goudonnet J.P. (1994): "A dark field photon scanning tunneling microscope under incoherent light illumination", Optics Comm., 107, 347
Chance R.R., Prock A., Silbey R. (1974): "Lifetime of an excited molecule near a metal mirror: energy transfer in the Eu^{3+} silver system", J. Chem. Phys., 60, 5, 2184
Chance R.R., Prock A., Silbey R. (1975): "Decay of an emitting dipole between two parallel mirrors", J. Chem. Phys., 62, 3, 771
Chance R.R., Prock A., Silbey R. (1975): "Comments on the classical theory of energie transfer", J. Chem. Phys., 62, 6, 2245
Chaumet P., Rahmani A., de Fornel F., Dufour J.P. (1998): "Evanescent light scattering: the validity of the dipole approximation", Phys. Rev. B, 58, 4, 2310
Chiacchiera S.M., Kosower E.M. (1992): "Deuterium exchange on micrograms of proteins by attenuated total reflection Fourier transform infrared spectroscopy on silver halide fiber", Anal. Biochem., 201, 43
Choo A.G., Jackson H.E. (1994): "Near field measurement of optical channel waveguides and directional couplers", Appl. Phys. Lett., 65, 8, 947
Clauss G., Kévorkian A., Persegol D., Rehouma F. (1994): "Guides optiques de profondeur variable: une nouvelle approche des capteurs à onde évanescente", J.N.O.G., Besançon
Cline J.A., Isaacson M. (1995): "Probe-sample interactions in reflection near-field scanning optical microscopy" Appl. Phys., 34, 22, 4869
Costa de Beauregard O. (1965):"Translational inertial spin effect with photons" Phys.Rev.139, 1443-1446
Cotton A., Mouton H. (1906): "Les ultramicroscopes. Les objets ultramicroscopiques", Masson, Paris
Courjon D., Sarayeddine K., Spajer M. (1989): "Scanning tunneling microscopy", Opt. Comm., 71, 23
Courjon D., Bainier C., Spajer M.: "Imaging of submicron index variation by scanning optical tunneling", J. Vac. Sci. Technol. B 10, 6, 2436.

Cros D., Auxemery P., Jiao X.H., Jarry B., Guillon P. (1990): "Un combineur de puissance à résonateur diélectrique dans la bande 75–110 Ghz", Ann. Télécommun., 45, 5-6, 288

Cros D., Guillon P. (1990): "Whispering gallery dielectric resonator modes for W-band devices", IEEE Trans. M.T.T., 38, 11,1667

Dakin J., Cushaw B. (1988): "Optical Fibre Sensors: Principles and Components", Artech House Inc., USA

Davies R.J., Pollard-Knight D. (1993): "An optical biosensor system for molecular interaction studies", Am. Biotechnol. Lab., 11, 52

Dawson P., de Fornel F., Goudonnet J.P. (1994): "Imaging surface plasmon propagation and edge interaction using a photon scanning tunneling microscope", Phys. Rev. Lett., 72, 18, 2927

Dawson P., Puygranier B.A.F., Cao W. , de Fornel F.(1999): "The interaction of surface plasmon polaritons with a silver film edge", Journal of Microscopy, 194, 2-3, 578

Dazzi A., Rahmani A., Jullien P., de Fornel F., Moretti P. (1994): "Observation de structures implantées en microscopie à effet tunnel optique", XIV èmes J.N.O.G., Besançon

Deutschmann R., Ertmer W., Wallis H. (1993): "Reflection beam splitter for multilevel atoms", Phys. Rev. A, 48, 6, R4023

Digonnet M. and Shaw H. J.(1983): "Wavelength multiplexing in single-mode fiber couplers". Appl. Optics, 22, 3

Drexhage K.H., Kuhn H., Schäffer F.P. (1968): "Wide angle interference and multipole nature of fluorescence and phosphorescence of organic dyes", Ber. Bunsenges. Phys. Chem., 72, 329

Drexhage K.H. (1970): "Influence of a dielectric interface on fluorescence decay time", J. Lumin., 1, 2, 693

Drexhage K.H. (1974): "Interaction of light with monomolecular dye layers", Progress in Optics, Wolf ed., vol. XII, North-Holland, Amsterdam, 163

Dubreuil N., Knight J.C., Leventhal D.K., Sandoghdar V., Hare J., Lefèvre V. (1995): "Eroded monomode optical fiber for whispering-gallery mode excitation in fused-silica microspheres", Optics Lett., 20, 8813

Ducloy M. (1997): "High-resolution reflection spectroscopy" in Atomic, Molecular and Cluster Physics, Narosa Publishing House, New Delhi

Durrant A.V., Hill K.E., Hopkins S.A., Usadi E. (1995): "A field-momentum approach the semiclassical theory of light forces on atoms", Journal of Modern Optics, 42, 1, 131

Dyer G. L. (1993): "Photon tunnelling microscopy : new instrument for old theory", Materials and Design, 14, 2

Egami C., Takeda K., Isai M., Ogita M. (1996): "Evanescent-wave spectroscopic fiber optic pH sensor", Optics Comm., 122, 122

Etourneau K. (1996): "Etude de l'interface coeur/gaine des fibres optiques polymères par excitation de modes de galerie", Thesis, Limoges, France

Facq P., Fournet P., Arnaud J. (1980): "Observation of tubular modes in multimode graded-index optical fibres", Electronics Lett., 16, 648

Facq P., de Fornel F., Jean F. (1984): "Tunable single mode excitation in multimode fibres", Electronics Letters, 20, 15, 613-614

Favennec P.-N. (1993): "L'implantation Ionique, pour la Microélectronique et l'Optique", Collection Technique et Scientifique des Télécommunications, Masson, Paris

Feit M.D., Fleck J.A. (1978): "Light propagation in graded-index optical fibers", Applied Optics, 17, 3990

Feit M.D., Fleck J.A. (1980): "The computation of modes properties in optical waveguide by a propagation beam method", Applied Optics, 19, 1154

Ferber J., Fischer U.C., Hagedorn N., Fuchs H. (1999): "Internal reflection mode scanning near-field optical microscopy with the tetrahedral tip on metallic samples", Appl. Phys. A, 69, 581

Ferdinand P. (1992): "Les capteurs à fibres optiques", Technique et Documentation, Lavoisier, Paris.

Ferrell T.L., Goudonnet J.P., Reddick R.C., Sharp S.L. (1991): "Photon scanning tunneling microscope", J. Vac. Scien. Technol., B9, 2, 525

Fillard J.P. (1996): "Near Field and Nanoscopy", World Scientific Publishing, Singapore

Fisher U.Ch., Dürig U.T., Pohl D.W. (1988): "Near-field optical scanning microscopy in reflection", Appl. Phys. Lett., 52, 4, 249

Fogret E., Fonteneau G., Rimet R., Lucas J. (1994): "Guides d'ondes plans confinés sur du verre de fluorures", J. N. O. G., Besançon, 1994

de Fornel F., Arnaud J. A., Facq P. (1983).: "Microbending effects on monomode light propagation in multimode fibers", J. Opt.Soc. Am.A, 73, 5, 661

de Fornel F., Ragdale C., Mears R. (1984): "Analysis of single mode fused tapered fibre couplers", I.E.E. Proceedings, 131, 4, 221

de Fornel F., Varnham M., Payne D. (1984): "Finite cladding effects in highly birefringent fibre taper-polarisers", Electronics Lett., 20, 10, 398

de Fornel F., Goudonnet J.P., Salomon L., Lesniewska E. (1989): "An evanescent optical microscope", SPIE Proceedings, 1139, 77

de Fornel F., Salomon L., Adam P., Bourillot E., Goudonnet J.P., Nevière M. (1992): "Resolution of the photon scanning tunneling microscope: influence of physical parameters", Ultramicroscopy, 42-44, 422

de Fornel F., Lesniewska E., Salomon L., Goudonnet J.P. (1993): "First images obtained in the near infrared spectrum with the photon scanning tunneling microscope", Optics Comm., 102, 1

de Fornel F., Salomon L., Adam P., Bourillot E., Goudonnet J.P. (1994): "Effect of the coherence of the source on the images obtained with a photon scanning tunneling microscope", Optics Lett., 19, 14

Garcia–Parajo M., Cambril E., Chen Y. (1994): "Simultaneous scanning tunneling microscope and collection mode scanning near-field optical microscope using gold coated optical fiber probes", Appl. Phys. Lett., 65, 12, 1498

Garret C.G.B., Kaiser W., Blond W.L. (1961): "Stimulated emission into optical whispering modes of spheres", Phys. Rev., 124, 6, 1807

Georges A.J.T., French R.R., Glennie M.J. (1995): "Measurement of kinetic binding constants of a panel of anti-saporin antibodies using a resonant mirror biosensor", J. Immunol. Methods, 183, 51

Georges A.J.T. (1999): "Measurement of the kinetics of biomolecular interaction using the IAsys resonant mirror biosensor", Current Protocol in Immunology, 33, 18.5.1

Gleyzes P., Boccara A.C., Bachelot R. (1995): "Near field optical microscopy using a metallic vibrating tip", Ultramicroscopy, 57, 318

Golden J.P., Anderson G.P., Rabbany S.Y., Ligler F.S. (1994): "An evanescent wave biosensor - Part II : fluorescent signal acquisition from tapered fiber optic probes", IEEE Transactions on Biomedical Engineering, 41, 6

Goodman J.W. (1972): "Introduction à l'optique de Fourier et à l'holographie", Masson, Paris

Gould (1992): Technical Document, Gould Inc Glen Burie (USA)

de Gramont A. (1945): "Vers l'Infiniment Petit", Gallimard, Paris

Gréco P., Hemers H., Rimet R., Kherrat R., Jaffrezic N. (1994): "Capteurs chimiques en optique intégrée sur verre", J.N.O.G., Besançon
Greffet J. J., Santenac A., Carminati R. (1995): "Surface profile reconstruction using near-field date", Optics Comm., 116, 20
Greffet J. J., Carminati R. (1997): "Image formation in near-field optics", Progress in Surface Science, 56, 3, 133
Grignand P. (1980): "Solutions of the exact time-dependent microbending equations", Optics Com., 33, 262
Grimaldi F. (1665): "Physico-mathesis de Lumine, Coloribus et Iridi", Florence
Guerra J.M. (1988): "Surface Measurements and Characterization", Proc.SPIE, 1009, 254
Guerra J.M. (1990): "Photon tunneling microscopy", Applied Optics, 29,26, 3741
Guerra J.M., Srinivasarao M., Stein R. (1993): "Photon Tunneling Microscopy of Polymeric Surfaces", Science, 262, 1393
Guerra J.M. (1993b): "Photon tunneling microscopy of diamond-turned surfaces", Appl. Optics, 32, 1, 24
Guerra J. (1995): "Super resolution through illumination by diffraction-born evanescent waves", Appl. Phys. Lett., 66, 26, 3555
Hall E.E. (1902): "The penetration of totally reflected light into the rarer medium", Phys. Rev., 15, 73,
Harootunian A., Betzig E., Isaacson M., Lewis A. (1986): "Super-resolution fluorescence near-field scanning optical microscopy"', Appl. Phys. Lett., 49, 11, 674
Harrick N.J. (1960): "Study of physics and chemistry of surface from frustrated total internal reflections", Phys. Rev. Lett., 4, 5, 224
Harrick N.J., du Pré F.K. (1966a): "Effective thickness of bulk materials and of thin films for internal reflection spectroscopy" , Appl. Opt., 5, 1739
Harrick N.J. (1966b): "Vertical double pass multiple reflection element for internal reflection spectroscopy", Appl. Opt., 5, 1
Harrick N.J. (1966c): "The rosette- a unioint multiple internal reflection element, Appl. Opt., 5, 1236
Harrick N.J. and Loeb G.I. (1973): "Multiple internal reflection fluorescence spectroscopy", Anal. Chem., 45, 687
Harrick N.J. (1979): "Internal Reflection Spectroscopy", Harrick Scientific Corporation, New York
Hartog A. (1983): "A distributed temperature sensor based on liquid-core optical fibers", J. of Light Techn., LT-1,3
Hecht E. (1974): "Optics" Addison-Wesley Publishing Company, Adelphi University
Hecht B., Bielefeld H., Inouye Y., Pohl D.W. (1997): "Facts and artefacts in near-field optical microscopy", J. Appl. Phys., 81, 6, 2492
Helmers H., Greco P., Rustard R., Kherret R., Bouvier G., Benech P. (1995): "Performance of a compact, hybrid optical evanescent-wave sensor for chemical and biological applications", Appl. Optics, 35, 4, 676
Helmers H., Benech P., Rimet R. (1996): "Integrated optical components employing slab waveguide for sensor application", IEEE Photonics Lett., 8, 1
Higgins D.A., Vanden Bout D.A., Kerimo J., Barbara P.F. (1996) "Polarization modulation near-field scanning optical microscopy of mesostructured materials", J. Phys. Chem., 100, 13 794
Hirschfeld T. (1965): "Attenuated total reflection, applications in the UV–visible region", Can. Spectrosc., 10, 128
Hirschfeld T. (1977): "Real time measurement and observation of viruses in biological fluids", in R.L. Whitman (ed.), Multidisciplinary Microscopy, Proc. SPIE, 104,16
Huard S. (1997): "Polarization of Light", Wiley, London

van Hulst N., Segerink F.B., Bölger B. (1992): "High resolution imaging of dielectric surfaces with an evanescent field optical microscope", Opt. Comm. 87, 212
van Hulst N., Segerink F.B., Achten F., Bölger B. (1992): "Evanescent-field optical microscopy: effect of polarization, tip shape and radiative waves", Ultramicroscopy, 42-44, 416
van Hulst N., Moers M.H.P., Noordman O.F.J., Tack R.G., Segerink F.B., Bölger B. (1993): "Near-field optical microscope using a silicon-nitride probe", Appl. Phys. Lett., 62, 461
Hunsperger R.G. (1985): "Integrated Optics: Theory and Technology", Springer-Verlag, Berlin
Imbert C. (1972): "Calculation and experimental proof of the transverse shift induced by total internal reflection of a circularly polarized light beam", Phys. rev. D, 5, 4, 787
Imbert C., Levy Y. (1975): "Déplacement d'un faisceau lumineux par reflexion totale: filtrage des états de polarisation et amplification", Nouv. Rev. d'Optique, 6, 5, 285
Inouye Y., Kawara S. (1994): "Near-field scanning optical microscope with a metallic tip", Optics Lett., 19, 3
Ito H., Sakaki K., Nakata T., Jhe W., Ohtsu M.(1995): "Optical potential for atom guidance in a cylindrical-core hollow fiber", Optics Comm., 115, 57
Ito H., Sakaki K., Ohtsu M., Jhe W. (1997): "Evanescent-light guiding of atoms through hollow optical fiber for optically controlled atomic deposition", Appl. Phys. Lett., 70, 19, 2496
Jackson H.E., Boyd J.T. (1991): "Raman and photon scanning tunnelling microscopy of optical waveguides", Optical and Quantum Electronics, 23, S901
Jahncke C.L., Paesler M.A., Hallen H.D. (1995): "Raman imaging with near-field scanning optical microscopy", Appl. Phys. Lett., 67, 17, 2483
Jalocha A., van Hulst N. (1995): "Dielectric and fluorescent samples imaged by scanning near field optical microscopy in reflection", Optics Comm., 119, 17
Jhe W., Othsu M., Hori H., Friberg S.R. (1994): "Atomic waveguide using evanescent waves near optical fiber", Jpn. J. Appl. Phys., 33, L1680
Jiang S., Tomita N., Ohsava H., Othsu M. (1991): "A photon scanning tunneling microscope using an AlGaAs Laser", Jpn. J. Appl. Phys., 30, 9A, 2107
Joannopoulos J.D., Meade J.D., Winn J. (1995): "Photonic Crystal", Princeton University Press, Princeton
Joindot I., Joindot M. (1996): "Les Télécommunications par Fibres Optiques ", Collection Technique et Scientifique du CNET, Dunod, Paris
Kaiser R., Lévy Y., Vansteenkiste N., Aspect A., Seifert W., Leipold D. and Mlynek J. (1994): "Resonant enhancement of evanescent waves with a thin dielectric waveguide", Optics Comm., 104, 234
Kallas N.C., Kallas R.H. (1997): "Déplacements Goos–Hänchen et Imbert à la réflexion totale sur l'interface séparant deux milieux relativement mobiles normalement au plan d'incidence", Can. J. Phys., 75, 677
Katz M., Katzir A., Schnitzer I., Bornstein A. (1994): "Quantitative evaluation of chalcogenide glass fiber evanescent wave spectroscopy", Appl. Opt., 33, 5888
Kawasaki B.S., Hill K.O., Lamont R.G. (1981): "Biconical-taper single mode fiber coupler", Optics Lett., 6, 327
Kawata S., Inouye Y., Sugiura T. (1994): "Near-field scanning optical microscope with a laser trapped probe", Jpn. J. Appl. Phys., 33, L1725
Kawata S., Tani T. (1996): "Optically driven Mie particules in an evanescent field along a channeled waveguide", Optics Lett., 21, 21, 1768
O'Keefe J.A. (1956): "Resolving power of visible light", J. Opt. Soc Am. 46, 359

Knight J.C., Dubreuil N., Sandoghdar V., Lefèvre-Seguin V., Raimond J.M., Haroche S. (1995): "Mapping whispering-gallery modes in microspheres with a near-field probe", Optics Lett., 20, 14, 1515

Knight J.C., Birks T.A., St. J. Russell P., Atkin D.M. (1996): "All silica single-mode optical fiber with photonic crystal cladding", Optics Lett., 21,1547

Koglin J., Fisher U.C., Fuchs H. (1996): "SNOM and simultaneous STM with the tetrahedral tip at resolution of 10nm", in "Optics at the nanometer scale. Imaging and storing with photonics fields" , M. Nieto-Vesperinas and N. Garcia, NATO ASI Series E: Applied Sciences, 319, Kluwer, Dordrecht

Krausch G., Wegscheider S., Kirsch A., Bielefeldt H., Meiners J.C., Mlynek J. (1995): "Near field microscopy and lithography with uncoated fiber tips: a comparison", Optics Comm., 119, 283

Krenn J.R., Gotschy W., Somitsch D., Leitner A., Aussenegg F.R. (1995): "Investigation of localized surface plasmons with photon scanning tunneling microscope", Appl. Phys. A, 6, 541

Kretschmann E., Raether H. (1968): "Plasma resonance emission in solids", Z. Naturforsch., A23, 2135

Khun H. (1970): "Classical aspects of energy transfer in molecular systems", J. Chem. Phys., 53, 1, 101 in "Fluorescence Spectroscopy", Plenum Press, New York

Lakowics J.R. (1992) "Topics in Fluorescence Spectroscopy", Plenum Press, New York

Lahrech A., Bachelot R., Gleyzes P., Boccara A.C. (1996): "Infra-red reflection mode near-field microscope using an apertureless probe with a resolution of $\lambda/600$", Optics Lett., 21, 17, 1315

Lamouche G., Lavallard P., Gacoin T. (1999): "Optical properties of dye molecules as a function of the surrounding dielectric medium", Phys. Rev. A, 59, 6, 4668

Landragin A., Courtois J.Y., Labeyrie G., Henkel C., Vansteenkiste N., Westbrook C.I., Aspect A. (1996): "Measurement of the van der Walls forces in an atomic mirror", Phys. Rev. Lett., 77, 8, 1464

Landragin A., Labeyrie G., Henkel C., Kaiser R., Vansteenkiste N., Westbrook C.I., Aspect A. (1996): "Specular versus diffuse reflection of atoms from an evanescent-wave mirror", Optics Lett., 21, 19, 1591

Lefèvre H., Simonpiétri P., Graindorge P. (1984): "High selectivity polarization splitting fiber coupler", S.P.I.E. Proceedings, 988, 63

Lefèvre-Seguin V., Knight J.C., Sandoghdar V., Weiss D.S., Hare J., Raimond J.M., Haroche S. (1995): "Very high whispering-gallery modes in silica microspheres for cavity-QED experiments", Advanced Series in Applied Phys. 3, R.K. Chang and A.J. Campillo Eds.

Leviatan Y. (1986): "Study of near-zone fields of a small aperture", J. Appl. Phys., 60, 5

Leviatan Y. (1988): "Electromagnetic coupling between two half-space regions separated by two perforated parallel conducting screens", IEEE Transaction on Microwave Theory and Techniques, 36, 1

Lévy Y., Zhang Y., Loulergue J.C. (1985): "Optical field enhancement comparison between long range surface plasma-waves indiced by resonant cavity", Optics Comm., 56, 155

Lewis A., Isaacson M., Harootunian A., Murray A. (1984): "Development of a 500A spatial resolution light microscope. I. Light is efficiently transmitted through $\lambda/16$ diameter aperture", Ultramicroscopy, 13, 227

Liao F., Boyd J. (1980): "Single-mode fiber coupler", Appl. Opt., 20, 2731

Lomer M. (1992): "Contribution à l'étude de modes de galerie optiques. Fabrication de guides optiques non linéaires par diffusion de plomb", Thesis, Limoges

Lotsch H.K.V. (1970a): "Beam displacement at total reflection: the Goos–Hänchen effect, I", Optik, 32, 116
Lotsch H.K.V. (1970b): "Beam displacement at total reflection: the Goos–Hänchen effect, II", Optik, 32, 189
Lotsch H.K.V. (1971a): "Beam displacement at total reflection: the Goos–Hänchen effect, III", Optik, 32, 319
Lotsch H.K.V. (1971b): "Beam displacement at total reflection: the Goos–Hänchen effect, IV", Optik, 32, 553
Lukosz W., Kunz R.E. (1977): "Light emission by magnetic and electric dipoles close to a plane interface. I. Total radiated power", J.O.S.A., 67, 12
Maitte B. (1981): "La lumière", Editions du Seuil, Paris
Marchman H.M., Novembre A.E. (1995): "Near field latent imaging with the photon tunneling microscope", Appl. Phys. Lett., 66, 24, 3269
Marcuse D. (1972): "Light Transmission Optics", Van Nostrand, New York
Marcuse D. (1973): "Coupled mode theory of round optical fiber", B.S.T.J., 52, 6, 817
Marcuse D. (1989): "Investigation of coupling between a single-mode fiber and an infinite slab"., J.Lightwave Technol., 7 122
Maréchal A., Françon M. (1960): "Diffraction, structure des images, influence de la cohérence de la lumière", Revue d'Optique Théorique et Instrumentale, Paris
Marti O., Bielefeldt H., Herminghaus S., Leiderer P., Mlynek J. (1993): "Near-field optical measurement of the surface plasmon field", Optics Comm., vol, 225
Massey G.A. (1983): "Microscopy and pattern generation with scanned evanescent waves", Appl. Optics, 23, 5, 658
Meixner A., Bopp M., Tarrach G. (1994): "Direct measurement of standing evanescent waves with a photon-scanning tunneling microscope", Applied Optics, 33, 34, 7995
Miller S.E. (1969): "Integrated optics: an introduction", Bell Syst. Techn. J., 48, 2059
Mirabella F.M. (1985): "Internal reflection spectroscopy", Applied Spectroscopy Reviews, 21, 1 and 2, 45
Moerner W.E., Orrit M. (1999): "Illuminating single molecules in condensed matter", Science, 283, 5408, 1670
Moers M., van Hulst N., Ruiter A., Bölger B. (1995): "Optical contrast in near field techniques", Ultramicroscopy, 57, 298
Monerie M. (1982): "Propagation in doubly clad single-mode fibers", IEEE Quantum Electron., QE-18, 535
Moyer P.J., Jahncke C.L., Paesler M.A., Reddick R.C., Warmack R.J. (1990): "Spectroscopy in the evanescent field with an analytical photon scanning tunneling microscope", Physics letters A, 145, 6-7, 343
Mukaiyama T., Takeda K., Miyazaki H., Jimba Y., Kuwata–Gonokami M. (1999): "Tight-binding photonic molecule modes of resonant bispheres", Phys. Rev. Lett., 82, 23, 4623
Nachet C.S. (1847): "Appareil destiné à permettre l'éclairage par une lumière oblique des objets que l'on observe au microscope", Compte Rendu de l'Académie des Sciences, XXIV, 976
Nachet (1979): "Catalogue de Fonds de 1854 à 1910", A. Brieux, Paris
Nevière M., Vincent P. (1993): "Diffraction gratings as components for photonscanning tunneling microscope image interpretation",in Near Field Optics, D. Pohl and D. Courjon, NATO ASI Series, 242, Kluwer, Dordrecht
Newton I. (1952): "Opticks", Dover, New York
Nha H., Jhe W. (1996): "Cavity quantum electrodynamics between parallel dielectric surfaces", Phys. Rev. A, 54, 4

Novotny L., Pohl D., Regli R. (1994): "Light propagation through nanometer sized structures: the two dimensional aperture scanning near field microscope", J. Opt. Soc. Am. A, 11, 1768

Novotny L., Hecht B, Pohl D.W. (1997): "Interference of locally excited surface plasmons", J. Appl. Phys., 81, 1798

Othsu M. (1998): "Near-Field Nano/Atom Optics and Technology", Springer, Tokyo

Oria M., Bloch D., Fichet M., Ducloy M. (1989): "Spectroscopie d'atomes au voisinage d'une interface" Bulletin de la SFP, 73, 18

Ostrowsky D.B. (1979): "Fiber and Integrated Optics", Plenum Press, New York

Pagnoux D. (1987): "Etude théorique et réalisation de coupleurs pour fibres optiques: méthode de fusion et étirage. Coupleurs à microoptique", Thesis, Limoges,

Painter O., Lee R.K., Schere A., Yariv A., O'Brien J.D., Dapkus P.D., Kim I. (1999): "Two-dimensional photonic band-gap defect mode laser", Science, 284, 1819

Panajotov K.P. (1994): "Polarization properties of a fiber-to-asymmetric planar waveguide coupler", J. Lightwave Technol., 12, 6, 983

Pangaribuan T., Yamada K., Jiang S., Ohsawa H., Ohtsu M. (1992): "Reproducible fabrication technique of nanometric tip diameter fiber probe for photon scanning tunneling microscope", Jpn. J. Appl. Phys., 9A, 1302

Parriaux O., Dierauer P. (1994): "Normalized expressions for the optical sensitivity of evanescent wave sensors", Optics Lett., 19, 7, 508

Parriaux O., Dierauer P. (1994): "Normalizd expressions for the optical sensitivity of evanescent wave sensors erratum", Optics Lett., 19, 20, 1665

Photonetics (1997): "Polarization splitting fiber couplers", Application note

Photonetics (1997): "Advanced fiber couplers", Application note,

Piednoir A., Creuzet F., Licoppe C., de Fornel F. (1992): "First specifications of a PSTM working in the infrared", Workshop on Optical Near Field, Arc et Senans, France

Piednoir A., Creuzet F., Licoppe C., Ortéga J.M. (1993): "Locally resolved infrared spectroscopy", Second International Conference on Near Field Optics, Raleigh, U.S.A

Pohl D.W. (1982): "Optical near-field microscope", European Patent Application No 0112401, 27/12/1982- US Patent 4, 604,520, 20/12/1983

Pohl D.W., Denk W., Dürig U.T. (1988): "Optical stethoscopy: image recording with resolution $\lambda/20$", Appl. Phys. Lett., 44, 651

Pohl D.W., Denk W., Dürig U.T. (1985): "Optical stethoscopy: image recording with resolution $\lambda/20$", in Micron and Submicron Integrated Circuit Metrology, Proc. SOIE, 565, 56

Pohl D.W. (1991): "Scanning near-field optical microscopy", in Advances in Optical and Electron Microscopy, T. Mulvey and C.J.R. Sheppard, eds., Academic Press, New York, 12, 243

Quartel J.C., Dainty J.C. (1999): "An infrared photon scanning tunneling microscope for investigations of near-field imaging", J. Opt. A: Pure Appl. Opt., 1, 517

Quincke G. (1866a): "Optische Experimental-Untersuchungen. I Über das Eindringen des total reflektierten Lichtes in das dünnere Medium", Ann. Phys.; 127, 1

Quincke G. (1866b): "Optische Experimental-Untersuchungen. II. Über die elliptische Plarisation des bei totaler Reflexion eingedrungen oder zurück-geworfenen Lichtes", Ann. Phys.; 127, 199

Ragdale C., Payne D., de Fornel F., Mears R. (1983): "Single-mode fused biconical taper fiber couplers", O.F.S., London

Rahmani A., de Fornel F. (1996): "Near-field optical probing of fluorescent microspheres using a photon scanning tunneling microscope", Optics Commun., 131, 253

Rather H. (1976): "Surface Plasmons in Smooth and Rough Surfaces and on Gratings", vol. 111 of Springer Tracts in Modern Physics, Spinger Verlag, Berlin

Reddick R.C., Warmack R.J., Ferrell T.L. (1989): "New form of scanning optical microscopy", Phys. Rev. B, 39, 767

Roberts A. (1987): "Electromagnetic theory of diffraction by a circular aperture in a thick, perfectly conducting screen", J.O.S.A. A, 4, 10

Roberts A. (1989): "Near-zone fields behind circular apertures in thick, perfectly conducting screens", J. Appl. Phys., 65, 5

Rogers J.K., Seiferth F., Vaez-Iravani M. (1995) "Near field probe microscopy of porous silicon: observation of spectral shifts in photoluminescence of small particles", Appl. Phys. Lett., 66, 3260

Roosen G., Imbert C. (1972): "Etude de la pression de radiation de l'onde évanescente par absorption de lumière résonnante", Optics Comm. 18, 3, 247

Rugar D., Hansma P.K. (1990): "Atomic force microscopy", Physics Today, 23

Saiki T., Mononobe S., Ohtsu M. (1995): "High resolution fluorescence imaging with enhanced sensitivity due to short range electromagnetic interaction in photon STM", Near Field optics-3 EOS Topical Meeting, 8, 127-128

Saleh B.E.A., Teich M.C. (1991): "Fundamentals of Photonics", John Wiley and Sons, New York

Salomon L. (1991): "Théorie et mise au point d'un microscope à effet tunnel photonique ", Thesis, Dijon

Salomon L., de Fornel F., Goudonnet J.P. (1991): "Sample-tip coupling efficiencies of the photon scanning tunneling microscope", J. Opt. Soc. Am. A, 8, 2009

Salomon L. (1992).: "Etude théorique et expérimentale de la profondeur de pénétration du champ frustré dans un microscope à effet tunnel optique", J. of Optics (Paris), 23, 49

Salomon L., de Fornel F. (1997): "Applications of the near-field optics to the caracterization of optoelectronic components" Annales des Télécommun., 52, 11-12,594

Salomon L., de Fornel F., Adam P.M. (1999): "Analysis of the near field and the far field diffracted by a metallized grating at and beyond the plasmon resonance", J. Opt. Soc.Am; A, 16, 11, 2695

Sandoghdar V., Treussart F., Hare J., Lefèvre-Seguin V., Raimond J.M., Haroche S. (1996): "Very low threshold whispering-tgallery-mode microsphere laser", Phys. Rev. A, 54, 3, R1777

Santenac A., Greffet J.J. (1995): "Study of the features of PSTM images by means of a perturbative approach", Ultramicroscopy, 57, 246

Savage C.M., Markstein S., Zoller P. (1993): "Atomic waveguides and cavities from hollow optical fibers " in "Fondamentals of quantum optics III", Ehlosky, Springer Verlag, 60

Savage C.M., Gordon D., Ralph T.C. (1995): "Numerical modeling of evanescent-wave atom optics", Phys. Rev. A, 52, 6, 4741

Schaafsma D.T., Mossadegh R., Sanghera J.S., Aggarwal I.D., Luce M., Generosi R., Perfetti P., Cricenti A., Gilligan J.M., Tolk N.H. (1999): "Fabrication of single-mode chalcogenide fiber probes for scanning near-field infrared optical microscopy", Opt. Eng., 38, 8, 1381

Schaffner M., Toraldo di Francia G. (1949): "Microonde evanescenti generate per diffrazione", Nuovo Cimento, 6, 2, 125

Sélényi P. (1913): "Sur l'existence et l'observation des ondes lumineuses sphériques inhomogènes", Comp. Rend., 157, 1408

Sharp S.L., Warmack R.J., Goudonnet J.P., Lee I., Ferrell T.L. (1993): "Spectroscopy and imaging using the Photon Scanning Tunneling Microscope", Acc. Chem. Res. 26, 377

Sheem S.K., Gaillorenzi T. (1979): “Single-mode fiber-optical power divider”, Opt. Lett., 4, 29
Sheem S.K., Taylor H.F., Moeller R.P., Burns W.K. (1981): “Propagation characteristics of single-mode evanescent field couplers”, Appl. Optics, 20, 6, 1056
Seifert W., Kaiser R., Aspect A., Mlynek J. (1994): “Reflection of atoms from a dielectric waveguide”, Optics Comm., 111, 566
Sigel M., Pfau T., Adams C.S., Kurtsiefer C., Seifert W., Heine C., Mlynek J., Kaiser R., Aspect A. (1993): “Optical elements for atoms : a beamsplitter and a mirror ” in “Fundamentals of Quantum Optics III ”, Ehlosky, Springer Verlag, 3
Simhi R., Gotshal Y., Bunimovich D., Sela B.-A., Katzir A. (1996): “Fiber-optic evanescent wave spectroscopy for fast multicomponent analysis of human blood”, Appl. Optics, 35, 19, 3421
Snyder A.W. (1969): “Asymptotic expressions for eigenfunctions and eigenvalues of a dielectric or optical waveguide”, IEEE Trans. Microwaves Theory Tech. MTT-17, 12, 1130
Snyder A.W. (1972): “Coupled mode theory”, J. Opt. Soc. Am., 62, 1267
Snyder A.W., Love J.D. (1983): “Optical Waveguide Theory”, Chapman and Hall, London
Stratton J.A. (1941): “Electromagnetic theory”, Mc Graw-Hill Inc., New York
Synge E. (1928): “Suggested method for extending microscopic resolution into the ultra-microscopic region”, Phil. Mag., 6, 357
Szczepanek P.S., Berthold J.W. (1978): “Side launch excitation of selected modes in graded-index optical fibers”, Appl. Optics, 17, 20, 3245
Tayeb G., Maystre D. (1997): “Rigorous theoretical study of finite-size two dimensional photonic crystals doped with microcavities”, J. Opt. Soc. Am. A, 14, 3323
Taylor A.M., A.M. Glover (1933a): “Studies in refractive index I and II”, J. Opt. Soc. Am., 23, 206
Taylor A.M., D.A. Durfee (1933b): “Studies in refractive index III”, J. Opt. Soc. Am., 23, 263
Taylor A.M., A.King (1933c): “Studies in refractive index IV”, J. Opt. Soc. Am., 23, 308
Temple P. A. (1981): “Total internal reflection microscopy”, Applied Optics, 20, 2656
Toraldo di Francia G. (1942a): “Alcuni fenomeni di diffrazione trattati mediante il principio dell’interferenza inversa”, Ottica, 7, 117
Toraldo di Francia G. (1942b): “Le onde evanescenti nella diffrazione”, Ottica, 7, 197
Toraldo di Francia G. (1952): “Super-gain antenna and optical resolving power”, Supplement al vol. IX del Nuovo Cimento, 3
Trautman J.K., Macklin J.J., Brus L.E., Betzig E. (1994): “Near field spectroscopy of single molecules at room temperature”, Nature , 369, 40
Treussart F., Hare J., Collot L., Lefèvre V., Weiss D. S., Sandoghdar V., Raimond J.M., S. Haroche (1994): “Quantized atom-field force at the surface of a microsphere”, Optics Lett., 19, 20, 1651
Tsai D.P., Jackson H.E., Reddick R.C., Sharp S.H., Warmack R.J. (1990): “Photon scanning tunneling microscope study of optical waveguides”, Appl. Phys. Lett., 56, 16,
Tsai D.P., Kovacs J., Wang Z., Moskovits M., Shalvev V.M., Suh J.S., Botet R. (1994): “Photon scanning tunneling microscopy images of optical excitations of fractal metal cilloid clusters”, Phys. Rev. Lett., 72, 4149
Udd E. (1995): “An overview of fiber-optic sensors”, Rev. Sci. Instrum., 66, 8,4015

Umbach A., Trommer D., Siefke A., Unterbrsch G. (1995): "50 Ghz operation of Waveguide Integrated Photodiode at 1.55μm", Proc. 21st. Eur. Conf. on Opt. Comm. (ECOC'95), Brussels

Vaez-Iravani M., Toledo-Crow R. (1992): "Amplitude, phase contrast, and polarization imaging in near-field scanning optical microscopy", Workshop on Optical Near Field, Arc et Senans, France

Vansteenkiste N., Aspect A., Courtois J.-Y., Desbiolles P., Jurczak C., Kaiser R., Landragin A., Senstock K., Vognolo P., Von Zanthier J., Westbrook C. (1994): "Ondes Évanescentes Exaltées pour Miroir à Atomes, Détection Non Destructive de Rebonds d'Atomes", J.N.O.G., Besançon, France

Vedrenne C., Arnaud J. (1982): "Whispering gallery modes of dielectric resonators", Proc. Inst. Elec. Eng., pt. H, 183

Vigoureux J.M., Payen R. (1974): "Interaction matière-onde évanescente de Fresnel. I. Radiation spontanée par un électron au voisinage d'un dioptre plan. Effet Cerenkov", J. Physique, 35, 617

Vigoureux J. M., Payen R. (1975): "Interaction matière-onde évanescente de Fresnel. II. Absorption par un atome au voisinage d'un dioptre plan", J. Physique, 36, 631

Vigué J. (1995): "Index of refraction of dilute matter in atomic interferometry", Phys. Rev. A, 52, 5, 3973

Wang Y., Chudgar M.H., Jackson H.E., Miller J.S., de Brabander G.N., Boyd J.T. (1992): "Characterization of Si_3N_4/SiO_2 optical channel waveguides by photon scanning tunneling microscopy", Proc. SPIE 1793, Integrated Optics and Microstructure

Wastiaux G.(1994): "La Microscopie Optique Moderne", Lavoisier, Paris

Weeber J.C., de Fornel F., Salomon L., Bourillot E., Goudonnet J.P. (1995): "Computation of the field diffracted by a dielectric grating: a comparison with experiments", Opt. Commun., 119, 23

Wickramasinghe H.K., Williams C.C. (1987): U.S. Patent 4 947 034 (28 April 1989); IBM Disclosure YO887-0949

Wickramashinge H.K. (1995): "Apertureless Near Field Microscope", Nato ARW on Near Field Optics, Miraflores, September 11-15

Wilks P.A., Hirschfeld T. (1967): "Internal Reflection Spectroscopy", Applied Spectroscopy Reviews, 1, 1, 99

Wolf E., Nieto-Vesperinas M. (1985): "Analyticalicity of the angular spectrum amplitude of scattered fields and some of its consequences", J. O. S. A. A, 2, 886

Xiao M. (1996): "Cutting off the diffraction: A numerical solution in scanning near-field optical microscopy", Appl. Phys. Lett., 69, 3125

Yeh C., Dong S.B., Oliver W. (1975): "Arbitrary shaped inhomogeneous optical fiber or integrateg optical waveguide", J. Appl. Phys., 46, 2125

Yeh P. (1985): "Resonant tunneling of electromagnetic radiation in superlattice structures ", J. Opt. Soc. Am. A, 2, 4, 568

Yariv A. (1997): "Optical Electronics in Modern Communications", Oxford University Press, New York

Yoshikawa H., Watanabe M, Ohno Y. (1988): "Distributed oil sensor using eccentriclly cladded fiber", OFS'88, 92, New Orleans

Zenhausern F., O'Boyle M.P., Wikramasinghe H.K. (1994): "Apertureless near-field optical microscope", Appl. Phys. Lett., 65, 1623

Zenhausern F., Martin Y., Wikramasinghe H.K. (1995): "Scanning interferometric apertureless microscopy: optical imaging at 10 angstrom resolution", Science, 269, 1083

Zhou Y., Laybourn P.J.R., Magill J.V., De La Rue R.M. (1991): "An evanescent fluorescence biosensor using ion-exchanged buried waveguides and the enhancement of peak fluorescence", Biosensors and Bioelectronics, 6, 595

Zhu S., Yu A.W., Hawley D., Roy R. (1986): "Frustrated total internal reflection: a demonstration and review" , Am. J. Phys., 7, 601

Zolatov E.M., Kiselyov V.A., Prokhorov A.M., Sacherbakov E.A. (1978): "Determination of characteristicsof diffused optical waveguide", Digest of Technical Papers, OSA Topical meeting on Integrated Optics, Salt Lake City, UT (16-18, january 1978)

Zolotaryov V.M. (1970): "Perturbed total internal reflection (PTIR) spectroscopy. Use of polarized light in compensative measurement", Opt. Spectr., 5, 519

Index

Printing (Computer to Film): Saladruck, Berlin
Binding: Lüderitz & Bauer, Berlin

Springer Series in
OPTICAL SCIENCES

New editions of volumes prior to volume 60

1 **Solid-State Laser Engineering**
By W. Koechner, 5th revised and updated ed. 1999, 472 figs., 55 tabs., XII, 746 pages

14 **Laser Crystals**
Their Physics and Properties
By A. A. Kaminskii, 2nd ed. 1990, 89 figs., 56 tabs., XVI, 456 pages

15 **X-Ray Spectroscopy**
An Introduction
By B. K. Agarwal, 2nd ed. 1991, 239 figs., XV, 419 pages

36 **Transmission Electron Microscopy**
Physics of Image Formation and Microanalysis
By L. Reimer, 4th ed. 1997, 273 figs. XVI, 584 pages

45 **Scanning Electron Microscopy**
Physics of Image Formation and Microanalysis
By L. Reimer, 2nd completely revised and updated ed. 1998,
260 figs., XIV, 527 pages

Published titles since volume 60

60 **Holographic Interferometry in Experimental Mechanics**
By Yu. I. Ostrovsky, V. P. Shchepinov, V. V. Yakovlev, 1991, 167 figs., IX, 248 pages

61 **Millimetre and Submillimetre Wavelength Lasers**
A Handbook of cw Measurements
By N. G. Douglas, 1989, 15 figs., IX, 278 pages

62 **Photoacoustic and Photothermal Phenomena II**
Proceedings of the 6th International Topical Meeting, Baltimore, Maryland,
July 31 - August 3, 1989
By J. C. Murphy, J. W. Maclachlan Spicer, L. C. Aamodt, B. S. H. Royce (Eds.),
1990, 389 figs., 23 tabs., XXI, 545 pages

63 **Electron Energy Loss Spectrometers**
The Technology of High Performance
By H. Ibach, 1991, 103 figs., VIII, 178 pages

64 **Handbook of Nonlinear Optical Crystals**
By V. G. Dmitriev, G. G. Gurzadyan, D. N. Nikogosyan,
3rd revised ed. 1999, 39 figs., XVIII, 413 pages

65 **High-Power Dye Lasers**
By F. J. Duarte (Ed.), 1991, 93 figs., XIII, 252 pages

66 **Silver-Halide Recording Materials**
for Holography and Their Processing
By H. I. Bjelkhagen, 2nd ed. 1995, 64 figs., XX, 440 pages

67 **X-Ray Microscopy III**
Proceedings of the Third International Conference, London, September 3-7, 1990
By A. G. Michette, G. R. Morrison, C. J. Buckley (Eds.), 1992, 359 figs., XVI, 491 pages

68 **Holographic Interferometry**
Principles and Methods
By P. K. Rastogi (Ed.), 1994, 178 figs., 3 in color, XIII, 328 pages

69 **Photoacoustic and Photothermal Phenomena III**
Proceedings of the 7th International Topical Meeting, Doorwerth, The Netherlands,
August 26-30, 1991
By D. Bicanic (Ed.), 1992, 501 figs., XXVIII, 731 pages

Springer Series in
OPTICAL SCIENCES

70 **Electron Holography**
By A. Tonomura, 2nd, enlarged ed. 1999, 127 figs., XII, 162 pages

71 **Energy-Filtering Transmission Electron Microscopy**
By L. Reimer (Ed.), 1995, 199 figs., XIV, 424 pages

72 **Nonlinear Optical Effects and Materials**
By P. Günter (Ed.), 2000, 174 figs., 43 tabs., XIV, 540 pages

73 **Evanescent Waves**
From Newtonian Optics to Atomic Optics
By F. de Fornel, 2001, 277 figs., XVIII, 268 pages

74 **International Trends in Optics and Photonics**
ICO IV
By T. Asakura (Ed.), 1999, 190 figs., 14 tabs., XX, 426 pages

75 **Advanced Optical Imaging Theory**
By M. Gu, 2000, 93 figs., XII, 214 pages

76 **Holographic Data Storage**
By H.J. Coufal, D. Psaltis, G.T. Sincerbox (Eds.), 2000
228 figs., 64 in color, 12 tabs., XXVI, 486 pages

77 **Solid-State Lasers for Materials Processing**
Fundamental Relations and Technical Realizations
By R. Iffländer, 2001, 230 figs., 72 tabs., XVI, 350 pages

78 **Holography**
The First 50 Years
By J.-M. Fournier (Ed.), 2001, 266 figs., XII, 460 pages

79 **Mathematical Methods of Quantum Optics**
By R.R. Puri, 2001, 13 figs., XIV, 285 pages